CONTROL AND DYNAMIC SYSTEMS

Advances in Theory and Applications

Volume 40

CONTRIBUTORS TO THIS VOULME

MASOUD AMIN-JAVAHERI
IZHAK BAR-KANA
ANTAL K. BEJCZY
AMIR FIJANY
ALLON GUEZ
BLAKE HANNAFORD
C. S. G. LEE
C. T. LIN
HOMAYOUN SERAJI
SUNIL K. SINGH
XIAOPING YUN

CONTROL AND DYNAMIC SYSTEMS

ADVANCES IN THEORY AND APPLICATIONS

Edited by
C. T. LEONDES

Department of Electrical Engineering
University of Washington
Seattle, Washington

VOLUME 40: ADVANCES IN ROBOTIC SYSTEMS
Part 2 of 2

ACADEMIC PRESS, INC.

Harcourt Brace Jovanovich, Publishers

San Diego   New York   Boston
London   Sydney   Tokyo   Toronto

Copyright © 1991 BY ACADEMIC PRESS, INC.
All Rights Reserved.
No part of this publication may be reproduced or transmitted in any form or
by any means, electronic or mechanical, including photocopy, recording, or
any information storage and retrieval system, without permission in writing
from the publisher.

Academic Press, Inc.
San Diego, California 92101

United Kingdom Edition published by
ACADEMIC PRESS LIMITED
24-28 Oval Road, London NW1 7DX

Library of Congress Catalog Card Number: 64-8027

ISBN   0-12-012740-7   (alk. paper)

PRINTED IN THE UNITED STATES OF AMERICA
91 92 93 94   9 8 7 6 5 4 3 2 1

CONTENTS

CONTRIBUTORS

Numbers in parentheses indicate the pages on which the authors' contributions begin.

Masoud Amin-Javaheri (285), *GMFanuc Robotics Corporation, Auburn Hills, Michigan 48057*

Izhak Bar-Kana (147), *Department of Electrical and Computer Engineering, Drexel University, Philadelphia, Pennsylvania 19104*

Antal K. Bejczy (315, 357), *Jet Propulsion Laboratory, California Institute of Technology, Pasadena, California 91109*

Amir Fijany (315, 357), *Jet Propulsion Laboratory, California Institute of Technology, Pasadena, California 91109*

Allon Guez (147), *Department of Electrical and Computer Engineering, Drexel University, Philadelphia, Pennsylvania 19104*

Blake Hannaford (1), *Department of Electrical Engineering, University of Washington, Seattle, Washington 98195*

C. S. G. Lee (33), *School of Electrical Engineering, Purdue University, West Lafayette, Indiana 47907*

C. T. Lin (33), *School of Electrical Engineering, Purdue University, West Lafayette, Indiana 47907*

Homayoun Seraji (205), *Jet Propulsion Laboratory, California Institute of Technology, Pasadena, California 91109*

Sunil K. Singh (105), *Thayer School of Engineering, Dartmouth College, Hanover, New Hampshire 03755*

Xiaoping Yun (259), *University of Pennsylvania, Department of Computer and Information Science, Philadelphia, Pennsylvania 19104*

PREFACE

Research and development in robotic systems has been an area of interest for decades. However, because of increasingly powerful advances in technology, the activity in robotic systems has increased significantly over the past decade. Major centers of research and development in robotic systems were established on the international scene, and these became focal points for the brilliant research efforts of many academicians and industrial professionals. As a result, this is a particularly appropriate time to treat the issue of robotic systems in this international series. Thus this volume and Volume 39 in this series are devoted to the timely theme of "Advances in Robotic Systems Dynamics and Control."

The first contribution to this volume, "Kinesthetic Feedback Techniques in Teleoperated Systems," by Blake Hannaford, is a particularly appropriate contribution with which to begin this second volume of this two volume sequence. As noted in this contribution, telemanipulation systems span an astonishing scale of ten orders of magnitude. As a result, in earlier times telemanipulation systems were designed to overcome barriers of distance between the operator and the manipulated object. However, recently teleoperation developments are aimed at overcoming the barriers of large differences of scale between the human operator and the manipulated object. Thus, teleoperated systems today span the scale from remote surgery to the Space Shuttle Remote Manipulation System (RMS) for satellite repair and other missions.

The next contribution, "Parallel Algorithms and Fault-Tolerant Reconfigurable Architecture for Robot Kinematics and Dynamics Computations," by C.S.G. Lee and C.T. Lin, presents powerful and robust computational techniques and architectures for the control of robot manipulators. The necessary goal is the development of algorithms of lower computational structures. In particular, the ultimate goal is the achievement of an order-of-magnitude and/or an order-of-complexity improvement in computational efficiency in robotics computations, in general, by taking advantage of parallelism, pipelining, and architectures, while at the same time maintaining efficiency and flexibility in the capability to solve robotic computational problems on the same architecture. In order to design a global architecture for a set of parallel robotic algorithms, the characteristics of these algorithms are identified according to six fundamental features. With the parallel robotics algorithms and a parallel computer architecture, a systematic mapping procedure to schedule the subtasks of the parallel algorithms onto the parallel architecture is presented. Because of the central importance to robotics of the issues presented in this contribution, it constitutes an essential element of these companion volumes on robotics.

The development of autonomous controllers, in general, and "intelligent" robots, in particular, has led to active research in "motion planning." The planning problem has been interpreted and solved in various ways by different researchers. It seems that many times planners completely ignore the dynamics of the system. This has led to the current trend of dividing the problem into smaller subproblems and solving each one separately. The three typical subproblems may be identified as task planning, trajectory planning, and trajectory tracking or motion control. The next contribution, "Trajectory Planning for Robot Control: A Control Systems Perspective," by Sunil K. Singh, provides an in-depth treatment of the central issues noted above, and as such is also an essential element of these companion volumes.

Among the modes of robotics control is adaptive control. The next contribution, "Simplified Techniques for Adaptive Control of Robotic Systems," by Izhak Bar-Kana and Allon Guez, which presents techniques in this area, is an important element of these companion volumes. In particular, in adaptive control techniques it is usually the case that such prior knowledge and conditions as the order of plant or system, the relative degree, inverse stability, stationarity, and external excitation are needed. In this contribution, rather powerful techniques for simple and robust adaptive controllers for nonlinear systems with unknown parameters with particular application for robot manipulators are presented.

The remarkable dexterity and versatility that the human arm exhibits in performing various tasks can be attributed largely to the kinematic redundance of the arm, which provides the capability of reconfiguring the arm without affecting the hand position. A robotic manipulator is called (kinematically) "redundant" if it possesses more degrees of freedom than necessary for performing various specific tasks. The next contribution, "Theory and Applications of Configuration Control for Redundant Manipulators," by Homayoun Seraji, is a rather comprehensive treatment of techniques in this broad area. Furthermore, control techniques for this complex problem area are presented which offer the possibility of efficient real-time control redundant manipulators.

The motions of robotic manipulators are either constrained or unconstrained while they perform tasks. For example, many robotic applications, such as assembly tasks, require constrained motion of manipulators. Two generic cases of constrained motion can be considered: (1) a single manipulator constrained by the environment and (2) multiple manipulators constrained with each other as well as constrained by the environment. The next contribution, "Nonlinear Feedback for Force Control of Robot Manipulators," by Xiaoping Yun, presents an in-depth treatment of issues and techniques in this major area of robotics. Numerous important results are presented. Not the least of these is that by the utilization of nonlinear feedback the simultaneous motion and force control of a constrained manipulator or two cooperative manipulators is converted into the design problem of decoupled linear subsystems, and this is a major result from an applied point of view.

A major challenge in effectively realizing advanced control schemes for robotic systems is the difficulty of implementing the kinematic and dynamic equations required for coordination and control in real time. While the total number of computations appears to be somewhat fewer than that of many scientific computations, implementations in real time imply that these computations must be repeated at high repetitive rates per second. This, then, results in an important computational problem in robotics control. It is these computational aspects of dynamic control techniques in robotics that are the main thrust of the next contribution, "Systolic Architectures for Dynamic Control of Manipulators," by Masoud Amin-Javaheri. The concepts, approach, and techniques presented in this chapter are general enough to be applied to a wide range of robotics control problems and their computational requirements.

In the next two contributions "Techniques for Parallel Computation of Mechanical Manipulator Dynamics Part I: Inverse Dynamics" and "Part II: Forward Dynamics," by Amir Fijany and Antal K. Bejczy, an in-depth treatment is presented for the solution of model-based control techniques in robotic systems, both for inverse dynamics (Part I) and for forward dynamics (Part II). Powerfully effective algorithms and systems architectures are developed and presented. The essential importance of these problems in robotic systems will make these two unique contributions a valuable source reference for workers in the field for years to come.

This volume is a particularly appropriate one as the second of a companion set of two volumes on advances in robotic systems dynamics and control. The authors are all to be commended for their splendid contributions, which will provide a significant reference source for workers on the international scene for years to come.

KINESTHETIC FEEDBACK TECHNIQUES
IN TELEOPERATED SYSTEMS

BLAKE HANNAFORD
Dept. of Electrical Engineering
University of Washington
Seattle, WA 98195

I: INTRODUCTION: OVERVIEW OF TELEOPERATION

Teleoperation, the ability to perform physical manipulations of objects from a distant control point, is the newest "tele" technology (coming after telegraphy, telephony, and television). Teleoperation was first reduced to practice by Goertz in the late 1940's.[1] Even at that time it was recognized that controlling the "slave" (remote) robot to track the position and orientation of a "master" manipulator held in the operator's hand was insufficient to effectively perform remote tasks. An essential feature of useful systems was the feedback* of force information to the operator arising from the interaction between the slave and its environment. The essential quality for effective remote manipulation is the replication of both force and incremental motion at the master and slave end effectors. A feedback system implementing this behavior is said to be "Kinesthetic". Kinesthesia is defined as

"The sensation of movement or strain in muscles, tendons, and joints."**

Thus, the effectiveness of kinesthetic remote manipulation comes from its ability to reproduce in the human nervous system the same kinesthetic sensations

---

*Designation of position as the forward command and force as the feedback variable is arbitrary as explained below. For the purposes of an initially descriptive common vocabulary, a reference system will be described having this configuration and in which the word force is used to indicate both force and torque, and position to indicate position and orientation.
** The Random House Dictionary, Random House, New York, 1978

CONTROL AND DYNAMIC SYSTEMS, VOL. 40

1

as would arise in directly manipulating an object.

The assessment of the quality or fidelity of remote kinesthesia is a major topic in its own right, beyond the scope of this chapter. Numerous studies have highlighted performance measurement methods and results[2,3,4,5,6,7,8]

The first kinesthetic remote manipulation systems, described above, consisted of identical master and slave manipulators. The master was different from the slave only in its base location, and in the attachment of a handgrip to the master at a point corresponding to the center of the slave gripper opening. The control systems for these "joint-based" teleoperators are designed to make sure that the joint angles of the master and slave manipulators correspond, and that their torques are opposed. Because of the identical kinematics, this control law ensures tracking between master and slave and feedback of slave contact force to the handgrip. The control systems for each joint are independent except for position dependent disturbance torques. Correctly coupling the feedback force to the contact force and the slave motion to the master motion defines kinesthetic correspondence between master and slave. The basic control architecture which derives the torques based on the position difference between the joints is referred to as the "classical master-slave teleoperator."

Constraining the master and slave systems to be identical can be costly. For example, if the work volume required to perform the task is much larger than the comfortable range of human manipulation, an unwieldy master is required. Ideally, the master should be kinematically optimized to the human operator and the slave to the task.[9] Teleoperators having dissimilar master and slave are said to posses "generalized teleoperation."[10] Generalized teleoperation requires that master joint motions be resolved in real time to some general representation (through forward kinematics of the master), and in turn that slave motions be resolved from the general representation (such as incremental motion vectors or frames) to joint increments (through the inverse kinematics or Jacobian models of the slave). The flexibility of modern computer systems used for these coordinate transforms means that alternate control modes can be implemented for testing and optimal task performance. In a recently implemented system,[11] one of ten distinct control and feedback modes can be independently selected for each task space axis. These options include for example position vs. rate control, force feedback or no force feedback, and complaint control of slave motion. The resulting number of possible combinations for the whole six axis system is thus equal to one million modes!.

No theory yet exists which can derive the optimal mode to use for a given task, we can only explore small regions of "mode space" with analysis and experiments. Even higher dimensionality results if we consider variations of continuous parameters such as gains and scales.

This chapter will consider a small slice of this rich space of possible teleoperation modes. First we will focus primarily on the effects of variations in two key parameters, the force scale and position scale, then we will describe two-port network models of the performance and dynamics kinesthetic remote manipulation systems.

I-1: Application domains

Although originally developed for handling dangerous materials in the nuclear industry from a distance, teleoperation is now a generic technology which can be applied to a wide variety of problems. Most of them still involve separation of the human operator from an inherently dangerous manipulation task.

In describing kinesthetic remote manipulation, we have referred to force (also refered to as "effort") as the feedback variable, and position ("flow") as the forward command. This way of thinking can be a convenient approach for system design. Such a design can be described as a "forward flow" design because the flow variable is transmitted from master to slave (and the effort variable is fed back). The essential quality for kinesthesia is that the efforts and flows at the handgrip and slave gripper closely correspond. Although the forward flow and the classical master-slave architectures are two ways of achieving remote kinesthesia, many others are possible. For example, an interesting property of kinesthetic remote manipulation systems is that they are bilateral - the kinesthetic correspondence applies if the slave robot is used as a master. Thus, the "forward effort" architecture is also valid and has been used.[12]

Rather than review applications by industry, we will instead illustrate the diversity of telemanipulation applications through the ranges of two key parameters in teleoperation system design: the position scale factor λ_p and the force scale factor λ_f. In the forward flow architecture, position (or velocity) commands sent from the master to the slave are multiplied by λ_p, and force information from the slave is multiplied by λ_f before being applied to the operator. In other architectures, λ_p and λ_f may not be explicit parameters in the control system, but they can be derived as described below.

Up until now, telemanipulation systems have been designed to overcome barriers of **distance** between the operator and the manipulated object. Recent teleoperation developments are aimed at overcoming the barriers of **large differences of scale** between the human operator and the manipulated object.

The position scale of telemanipulation systems has today spanned an astonishing 10 orders of magnitude (Figure 1). In systems which have been implemented to date, λ_p has varied from approximately 10 in the case of the Space Shuttle Remote Manipulation System (RMS)*, to about 10^{-9} in the case of recent work teleoperating the scanning, tunneling microscope (STM) and atomic force microscope (AFM).[13,14]

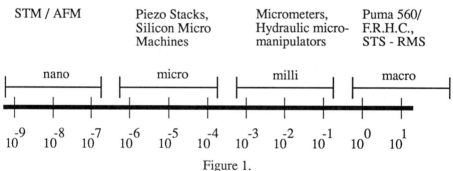

Figure 1.
Telemanipulation systems have been developed which scale the operators position commands over a wide range.

---

*Only resolved rate control has been used to control the RMS to date. The λ_p estimate is based on the approximate ratio of the size of the RMS to the size of the human arm.

To define some terms, ranges can be designated on the λ_p scale defining "macro-telemanipulation," $10^0 \leq \lambda_p < 10^3$; "milli-telemanipulation," $10^{-3} \leq \lambda_p < 10^0$; "micro-telemanipulation," $10^{-6} \leq \lambda_p < 10^{-3}$; and "nano-telemanipulation," $10^{-9} \leq \lambda_p < 10^{-6}$. Although these definitions haven't yet been standardized, they are beginning to achieve common usage, especially in Japan. While the specific scales associated with each range are open to debate, clearly the four terms should be separated by a factor of 10^3 and each span a range of 10^3. One difficulty with consistent terminology is that a typical robot manipulator has two relevant scales usually separated by a factor of about 1000. For example, an industrial robot may have a work volume radius of about 1 meter, and a position resolution at the end effector of about 1 mm. The terminology in this chapter is based on the position scale λ_p and thus avoids this ambiguity.

Nano-telemanipulation has leapt to attention with the STM and AFM applications in which a human controls a probe tip with atomic scale resolution. Hunter[15] has developed a six degree of freedom nano-manipulator to kinesthetically manipulate individual muscle fibers (carefully cultured together with the apparatus) during muscle physiology experiments. His system has focused primarily on delivering controlled position increments to muscle fiber preparations for nano-biomechanical tissue characterization.

Micro-telemanipulation is an area of opportunity raised by the recent development of small mechanical systems fabricated on silicon by microscopic photo-lithography and processes from the VLSI industry, but this author has not yet seem kinesthetic teleoperation implemented at this scale.

Milli-telemanipulation is widely expected to have significant impact in the field of micro-surgery,[16] micro-neurography, and electrophysiology. Additional applications under exploration at the "milli" scale include: injection of genetic material into cells, embryological research, in-vitro fertilization, and electronic assembly. Today, all of these applications are performed either with micrometer drive or hydraulic reduction drive equipment or manually.

By virtue of their huge mechanical reduction of the operator's motion through high-pitch lead screws, micrometer drives are completely rigid with respect to the delicate tissues they are manipulating. Direct human manipulation at these scales while remarkably capable, (in retinal surgery, human surgeons can "routinely" make repeatable controlled incisions as small as 200 microns[17]), is limited to a few individuals and like the micrometer drive is insensitive to the small forces generated by the tissues.

The other crucial design or operating parameter is λ_f, the force scale factor. Nominally, λ_f is the inverse of λ_p. However, the two can be varied independently to effectively adjust the impedances of the operator and task. If λ_f is plotted against λ_p, we can visualize a space in which different kinesthetic feedback applications reside (Figure 2).

The mechanical impedance felt by the operator of such a device can be derived in terms of the environment impedance Z_e. To a first approximation, it can be shown (section II-3) that the felt impedance is

$$Z_f = \lambda_p \lambda_f Z_e \tag{1}$$

where Z_e is the mechanical impedance of the environment.
The line

$$\lambda_p = \frac{1}{\lambda_f} \tag{2}$$

defines the locus of scale factors for which the operator's perception of the environment mechanical impedance is unscaled. Operation above the line, i.e.

$$\lambda_p > \frac{1}{\lambda_f} \tag{3}$$

increases the perceived impedance relative to the actual load and operation below the line reduces the perceived mechanical impedance.

Similarly, power gain across the teleoperator is (section II-4)

$$\frac{P_{load}}{P_{operator}} = \frac{\lambda_p}{\lambda_f} \tag{4}$$

Thus, the line

$$\lambda_f = \lambda_p \tag{5}$$

defines the locus of unity power gain (passivity). Operation above the line, i.e.

$$\lambda_f > \lambda_p \tag{6}$$

attenuates power from the operator and operation below the line amplifies

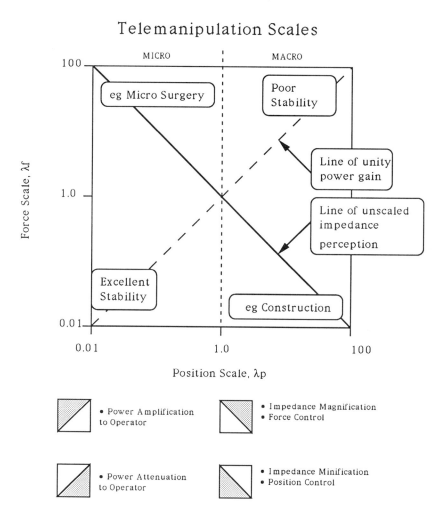

Figure 2.
Properties of scaled telemanipulation as a function of position scale λ_p, and the force scale λ_f

power from operator to environment.

This ability to amplify or attenuate energy flows and modulate perceived task impedance raises important opportunities. Clearly power attenuation between surgeon and tissue in micro-telemanipulation is a desirable trait. Similarly, power amplification is needed for the domain of macro-telemanipulation for construction. Normally, one might consider unscaled impedance perception a desirable operating condition. However, one of the open research questions in this technology is, what is the mechanical impedance of the relevant environments, such as biological tissues, at small scales (e.g. 100 microns)?. In other words, if a cell membrane is displaced by a few microns, what is the resulting force? This is of interest because of the generally exponential stress strain characteristic of biological tissues[18,19] which predicts vanishing mechanical impedance for microscopic deflections. Some recent work has been done on the mechanics of cell membranes which indicates a stiffness in the range of 0.2 to 0.5 N/M.

II: NETWORK APPROACHES TO BILATERAL CONTROL

II-1: Introduction

Recently, the study of electrical networks has provided tools for the analysis of bi-lateral manipulation systems. Network models of both electrical and mechanical systems are part of a closed graph of model classes (Figure 3). All physical systems can be represented by all of these model types but the optimal model type depends on the problem to be solved, the level of representation accuracy, and the system complexity. New models are most easily derived by traversing from an established model to an adjacent node. Network models of robotic systems are being developed by generalizing rigid body analysis.

Network concepts evolved in parallel with classical field descriptions and rigid body dynamic descriptions of physical interactions. Early network pioneers such as Kirchoff recognized that networks were a reticulated case of fields and fluxes.[20] An early example of the use of network ideas for electromechanical systems is the work of Kron,[21] who developed tensor

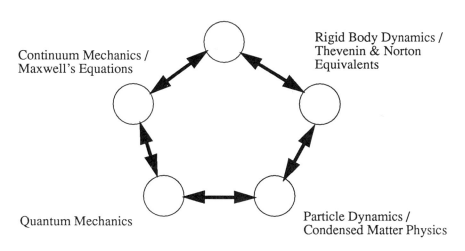

Mechanical Network /
Electrical Network

Rigid Body Dynamics /
Thevenin & Norton
Equivalents

Continuum Mechanics /
Maxwell's Equations

Quantum Mechanics

Particle Dynamics /
Condensed Matter Physics

Figure 3.
A network of descriptions of physical systems. The various levels of descriptions
of electrical and mechanical systems can be arranged in a closed graph as opposed
to a hierarchy.

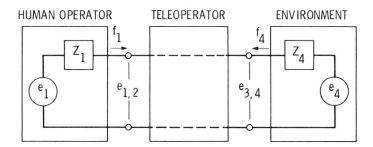

Figure 4.
Two-port model of a teleoperator with kinesthetic force feedback. The network is
expressed in terms of the variables effort and flow which represent force (or torque)
and velocity (angular velocity).

descriptions of rotating electrical machines.

A robot in contact with the environment exchanges mechanical energy with the environment through a single port - namely the contact point. This interaction can be modeled with two interconnected one-port networks representing the robot and environment. This idea was used by Hogan[22] to define the impedance control strategy.

A teleoperator with kinesthetic force feedback is a system which exchanges mechanical energy at two ports; the slave/environment contact point, and the operator/master contact point. This type of system can be modeled as a two-port network connected to one-port networks representing the human operator and the environment.

A network model of a one-dimensional system is shown in Figure 4. Because we will frequently be interchanging between mechanical, electrical, and informational quantities, it will be convenient to adopt Paynter's "effort" and "flow" terminology. Effort and flow are generalized physical quantities whose product is the rate of energy transmission, power. In mechanical systems, effort and flow are force (torque) and velocity (angular velocity), and in electrical systems, they are voltage and current. Since our initial example sends position/velocity commands to the slave and force back to the master, we can designate this a "forward flow" system. Note that this implies a choice of causality[20] direction because the physical efforts and flows at the master and slave ports are transformed to a bi-directional information flow in the communication link.

Six possible matrix representations for two ports arise from the 6 ways to choose two independent variables from four variables. Of these, the impedance (Z), admittance (A), and hybrid (H) parameters are the most commonly used. As detailed in Chua et.al.[23] all six are projections of the matrix equation:

$$ME+NF = 0 \tag{7}$$

where M and N are 2 x 2 matrices of functions of s, and E and F are vectors representing efforts and flows at the ports.

Two port network ideas are now being applied to actual system designs. Fukuda[24] derived a bilateral controller for a micro-teleoperator which was similar two the general matrix formulation (above). Jansen and Herndon[25] have recently used various two-port ideas in their full scale teleoperation design. Goldenberg and Bastas[26] make extensive use of the Hybrid two-port model (below) in their system design and preliminary performance experiments.

II-1.1: Impedance Matrix Design

Paynter[20] has pointed out a history of applications of two-port impedance matrices to transducers and energy conversion devices starting with Poincare.

Raju[27,28] has recently analyzed the classical, position error controlled teleoperator by designating the two flows (velocities) as independent variables, and the efforts (torques) as dependent. In this case, the control law is naturally characterized by a matrix Z whose elements relate motor torques to the two joint positions:

$$\begin{bmatrix} e_{in} \\ e_{out} \end{bmatrix} = \begin{bmatrix} z_{11} & z_{12} \\ z_{21} & z_{21} \end{bmatrix} \begin{bmatrix} f_{in} \\ f_{out} \end{bmatrix} . \tag{8}$$

His control system allowed for global stability for all passive loads and operator models and allowed the operator to adjust the effective Z matrix of the teleoperator for optimal task performance. The results were experimentally verified in a one degree of freedom laboratory system with a non-linear task.

II-1.2: Scattering Matrix Design

Anderson[29] formulated bi-lateral teleoperation in terms of the scattering theory developed for the study of transmission lines and distributed networks. The fundamental relation is the scattering operator (a matrix for LTI systems):

$$\begin{bmatrix} e_{in}-f_{in} \\ e_{out}-f_{out} \end{bmatrix} = \begin{bmatrix} S_{11} & S_{12} \\ S_{21} & S_{21} \end{bmatrix} \begin{bmatrix} e_{in}+f_{in} \\ e_{out}+f_{out} \end{bmatrix} . \tag{9}$$

The main result of this work is to use the concept of passivity to guarantee stability of a bi-lateral teleoperator. If the three main blocks of the system: the master, the communication system, and the slave, can all be modeled as passive 2-ports, **and** the environment and human operator are assumed to be passive, then stability of the system is guaranteed. Anderson's central idea in this work was to define a control law for the communication link 2-port which is passive in spite of any time delays it contains. The resulting system is stable, but suitable performance, beyond the ability of the slave to respond to environmental forces, was not shown.

Anderson's later analysis of robots and teleoperators[30] developed "Hilbert Networks" as a general representation of multidimensional dynamic systems such as robotic arms and teleoperators. This work ramified the network idea to parallel and serial link manipulator kinematic and dynamic analysis, and to such subtle multi-dimensional problems as multiple points of contact (with friction) between manipulator and environment.

II-1.3: Hybrid Matrix Design

The dynamic loop of kinesthetic force feedback has also been modeled with the hybrid two-port model[31] based on the hybrid two-port model of network theory.[32,33,34,23] Like the impedance and scattering matrices, the hybrid matrix is a useful linear representation of networks which exchange energy with other networks at two distinct pairs of terminals - ports.

The hybrid two port model relates forces and velocities at the input and output ports with a 2 x 2 matrix H. The elements of H, which are frequency dependent, are well understood quantities such as gains, impedances, and admittances. The 2 port hybrid parameters can be derived from the system architecture and thus can be used to relate components or parameters of the system to overall system performance.

The hybrid two-port model is ideally suited to the "forward flow" architecture depicted in Figure 5, although like all of the two port models it is inherently a black box description and not a control architecture.

The defining equation of the hybrid two port model is

$$\begin{bmatrix} F_1 \\ V_2 \end{bmatrix} = \begin{bmatrix} h_{11} & h_{12} \\ h_{21} & h_{21} \end{bmatrix} \begin{bmatrix} V_1 \\ F_2 \end{bmatrix} \quad . \tag{10}$$

Where F is the force (effort) at the port indicated by the subscript and V is the velocity (flow). It is shown below that H can be described by

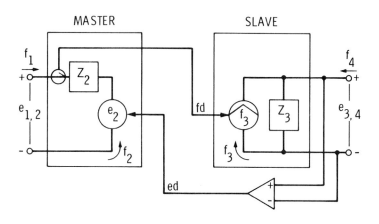

Figure 5.
Two port model of the forward flow teleoperator. This architecture has been used, for example in the recent JPL system (9,10,11).

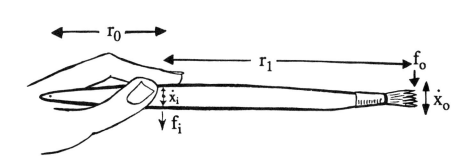

Figure 6.
An artist's long handled paintbrush can be viewed as a two-port variable transformer teleoperator and modeled with the hybrid 2-port network (see text).

$$H = \begin{bmatrix} Z_2 & \lambda_f \\ -\lambda_p & \dfrac{1}{Z_3} \end{bmatrix} \tag{11}$$

where Z_2 and Z_3 are as defined in Figure 5.

A simple example illustrates the utility of this approach. Consider the artist's paintbrush (Figure 6). The artist varies his/her grip on the brush (varying the effective lever arm) in order to achieve specific modulation of the output impedance of the brush tip. With a few approximations, the H matrix for the brush can be written

$$\begin{bmatrix} f_i \\ \dot{x}_o \end{bmatrix} = \begin{bmatrix} \dfrac{\rho(r_0+r_1)^2 s}{2} & \dfrac{r_0+r_1}{r_0} \\ \dfrac{r_0+r_1}{r_0} & \dfrac{r_1 s}{k} \end{bmatrix} \begin{bmatrix} \dot{x}_i \\ f_o \end{bmatrix} \tag{12}$$

where ρ is the linear density of the brush, k is related to the elastic modulus of the brush, s is the laplace transform variable, and r_0 and r_1 are described in Figure 6. Thus, for example,

$$\dot{x}_o = h_{22}f_o\Big|_{\dot{x}_i=0} \tag{13}$$

$$f_i = h_{11}\dot{x}_i\Big|_{f_o=0} \tag{14}$$

h_{11} represents the inertial load seen by the fingers when the brush is free to move without contact. h_{22} is the compliance of the brush as if griped in a vise. h_{12} and h_{21} represent mechanical advantage or "ideal transformer" behavior determined by the length of the brush (r_0+r_1) and the location of grasp (r_0). We can show that the effective mechanical impedance of the artist and brush, "seen" by the canvas, Z_{AB}, varies as

$$Z_{AB} = Z_A \left(\frac{r_0}{r_0+r_1} \right)^2 \tag{15}$$

where Z_A is the mechanical impedance of the artist's hand. As the grip is varied,

$$\text{as } r_0 \to 0 \text{ then } Z_{AB} \to 0$$

$$\text{as } r_1 \to 0 \text{ then } Z_{AB} \to Z_A$$

in the former case, $Z_{AB} \to 0$ is ideal for controlling forces applied to the canvas/paint, while $Z_{AB} \to Z_A$ gives the highest impedance available to the artist for precise position control.

II-2: Hybrid Models

II-2.1: Hybrid 2-Port Model of Teleoperation

We consider that the design goal is to achieve so-called telepresence in which the operator receives feedback from the manipulator to provide kinesthetic information vital to manipulation. This process can be modeled for a single degree-of-freedom as shown in Figure 4. The task of the teleoperator then is to duplicate the effort and flow of the environment at the hand controller and at the same time, to reproduce the effort and flow of the human operator at the manipulator tip. This constrains these efforts and flows to be identical. The dashed lines in the figure represent an ideal teleoperator, i.e. one which has so little distortion or frequency dependence that it is equivalent to direct manipulation of the environment by the operator.

An alternate formulation of the ideal teleoperator is one in which the effort and impedance of the input port are controlled to exactly match the effort and impedance of the environment and vice versa.

The notion of an "ideal" teleoperator is of course somewhat subjective. As detailed above, by manipulation of λ_p and λ_f we can modulate the task impedance and the apparent impedance of the tool tip as desired. The paintbrush is an example of a passive teleoperator which has this property. Thus a specific application may be optimized through this "distortion" of the actual impedances. However, this deviation from the "ideal" response (unity power gain and impedance transformation) is always performed in the context of a specific task. If forced to choose a single fixed configuration for all tasks, the "ideal" teleoperator described above would be chosen because it embodies no assumptions about the task.

Especially when output force sensing is available, it is convenient to choose as independent variables e_{out}, the output effort (force), and f_{in}, the input flow (velocity). In this case, the dependent variables are related to the independent variables by the the matrix H, the hybrid 2-port parameters:

$$\begin{bmatrix} e_{in} \\ f_{out} \end{bmatrix} = \begin{bmatrix} h_{11} & h_{12} \\ h_{21} & h_{21} \end{bmatrix} \begin{bmatrix} f_{in} \\ e_{out} \end{bmatrix} \, . \tag{16}$$

where e_k and f_k, $k = \{in, out\}$ indicate the "effort" and "flow" nomenclature of Paynter[20] at the input and output ports. From (2) we can obtain each h parameter by constraining one of the independent variables to zero. The resulting h parameter definitions in the different problem domains are illustrated in Table I.

Table I. Definitions of Hybrid 2-Port Parameters			
h Parameter	Electrical	General	Mechanical
h_{11}	$\left. \dfrac{v_1}{i_1} \right\|_{v_2=0}$	$\left. \dfrac{e_1}{f_1} \right\|_{e_2=0}$	$\left. \dfrac{f_{in}}{\dot{x}_{in}} \right\|_{f_{out}=0}$
h_{12}	$\left. \dfrac{v_1}{v_2} \right\|_{i_1=0}$	$\left. \dfrac{e_1}{e_2} \right\|_{f_1=0}$	$\left. \dfrac{f_{in}}{f_{out}} \right\|_{\dot{x}_{in}=0}$
h_{21}	$\left. \dfrac{i_2}{i_1} \right\|_{v_2=0}$	$\left. \dfrac{f_2}{f_1} \right\|_{e_2=0}$	$\left. \dfrac{\dot{x}_{out}}{\dot{x}_{in}} \right\|_{f_{out}=0}$
h_{22}	$\left. \dfrac{i_2}{v_2} \right\|_{i_1=0}$	$\left. \dfrac{f_2}{e_2} \right\|_{f_1=0}$	$\left. \dfrac{\dot{x}_{out}}{f_{out}} \right\|_{\dot{x}_{in}=0}$

Definitions of symbols in Table I: v_j is the voltage at port j, i_j is the current, e_j is generalized "effort", f_j, generalized "flow," $f_{in/out}$ the force, and $\dot{x}_{in/out}$ the velocity at the teleoperator inputs and outputs.

In terms of these definitions, it is clear that the h matrix representing the ideal teleoperator is,

$$h_{ideal} = \begin{bmatrix} 0 & 1 \\ -1 & 0 \end{bmatrix} . \tag{17}$$

This matrix is then a standard by which teleoperator systems can be judged, or by which different configurations of a teleoperator system can be compared. A similar analysis using the Z matrix approach, in which the two flows are the independent variables, gives infinity for each element of Z as a representation of the ideal teleoperator.

Anderson[29] showed the relation between the scattering matrix and the hybrid matrix (see below) to be:

$$S(s) = \begin{bmatrix} 1 & 0 \\ 0 & -1 \end{bmatrix} (H(s)-I)(I+H(s))^{-1} \tag{18}$$

Applying this to the H matrix representation of the "ideal" teleoperator, we get:

$$S_{ideal}(s) = \begin{bmatrix} 0 & 1 \\ 1 & 0 \end{bmatrix} \tag{19}$$

II-2.2: Derivation of H Matrix for two current architectures

Two commonly used bi-lateral control architectures are the "classical master slave" (Figure 7), and the "forward flow" architecture. The control law for the "classical master slave system" is

$$e_{2a} = e_{3a} = (f_1 + f_4)G \tag{20}$$

By solving loop equations for the master and slave sides,[31] we can compute the two-port model for the teleoperator as:

$$h = \begin{bmatrix} Z_{2a}+G \left[1-\dfrac{G}{Z_{3a}+G} \right] & \dfrac{G}{Z_{3a}+G} \\ \dfrac{-G}{Z_{3a}+G} & \dfrac{1}{Z_{3a}+G} \end{bmatrix} \tag{21}$$

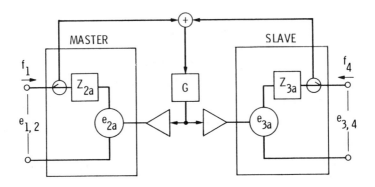

Figure 7.
Network model of the classical master slave control system for kinematically correspondent teleoperators.

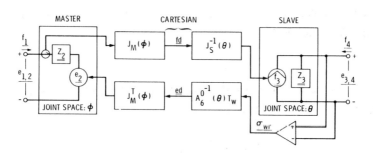

Figure 8.
Network model of the 6 axis forward flow telemanipulation system.

We now consider the "forward flow" architecture (Figure 5) which has been implemented in the JPL FRHC - PUMA generalized bilateral teleoperator.[10,35] It is interesting to point out that there is nothing about this architecture that mandates the designation of "forward" to the flow path. In fact the master can be moved readily by human manipulation of the slave. This "forward effort" mode has been used in the JPL "force reflecting" trigger[12] in which finger force is sensed as the command to a gripper whose position is fed back through a lead screw drive mechanism.

Using the notation established in the figure, it is straightforward to calculate the H matrix for the basic forward flow architecture:

$$h_{11} = \left. \frac{e_{1,2}}{f_1} \right|_{e_{3,4} = 0} = Z_2 \tag{22}$$

$$h_{21} = \left. \frac{f_4}{f_1} \right|_{e_{3,4} = 0} = -\frac{f_3}{f_1} = -1 \tag{23}$$

$$h_{12} = \left. \frac{e_{1,2}}{e_{3,4}} \right|_{f_1 = 0} = 1 \tag{24}$$

$$h_{22} = \left. \frac{f_4}{e_{3,4}} \right|_{f_1 = 0} = \frac{1}{Z_3} \; . \tag{25}$$

In comparing this model to the ideal system, note that the equivalent impedances, Z_2 and Z_3 are typically a function of the controller gains as well as the mechanism impedances. Thus the system deviates from the ideal response only to the extent that the effects of the mechanism impedances are not entirely canceled by feedback control.

II-2.3: Extension to the Six D.O.F. Case.

To extend the above analysis to the case of a teleoperator with multiple mechanical degrees-of-freedom, there are two possible approaches. First, as in conventional master slave teleoperators, the master and slave can be kinematically identical and multiple independent control systems can be connected between the corresponding joints. A current application of this concept is the coupling of human exoskeletal controllers with anthropometric arms and hands.

In this case, many degrees of freedom will be coordinated by and redundancies will be resolved by the human nervous system. In applications where the master and slave are dissimilar, the forward flow system can be expanded as in Figure 8.

Each input flow (velocity) and effort (force) becomes a six vector which is transformed to Cartesian coordinates through the appropriate transformations. Note that in the "forward flow" implementation, slave force is sensed at the wrist, first in terms of load cell strain σ_w which is transformed to wrist coordinates through the transformation T_w. Forces and torques in wrist space are then transformed to Cartesian space by $A_6^{0^{-1}}(\theta)$. Which is the transform from wrist to base coordinates. This force feedback is multiplied by $J_m^T(\phi)$, the Jacobian matrix transpose for the master, and then drives the master motors. Forward velocity commands are transformed into Cartesian space by $J_m(\phi)$, and to slave joint space through $J_s^{-1}(\theta)$, the inverse of the slave Jacobian matrix. The mechanism impedances, Z_2 and Z_3, become matrices.

As in equations (5-8),

$$H_6 = \begin{bmatrix} Z_2(\phi) & J_m^T(\phi)A_6^{0^{-1}}T_w \\ J_m(\phi)J_s^{-1}(\theta) & Z_3^{-1}(\theta) \end{bmatrix} \tag{26}$$

where ϕ represents the joint configuration of the master, and θ the joint configuration of the slave. Problems due to conditioning in the neighborhood of singularities are revealed in the explicit formulation and their effects can be related to teleoperator fidelity through the hybrid parameters. Note that effort (force) feedback is not transformed by the slave Jacobian because it is never represented in joint space.

II-2.4: General Master Slave Architecture

Consideration of the Forward Flow teleoperator architecture raises some interesting questions. First, if the system can be operated in either direction, which direction, if any, is optimal? Second, although physically realized systems which implement the forward flow architecture tend to have very stiff mechanisms on the effort sensing side, and compliant ones on the flow sensing side, is there any basis for selecting these characteristics? Finally, although the forward flow architecture does come close to realizing the ideal response, the interaction between operator and environment which provides the kinesthetic sensation takes place via a "long loop" formed by the forward flow command and the returning effort feedback. Although this works well in current systems, recent simulation and experimentation[8,36] has shown that it is extremely sensitive to even small time delays.

As a result of these factors, a new, general architecture, "bilateral impedance control," is proposed for teleoperators. This architecture (Figure 9) features identical models for master and slave sides which can be fit to actual hardware by parameter variation. A local control law which enforces a desired impedance is implemented on each side. Each side of the teleoperator consists of a general machine built around an actuator with effort source, e_{ia}, (where i indicates the machine number and a stands for "actuator") and series mechanical impedance, Z_{ia}. Each machine's controller implements the impedance law and thus has two inputs, e_{id}, and Z_{id}. The combined properties of the mechanism and control law thus yield effective efforts and impedances, e_i and Z_i. The actuator effort e_{ia} is given by the control law,

$$e_{ia} = G(e_{id} - e_{out} - Z_{id}f_{out}) \tag{27}$$

for example, in the case of the slave actuator, designated by the subscript 3,

$$e_{3a} = G(e_{3d} - e_{3,4} - Z_{3d}f_3) \tag{28}$$

where G is a controller gain which may be frequency dependent.

For bilateral impedance control, we can derive[31] the complete H matrix in terms of the actuator impedances, gains, and estimator transfer functions:

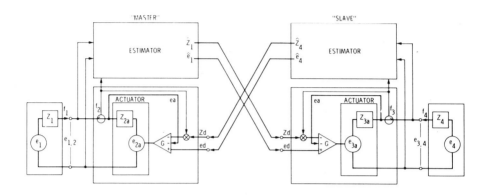

Figure 9.
Bilateral Impedance Control. A new teleoperator architecture which is symmetrical
with respect to master and slave and does not embody assumptions about their mechanical properties.

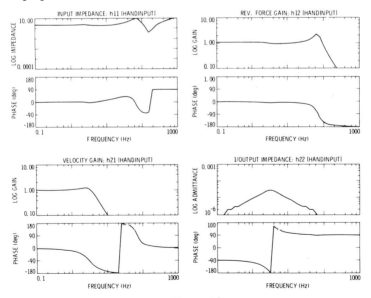

Figure 10.
Frequency dependent H matrix description of a forward flow teleoperator computed
by a SPICE simulation of a detailed network model.

$$
H = \begin{bmatrix} \dfrac{Z_{2a}}{1+G} & \left[\dfrac{G}{1+G}\right]\dfrac{\hat{e}_4}{e_4} \\[2em] -\dfrac{\hat{e}_1\left[Z_1+\dfrac{Z_{2a}}{1+G}\right]}{e_1\left[\hat{Z}_1+\dfrac{Z_{3a}}{G}\right]} & \dfrac{1}{\dfrac{G}{1+G}\left[Z_{max}+\dfrac{Z_{3a}}{G}\right]} \end{bmatrix} . \tag{29}
$$

where the quantities $\hat{e}_4$ and $\hat{Z}_1$ are outputs of the estimators, and Z_{max} is the maximum representable impedance.

Thus as the estimates approach their actual quantities, e.g. $\hat{e}_1 \rightarrow e_1$, for small values of the actuator impedance to gain ratios, and, in the case of h_{22} assuming Z_{max} is large, we have,

$$
H \rightarrow \begin{bmatrix} 0 & 1 \\ -1 & 0 \end{bmatrix} . \tag{30}
$$

Bi-lateral impedance control depends on the existence of an estimator which is capable of identifying the impedance of the environment and of the human operator. In general, this is a very difficult problem because of numerical conditioning problems and noise. Although ideal response (equations 26 and 27) may be difficult to achieve in general, making assumptions about the environment can extend the usefulness of this approach. For example, the estimators can be assisted by an intelligent system with a reduced set of impedance vectors ($\hat{Z}_i$). The task of the estimator is then to classify the effort and flow sensor readings into one of the $\hat{Z}_i$. If the manipulation environment is man-made, the $\hat{Z}_i$ may correspond to known objects based on design properties.

In the case of the human operator estimator, the $\hat{Z}_i$ would correspond to pre-defined manipulation states, that is levels of impedance corresponding to different manipulation sub-tasks. For example, fine position control (high mechanical impedance), free motion (medium), and force control (low). It

would be desirable to identify these properties separately for different directions
so that hybrid[37] control strategies can be transmitted to the slave.

II-2.5: SPICE Model

Simulation is a valuable tool to provide a bridge between theoretical ideas
and the physical system. We have chosen to simulate bilateral teleoperation by
transforming the system to an equivalent electrical network and simulating that
network using SPICE, the circuit simulation system from the University of California, Berkeley.[38] Although there are several possible mappings from mechanical to electrical systems, effort (force) to voltage, and flow (velocity) to current
has several advantages[36] and was used throughout. Each component of the H
matrix is a complex function of frequency. With the SPICE simulation, the frequency dependence of the H matrix can be calculated by using the built in AC
analysis function of SPICE. The AC analysis applies a sinusoidal signal to one
of the independent variables (a voltage or current source), and observes the
magnitude and phase of one of the dependent variables.

The SPICE model consists of a set of files containing SPICE input "cards"
describing each of the main components of the system: the human operator,
the hand controller, the communication channel, the manipulator, and the
environment. For reasons of space, the SPICE model is not included here, but
a full circuit diagram and input "deck" are given in.[39]

When assembled together in a matrix (Figure 10), the Bode plots
representing the h parameters completely describe the frequency dependence of
the teleoperator. Each describes a measurable quantity characteristic of the
system's mechanical properties.

h_{11}, the input impedance (Figure 10, upper left) is relatively flat out to
about 30 Hz, but then shows strong frequency dependencies. Experimenting
with variations of model parameters showed that the resonant peak at about 60
Hz was due to the natural frequency of the hand controller cable/handle system.

h_{12}, the reverse force gain (Figure 10, upper right) shows flat behavior out
to about 80 Hz, but then drops sharply for higher frequencies. This effect is

also due to the hand controller cable drive.

h_{21}, the velocity gain (Figure 10, lower left) rolls off at a much lower frequency, beginning at about 3 Hz. This is of course limited by the slave manipulator dynamics.

h_{22}, the inverse of the output impedance, (Figure 10, lower right) shows a peak at the manipulator resonant frequency. Recall that the ideal h_{22} is zero, i.e. infinite output impedance (measured with zero input velocity). The peak at 3 Hz represents the manipulator's ability to absorb energy at its resonant frequency.

The frequency dependent h parameters thus illustrate how dynamic properties of the master and slave mechanisms effect teleoperator performance. With a suitable model and simulation the designer can thus observe the effects of mechanism parameters such as inertia and compliance on teleoperator performance. For example, the simulation indicates that force frequency response to the operator can be extended by attention to the dynamics of the hand controller cable drive.

II-2.6: Impedance Scaling

We can use the hybrid two port model to calculate the impedance perceived by the operator as a function of the environment impedance, Z_4 and the hybrid parameters as follows.

The impedance felt by the operator, Z_{felt} is

$$Z_{felt} = \frac{e_1}{f_1} \tag{31}$$

From the definition of the H parameters,

$$e_1 = h_{11}f_1 + h_{12}e_4 \tag{32}$$

The environment impedance constrains the output so that

$$e_1 = h_{11}f_1 + h_{12}(-f_4 Z_4) \tag{33}$$

Using

$$f_4 = h_{21}f_1 + h_{22}(-f_4 Z4) \tag{34}$$

we get

$$f_4 = -h_{21}\frac{f_1}{1+Z_4 h_{22}} \tag{35}$$

thus

$$e_1 = h_{11}f_1 - \frac{h_{12}h_{21}}{1+Z_4 h_{22}}Z_4 f_1 \tag{36}$$

Finally,

$$Z_{felt} = h_{11} - \left[\frac{h_{12}h_{21}}{1+Z_4 h_{22}}\right]Z_4 \tag{37}$$

In section I-1, we considered the problem of impedance scaling as a function of the two scale parameters, $\lambda_f = h_{12}$, and $\lambda_p = -h_{21}$. We can define the "ideal scaled" teleoperator as one in which $h_{11} = h_{22} = 0$. In this case we get the result for the felt impedance described in section I-1.

II-2.7: Energy Flows in Force Reflecting Teleoperation

We have seen above (sec. I-1) that through varying the scale factors for position and force transmission, we can achieve desirable modulation of power flows through the teleoperator. We can model the flows of energy by considering the product of effort and flow at the teleoperator ports, and the dissipative (in phase) power in the impedances. Relationships based on conservation of energy are inherently steady state and can thus safely ignore the quadrature component. One way to view the flow of mechanical energy in the kinesthetic telemanipulation system is to derive the power flowing into the input port, P_1, in terms of the effort and flow at the output, F4,f4.

$$P_1 = f_1 e_2 + Z_2 f_1^2 \tag{38}$$

$$P_1 = f_1 \lambda_f e_4 + Z_2 f_1^2 \tag{39}$$

$$P_1 = \frac{\lambda_p}{\lambda_f}f_3 e_4 + Z_2\frac{f_3^2}{\lambda_p^2} \tag{40}$$

$$f_3 = \frac{e_4}{Z_3} - f_4 \tag{41}$$

$$P_1 = \frac{\lambda_p}{\lambda_f}\left[\frac{e_4^2}{Z_3}-e_4f_4\right] + \frac{Z_2}{\lambda_p^2}\left[\frac{e_4}{Z_3}-f_4\right]^2 \tag{42}$$

$$P_1 = \frac{\lambda_p}{\lambda_f}\left[\frac{e_4^2}{Z_3}-e_4f_4\right] + \frac{Z_2}{\lambda_p^2}\left[\frac{e_4^2}{Z_3^2}-2e_4\frac{f_4}{Z_3}+f_4^2\right] \tag{43}$$

$$P_1 = -e_4f_4\left[\frac{\lambda_p}{\lambda_f}+\frac{2Z_2}{\lambda_p^2 Z_3}\right] + \frac{e_4^2}{Z_3}\left[\frac{\lambda_p}{\lambda_f}+\frac{Z_2}{\lambda_p^2 Z_3}\right] + f_4^2\frac{Z_2}{\lambda_p^2} \tag{44}$$

$$P_1 = -P_4\,Q_1 + \frac{e_4^2}{Z_3}\,Q_2 + f_4^2 Z_2\,Q_3 \tag{45}$$

where P_4 is the output power. The three dimensionless terms in Eq. (45), $Q_{1,2,3}$ have the following interpretations: Q_1 which multiplies the output power, consists of the scale ratio ($\frac{\lambda_p}{\lambda_f}$) and the ratio of the circuit impedances with Z_3 reflected into the master circuit by λ_p^2. Q_2 reflects the "shunt loss" of the output port back to the input, and Q_3 reflects the "series loss" of the input port to the output flow variable, f_4. Alternatively, we can keep the last term of Eq. (41) expressed in terms of the input flow, and get

$$P_1 = -P_4\frac{\lambda_p}{\lambda_f} + \frac{e_4^2}{Z_3}\frac{\lambda_p}{\lambda_f} + Z_2 f_1^2 \tag{46}$$

This simpler form shows the output power and shunt loss scaled by the scale ratio and the series loss. These flows of energy are depicted in Figure 11.

The energy balance can be expressed in terms of the H parameters as

$$P_1 = -P_4\left[-\frac{h_{21}}{h_{12}}+\frac{2h_{11}h_{22}}{h_{21}^2}\right] + e_4^2 h_{22}\left[\frac{h_{21}}{h_{12}}+\frac{h_{11}h_{22}}{h_{21}^2}\right] + f_4^2\frac{h_{11}}{h_{21}^2} \tag{47}$$

Thus for the "ideal scaled" teleoperator, we obtain the relationship derived in section I-1.

II-2.8: Stability of Teleoperation

The two port hybrid model can similarly be used to analyze the stability of teleoperation. Thus, the tradeoff between stability and performance can be analyzed. The forward flow description can be viewed as a feedback loop in which the environment impedance and human operator impedance couple signals between the effort and flow control systems. The open loop gain of this loop can be expressed in terms of the hybrid parameters and the human operator and environment impedances[40] as:

$$G_l(H,s) = \frac{-h_{12}h_{21}Z_e}{(Z_{ho}+h11)(1+h_{22}Z_e)} \tag{48}$$

Thus, we can analyze stability using classical techniques such as Routh Horowitz, Root Locus, and Nyquist methods.

Stability will not be considered in detail here, but a wide variety of approaches have been used including Lyapunov stability theory[29] and passivity theory.[28,29] Experimental studies[8,36] have confirmed effects predicted by the theory such as the effect of human operator grasp strength on impedance.

III: CONCLUSIONS

Kinesthetic feedback techniques will be essential for the foreseeable future in effectively operating robots in low-repetition rate, high value operations. The alternative of fully or partially autonomous sensor based intelligent control is not so far economically justifiable for low repetition rate, unstructured tasks because of the difficulty and cost of software development. Even when such systems are available, in remote operation, teleoperation capability must be provided as a backup contingency. This chapter has described a variety of network methods for the analysis of remotely operated robots with kinesthetic feedback.

Many challenging areas remain for the development of kinesthetic remote manipulation. For example, stable and sensitive force feedback from micromanipulators will be a challenge to sensor and control development as detailed above.

Secondly, when remote operation is conducted under time delays, stability is compromised and the value of force feedback to the operator comes into question. A recent study[8] experimentally explored these two effects of time delay.At delays below about 250 ms., stability is compromised but the force information is still comprehensible to the operator. At further delays, even if stability can be maintained* it is very difficult for the operator to make use of the force information because of the loss of the kinesthetic correspondence with direct performance of the original task.

Finally, while network models lend themselves well to questions of force reflection fidelity and stability, full understanding of remote task performance requires models of the whole manipulation process. This model must include the operator's cognitive functioning as he/she accomplishes various stages of the task. A recent step in this direction has been taken through the use of Hidden Markov Models[41] to simulate and analyze data from experimental performance of remote manipulation tasks of medium complexity.

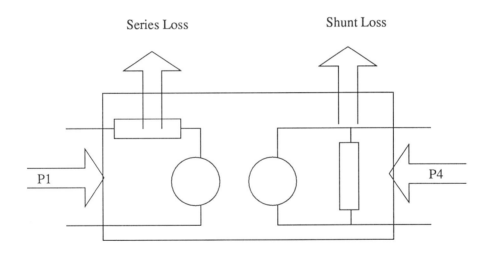

Figure 11.
Flows of mechanical energy and losses in a teleoperation system with kinesthetic feedback.

---

* As for example in[30]

References

1. R.C. Goertz and W.M. Thompson, "Electronically Controlled Manipulator," *Nucleonics*, pp. 46-47, Nov. 1954.

2. D.A. Kugath, *Experiments Evaluating Compliance and Force Feedback Effect on Manipulator Performance,* General Electric Corporation, NASA-CR-128605, Philadelphia, Pa, August, 1972.

3. J.W. Hill and J.K. Salisbury, *Study to Design and Develop Remote Manipulator Systems, Annual Report,* SRI International, Menlo Park, CA, Nov. 1977.

4. J.V. Draper, W.E. Moore, and J.N. Herndon, "Effects of Force Reflection on Servomanipulator Task Performance," *Proceedings of USDOE/FRG Specialists' Meeting on Remote Systems Technology*, Oak Ridge, TN, June 1-3, 1987.

5. B. Hannaford, "Task Level Testing of the JPL-OMV Smart End Effector," *Proceedings of the JPL - NASA Workshop on Space Telerobotics*, vol. 2, pp. 371-380, JPL Publication 87-13, Pasadena, CA, July 1, 1987.

6. B. Hannaford and L. Wood, "Performance Evaluation of a 6 Axis High Fidelity Generalized Force Reflecting Teleoperator," *Proceedings JPL/NASA Conference on Space Telerobotics*, JPL Publication 89-7, Pasadena, CA, January 1989.

7. B. Hannaford, L. Wood, B. Guggisberg, D. McAffee, and H. Zak, *Performance Evaluation of a Six-Axis Generalized Force-Reflecting Teleoperator,* JPL Publication 89-18, Pasadena, CA 91109, June 15, 1989.

8. B. Hannaford and W.S. Kim, "Force Reflection, Shared Control, and Time Delay in Telemanipulation," *Proceedings, IEEE Intl. Conference on Systems, Man, & Cybernetics*, Cambridge, MA, Nov. 1989.

9. A.K. Bejczy and J.K. Salisbury, "Kinesthetic Coupling Between Operator and Remote Manipulator," *Proceedings: ASME International Computer Technology Conference*, vol. 1, pp. 197-211, San Francisco, CA, Aug 12-15, 1980.

10. A.K. Bejczy and Z. Szakaly, "Universal Computer Control System for Space Telerobotics," *Proceedings of the IEEE Conference on Robotics and Automation*, vol. 1, pp. 318-324, Raleigh, N.C., 1987.

11. A.K. Bejczy, B. Hannaford, and Z. Szakaly, "Multi-Mode Manual Control in Telerobotics," *Proceedings of Romansy '88*, Udine, Italy, September 12-15, 1988.

12. P. Fiorini, B. Hannaford, B. Jau, E. Kan, and A. Bejczy, "Hand Trigger System for Bi-Lateral Gripping Control in Teleoperation," *Proceedings of the IEEE International Conference on Robotics and Automation*, vol. 1,

pp. 586-592, April 1987.

13. R.L. Hollis, S. Salcudean, and D.W. Abraham, *Toward a Tele-Nanorobotic Manipulation System with Atomic Scale Force Feedback and Motion Resolution*, pp. 115-119, Napa Valley, CA, Feb. 1990.

14. Y. Hatamura and H. Morishita, "Direct Coupling betweeen Nanometer World and Human World," *Proceedings IEEE Micro Electro Mechanical Systems Conference*, pp. 203-208, Napa Valley, CA, Feb. 1990.

15. I. Hunter, S. Lafontaine, P.M. Nielson, and P.J. Hunter, "Manipulation and Dynamic Mechanical Testing of Microscopic Objects using a Tele-Micro-Robot System," *Proceedings, IEEE Intl. Conf. on Robotics and Automation*, pp. 1553-58, Scottsdale, 1989.

16. S. Charles, R.E. Williams, and W. Hamel, "Design of a Surgeon-Machine Interface for Teleoperated Micro-Surgery," *In Press: Proceedings IEEE EMBS*, Seattle, Nov. 1989.

17. S. Charles and R.E. Williams, "Measurement of Hand Dynamics in a Microsurgery Environment: Preliminary Data in the Design of a Bimanual Telemicro-operation Test Bed," *Proceedings of the NASA Conference on Space Telerobotics*, JPL Publication 89-7, Pasadena, CA, January, 1989.

18. Y.C. Fung, *Biomechanics,* Springer Verlag, 1981.

19. J.M. Winters, *Generalized Analysis and Design of Antagonistic Muscle Models: Effect of Nonlinear Properties on the Control of Human Movement; Ph.D. Dissertation,* University of California, Berkeley, July, 1985.

20. H.M. Paynter, *Analysis and Design of Engineering Systems,* MIT Press, Cambridge, 1961.

21. G. Kron, *Tensor Analysis of Networks,* Wiley, 1939.

22. N. Hogan, "Impedance Control: An Approach to Manipulation: Part I - Theory," *Journal of Dynamic Systems, Measurement, and Control*, vol. 107, no. 1, pp. 1-7, 1985.

23. L.O. Chua, C.A. Desoer, and E.S. Kuh, *Linear and Non-linear Circuits,* McGraw-Hill, New York, 1987.

24. T. Fukuda, K. Tanie, and T Mitsuoka, "A New Method of Master Slave Type Teleoperation for a Micro-Manipulator," *Proceedings IEEE Micro Robots and Teleoperators Workshop*, Hyannis, MA, Nov. 1987.

25. J.F. Jansen and J.N. Herndon, "Design of a Telerobotic Controller with Joint Torque Sensors," *Proceedings of IEEE Intl Conf. on Robotics & Automation*, pp. 1109-1115, Cincinnati, OH, May, 1990.

26. A.A. Goldenberg and D. Bastas, "On the Bilateral Control of Force Reflecting Teleoperation," *To Appear: Proceedings IFAC '90*, August 1990.

27. T.B. Sheridan and Staff, "MIT Research in Telerobotics," *Proceedings of the Workshop on Space Telerobotics*, vol. 2, pp. 403-412, JPL publication 87-13, July 1, 1987.

28. G.J. Raju, "Operator Adjustable Impedance in Bilateral Remote Manipulation," *Ph.D. Dissertation, Department of Mechanical Engineering M.I.T.*, September, 1988.

29. R.J. Anderson and M.W. Spong, "Bilateral Control of Teleoperators with Time Delay," *Proc. IEEE Intl. Conf. Systems, Man, & Cybernetics*, vol. 1, p. 131, Beijing, China, Aug. 1988.

30. R.J. Anderson, "A Network Approach to Force Control in Robotics and Teleoperation, Ph.D. Dissertation," *Department of Electrical Engineering, University of Illinois*, Urbana-Champaign, 1989.

31. B. Hannaford, "A Design Framework for Teleoperators with Kinesthetic Feedback," *IEEE Transactions on Robotics and Automation*, vol. 5, no. 4, pp. 426-434, 1989.

32. J. Choma, *Electrical Networks,* John Wiley & Sons, New York, 1985.

33. J.B. Murdoch, *Network Theory,* McGraw-Hill, New York, 1970.

34. C.A. Desoer and E.S. Kuh, *Basic Circuit Theory,* McGraw-Hill, New York, 1969.

35. A.K. Bejczy and M. Handlykken, "Experimental Results with a Six-Degree-of-Freedom Force Reflecting Hand Controller," *Proceedings of the 17th Annual Conference on Manual Control*, Los Angeles, CA, June 1981.

36. B. Hannaford and R. Anderson, "Experimental and Simulation Studies of Hard Contact in Force Reflecting Teleoperation," *Proc. IEEE Conference on Robotics and Automation*, pp. 584-589, April, 1988.

37. M.H. Raibert and J.J. Craig, "Hybrid Position/Force Control of manipulators," *ASME J. of Dynamic Systems, Measurement, and Control*, vol. 102, pp. 126-133, 1981.

38. A. Vladimirescu, K. Zhang, A.R. Newton, D.O. Pederson, and A. Sangiovanni-Vincentelli, *SPICE version 2G.5 User's Guide,* University of California, Department of Electrical Engineering and Computer Science, Berkeley CA 94720, August 1981.

39. B. Hannaford and P. Fiorini, "A Detailed Model of Bilateral (Position/Force) Teleoperation," *Proceedings International Conference on Systems, Man, and Cybernetics*, pp. 117-121, Bejing, August, 1988.

40. B. Hannaford, "Stability and Performance Tradeoffs in Bi-Lateral Telemanipulation," *Proceedings of IEEE Intl. Conf. on Robotics & Automation*, vol. 3, pp. 1764-1767, Scottsdale, AZ, May 1989.

41. B. Hannaford and P. Lee, "Hidden Markov Model Analysis of Force/Torque Information in Telemanipulation," *Proceedings 1st International Symposium on Experimental Robotics*, Montreal, June 1989.

Parallel Algorithms and Fault-Tolerant Reconfigurable Architecture for Robot Kinematics and Dynamics Computations

C. S. G. Lee and C. T. Lin

School of Electrical Engineering
Purdue University
West Lafayette, Indiana 47907

ABSTRACT

The computations of kinematics, dynamics, Jacobian, and their corresponding inverses are six essential problems in the control of robot manipulators. Efficient parallel algorithms for these computations are discussed and analyzed, and their characteristics are identified based on type of parallelism, degree of parallelism, uniformity of the operations, fundamental operations, data dependencies, and communication requirement. It is shown that most of the algorithms for robotic computations possess highly regular properties and some common structures, especially the linear recursive structure. They are well-suited to be implemented on a single-instruction-stream multiple-data-stream (SIMD) computer with reconfigurable interconnection networks. A reconfigurable, dual-network, SIMD (DN-SIMD) machine with internal direct feedback that best matches these characteristics has been designed. To achieve high efficiency in the computations of robotic algorithms on the proposed parallel machine, a Generalized Cube interconnection network is proposed and investigated. A centralized network switch control scheme is also developed to support the pipeline timing of this machine. Moreover, to maintain a high reliability in the overall system, a fault tolerant Generalized Cube network is designed to improve the original network. With this improvement, it is shown that the proposed parallel architecture performs correctly even under a single fault condition including a processing element fault

This work was supported in part by the National Science Foundation under Grant CDR 8803017 to the Engineering Research Center for Intelligent Manufacturing Systems and in part by a grant from the Ford Fund.
Any opinions, findings, and conclusions or recommendations expressed in this article are those of the authors and do not necessarily reflect the views of the funding agencies.

or a network component fault. The use of the DN-SIMD machine is illustrated through an example of the inverse dynamics computation. A systematic procedure to map these computations to the proposed machine is presented. A new scheduling problem for SIMD machines is investigated and a heuristic algorithm, called neighborhood scheduling, that reorders the processing sequence of subtasks to reduce the communication time is described.

1. Introduction

Robot manipulators are highly nonlinear systems and their dynamic performance depends on the efficiency of computing such tasks as coordinate transformation between the joint-variable space and the Cartesian space, generalized forces/torques to drive the joint motors, the manipulator inertia matrix for model-based control schemes, and the Jacobian matrix which relates the joint velocity in the joint-variable space to the Cartesian space. These are the basic computations for the control of robot manipulators. They are equivalent to the computations of robot arm kinematics, dynamics, Jacobian, and their corresponding inverses. These six basic robotic computations are required at various stages of robot arm control and computer simulation of robot motion, and reveal a basic characteristic and common problem in robot manipulator control — *intensive computations with a high level of data dependency*. They have become major computational bottlenecks in the control of robot manipulators. Despite their impressive speed, conventional general-purpose uniprocessor computers cannot efficiently handle the kinematics and dynamics computations at the required computation rate because their architectures limit them to a mostly serial approach to computation. Furthermore, less efficient, serial computational algorithms must be used to compute these robotics computations on a uniprocessor computer. Consequently, the quest for real-time robot arm control and motion simulation rests on the study and development of parallel algorithms of lower computational complexity with faster computational structures. The ultimate goal is to achieve an *order-of-magnitude* and/or an *order-of-complexity* improvement in computational efficiency in these robotics computations by taking advantage of parallelism, pipelining, and architectures.

A common feature of today's research on robotic computational problems is that a specific problem, mostly the inverse dynamics or the inverse kinematics, is studied at a time, and usually an algorithmically-specialized architecture or processor is developed for that particular algorithm [5,11]. Obviously, this specialized architecture can make the most use of the parallel properties of the algorithm. However, most advanced robot control schemes always require to solve a combination of some or all of the six basic robotic computations. One solution for this problem is to wire these specialized architectures or processors together. This method is inflexible because the combination of these components is dedicated to a particular control scheme and cannot be used efficiently for another scheme. Another solution is to connect the architectures or processors to a bus as peripherals of a general-purpose computer. This is more flexible, but the bus becomes a bottleneck and time is wasted in data movements between different computational processes. Another possible solution is focussed on partitioning the original algorithm/task into a set of subtasks with precedence relationship and then developing efficient scheduling algorithms to map these subtasks onto a general-purpose multiprocessor system. This solution is much more flexible because most computational algorithms can be represented by directed task graphs. However, this approach may result in ignoring some inherent parallelism in robotics algorithms.

This paper focuses on designing a machine which provides the flexibility needed to solve these robotic computational problems on the same architecture while maintaining high efficiency by taking into account the inherent parallelism of robotic algorithms. To exploit their inherent parallelism, our approach is first to characterize the set of parallel robotic algorithms based on the six specified characteristics and features, such as type of parallelism, degree of parallelism, uniformity of the operations, fundamental operations, data dependency, and communication requirement. Our analysis shows that machines operating in the single-instruction-stream multiple-data-stream (SIMD) mode are the most efficient and suitable for our robotic algorithms. By fully considering the common characteristics and inherent parallelism of the robotic algorithms, a medium-grained, reconfigurable, dual-network, SIMD (DN-SIMD) machine with internal direct feedback has been designed for the computation of these kinematic and dynamic computational tasks. To achieve high efficiency by solving the communication problem in the proposed parallel machine, the Generalized Cube network is used as the interconnection networks in our structure. A centralized switch control scheme is developed to reconfigure the network easily and quickly. To assure high reliability in the proposed DN-SIMD machine, a fault tolerant Generalized Cube network is designed by adding two extra stages to the original Generalized Cube network. The fault tolerance capability of the new system is discussed under a well-defined fault model in which both the network component failure and the processing element (PE) failure are considered. General rules were developed for both failures by modifying the switch control matrix and the extra stage control arrays to make the least changes in the required connection types in the network such that the required permutation can still be performable in one pass. Our design makes the system function normally even under the single-fault situation. A systematic mapping procedure has been developed for scheduling these robotic computational tasks onto the proposed DN-SIMD machine. This procedure builds a task table which contains the subtask assignment from the original parallel algorithm. The task table is then used as input to the neighborhood scheduling algorithm which reorders the processing sequence of the subtasks into a rescheduled task table to reduce the communication time. Finally, the subtasks in this rescheduled task table are mapped onto the proposed DN-SIMD machine and a control table which describes the control sequence in the machine is produced. As an example, the implementation of the inverse dynamics computation is presented to show the use of the DN-SIMD machine through the systematic mapping procedure under either normal or faulty situation.

2. Characteristics of Basic Robotic Parallel Algorithms

In dealing with the problem of mapping parallel algorithms onto parallel architectures, three different approaches can be distinguished. In the first approach, a good target parallel architecture with a well-chosen interconnection network is given, and the parallel algorithm is reformulated to match this target architecture by taking the features of the target architecture into consideration. It is obviously less algorithmically-featured concerned but more architecturally-featured concerned. In the second approach, an algorithm is first reformulated to exploit the parallelism in the problem and then an algorithmically-specialized parallel architecture is designed to fully support the algorithm. Obviously, this results in a special-purpose architecture designed for the particular algorithm. Thus, it is fully algorithmically-featured concerned. The third approach is between the above two extreme cases. That is, to design a good parallel algorithm as well as a good parallel architecture simultaneously by considering the tradeoff of their features. This scheme is suitable for designing a *sub-special* parallel architecture for a fixed set of parallel algorithms. This is the approach we choose for our robotic computational tasks.

A key factor to the design of a parallel architecture for a group of algorithms is the understanding of their architectural requirements, and this requires us to identify the characteristics of these algorithms. This identification is usually helpful because the algorithms from a given application area such as robotics often possess an identifiable structure. In order to examine the characteristics of the six basic robotic algorithms, a set of features which have the greatest effects on the execution of parallel algorithms is defined for robotics application [1-3].

■ **Type of parallelism.** Two levels of parallelism, the job-level and the task-level, can be identified. In the job-level parallelism, the original algorithm is reformulated to a parallel processable form which is often amenable to the SIMD implementation. In this level, the variables carrying the same kind of information but with different indices (e.g., for different links or joints of a manipulator) are processed parallelly through an identical computational procedure but with different set of data. In the job-level parallelism, the required number of processors depends on the number of degrees of freedom of the manipulator. In the task-level parallelism, the original algorithm is decomposed into multiple subtasks which can be processed concurrently, this implies multiple-instruction-stream multiple-data-stream (MIMD) operations.

Furthermore, for this level of parallelism, a subtask usually performs the same computation for different set of data, and hence the operation can be pipelined. An advantage of this task-level parallelism is that the required number of processors is independent of the number of degrees of freedom of the manipulator.

- **Degree of parallelism (Granularity).** Three levels of granularity are distinguished. They are *large-grain, medium-grain,* and *fine-grain granularities.* In the *large grain granularity*, the parallelism is performed at the algorithmic level. That is, only the parallelism between different segments or subtasks is considered. For the *medium grain granularity*, the concurrency is considered at the operation level and the parallelism is performed based on some basic mathematical operations such as vector cross product and matrix-vector multiplication. If we consider the implementation of parallelism within the basic arithmetic operations, then the *fine grain granularity* is achieved. Different degrees of parallelism often imply different synchronization requirements. The finer the granularity is, the more frequent synchronization is required.

- **Uniformity of operations.** A robotic algorithm is said to possess uniformity of operations if the required computations for some set of variables (e.g., the joint variables) are uniform. An algorithm with operation uniformity can be implemented on an SIMD machine with higher efficiency.

- **Fundamental operations.** Algorithms in an application area usually perform similar mathematical operations. The identification of basic operations performed in the algorithm will dictate the processor capabilities needed.

- **Data dependency.** Three kinds of data dependency are classified for robotic algorithms: *local neighborhood dependency, special type dependency,* and *global dependency.* The local dependency means that the required operands in an operation come from its neighborhood; for example, from the results of last operation or using the same operands of last operation. The special type dependency is defined for some special equation or problem. There are some special types of data dependency that are peculiar and inherent to the robotics algorithms. Among them, the homogeneous (or hetero-homogeneous) linear recursive type of dependency which describes the data dependency in a homogeneous (or hetero-homogeneous) linear recursive equation appears most frequently. This linear recurrence structure plays a major role in the robotics algorithms because the variables of a joint are usually related to the corresponding variables of its adjacent joint due to the robot's serial link structure. Other special types of data dependency are defined for some well-known problems; for example, system of linear equations and Column-Sweeping algorithm for a triangular linear system. The global dependency means that the results of some operations may be required by other operations or equations that may appear in other places of the algorithm. Since few algorithms possess absolutely one kind of data dependency, we can just identify whether an algorithm is local data dependency oriented or not. The data dependency in an algorithm usually dictates memory organization, data allocation, and communication requirements.

■ **Communication requirement.** The communication requirement decides the required interconnection type between processors or between processor and memory. Three types of interconnection are considered: *one-to-one*, *permutation* and *broadcast* connections. Of course, the exact required interconnection type for each computation in an algorithm depends on many factors such as task assignment of each processor, data allocation in the memories, and data dependency of each computation. Hence, the exact required interconnection type can only be decided at the time of the algorithm-architecture mapping process. In examining the features of robotics parallel algorithms, only rough connection requirements can be observed.

Based on the above set of features, each of the six basic robotic algorithms has been carefully examined and analyzed to find the common features and characteristics among them. The final results are presented in Table 4, which are useful for better understanding of the robotic computations and for designing a suitable parallel architecture for their computations.

2.1. Inverse Dynamics Problem

The inverse dynamics problem can be stated as: Given the joint positions and velocities as $\{ q_i(t), \dot{q}_i(t) \}_{i=1}^{n}$ which describe the state of the manipulator at time t, expressed in the base (or reference) coordinate system, together with the desired joint accelerations $\{ \ddot{q}_j(t) \}_{j=1}^{n}$ at that time, solve the dynamic equations of motion for the joint torques $\{ \tau_j(t) \}_{j=1}^{n}$. Among various methods for computing the applied generalized forces/torques, the one based on the Newton-Euler (NE) equations of motion is the most efficient [4]. Since this method has been shown to possess the time lower bound of $O(n)$ running on uniprocessor computers, where n is the number of degrees-of-freedom of the manipulator, further substantial improvements in computational efficiency appear unlikely. Nevertheless, some improvements could be achieved by taking advantage of particular computation structures [5], customized algorithms/architectures for specific manipulators [6,7], parallel computations [8,9], and scheduling algorithms for multiprocessor systems [10-13].

The NE equations of motion are very efficient in evaluating the inverse dynamics whether they are formulated in the base coordinate frame or in the link coordinate frames. Lee and Chang [9] have shown that, when expressed in the base coordinate system, the NE equations are in a homogeneous linear recursive form (see Table 1) which is more suitable for parallel processing on an SIMD computer. One possible approach to parallelly process these equations in Table 1 is to decompose them into some computational processes, each of which calculates the kinematic or dynamic variables for all the joints such as angular velocities, angular and linear accelerations, joint forces and moments. For example, equations (1-a) to (1-i) in Table 1 are treated as consisting of nine computational processes. Then each one of these computational processes is assigned to one processor and different processes are processed concurrently. This type of parallelism is the task-level parallelism. However, if we observe the structure of these equations more carefully, one can find that equations (1-a), (1-b), (1-c), (1-g), and (1-h) are all in HLR form, and solving a linear recursive equation on a uniprocessor computer has a time complexity of $O(n)$ [9]. This time complexity can be reduced to $O(\lceil \log_2 n \rceil)$ if the recursive doubling technique is applied to the n processors [9]. Moreover, due to the data precedency, some processor(s) will be idle for waiting the required data calculated by other processors. For example, not until the processor which is assigned to compute the angular velocities ω_i, $i = 1, 2, \cdots, n$, completes its work, no other processors can start initiating their computations. This means that the distribution of tasks among processors will lead to low processor utilization. So a better parallelization approach is to perform the job-level parallelism. That is, n processors collaborate to compute each equation in Table 1. Hence, final results of each equation are produced at the same time and no data precedency problem needs to be worried about as long as these equations are processed in order. The above parallelization is considered at the algorithmic level; that is, at the large grain granularity. Moreover, since the procedures to calculate the variables with the same kinematic or dynamic meaning but for different links, e.g., ω_i, $i = 1, 2, \cdots, n$, are identical, the concurrency can be achieved at each basic operation (i.e., each matrix-vector operation). This means that the NE equations of motion possess the uniformity of operations. So, as far as the degree of parallelism is concerned, these NE equations can be processed with parallelism at the medium grain granularity. Furthermore, parallelism at the fine grain granularity is also possible because the fundamental operations of the NE equations are some basic matrix-vector operations with fixed data size (e.g., 3×3 matrix or 3×1 vector), so a special-purpose processor can be designed to perform these operations parallelly.

Considering their data dependencies, a number of the NE equations are in homogeneous linear recursive form and so possess the so called dependency of HLR type. A rough observation indicates that the most important communication requirement is permutation and no broadcast capability is necessary because there

Table 1. Newton-Euler Equations of Motion Referenced to the Base Coordinate System.

Forward (or Outward) Equations: $i = 1, 2, \cdots, n$

$$\omega_i = \omega_{i-1} + \dot{q}_i z_{i-1} (1 - \lambda_i) \tag{1-a}$$

$$\dot{\omega}_i = \dot{\omega}_{i-1} + (\ddot{q}_i z_{i-1} + \omega_{i-1} \times \dot{q}_i z_{i-1})(1 - \lambda_i) \tag{1-b}$$

$$\dot{v}_i = \dot{\omega}_i \times p_i^* + \omega_i \times (\omega_i \times p_i^*) + \dot{v}_{i-1} + (2\omega_i \times \dot{q}_i z_{i-1} + \ddot{q}_i z_{i-1})\lambda_i \tag{1-c}$$

$$a_i = \dot{\omega}_i \times s_i + \omega_i \times (\omega_i \times s_i) + \dot{v}_i \tag{1-d}$$

Backward (or Inward) Equations: $i = n, n-1, \cdots, 1$

$$F_i = m_i a_i \tag{1-e}$$

$$N_i = I_i \dot{\omega}_i + \omega_i \times (I_i \omega_i) \tag{1-f}$$

$$f_i = F_i + f_{i+1} \tag{1-g}$$

$$n_i = n_{i+1} + p_i^* \times f_{i+1} + (p_i^* + s_i) \times F_i + N_i \tag{1-h}$$

$$\tau_i = (n_i^T z_{i-1})(1 - \lambda_i) + (f_i^T z_{i-1})\lambda_i + b_i \dot{q}_i \tag{1-i}$$

where all the variables are referenced to the base coordinate frame, and the initial conditions are:

$\omega_0 = \dot{\omega}_0 = 0.$

$\dot{v}_0 = g = (g_x, g_y, g_z)^T$ to include gravity, and $|g| = 0.98062 \, m/s^2$.

$f_{n+1} = f_e$ = external force f_e exerting on link n.

$n_{n+1} = n_e$ = external moment n_e exerting on link n.

is no constant or variable which is common to all the manipulator links. Moreover, the required permutation capabilities are the one which can provide the recursive doubling technique to solve HLR equations and the one which connects a processor to its corresponding memory module (straight connection) or to its neighboring processor and corresponding memory module (nearest neighborhood connection) because some joint variables relate to the variables of its own joints or its adjacent joint. For example, in Table 1, $\mathbf{a}_i$ in equation (1-d) is a function of $\boldsymbol{\omega}_i$, $\dot{\boldsymbol{\omega}}_i$, $\mathbf{s}_i$, and $\dot{\mathbf{v}}_i$, all from its own link i; $\dot{\boldsymbol{\omega}}_i$ in equation (1-b) is a function of $\boldsymbol{\omega}_{i-1}$, $\mathbf{z}_{i-1}$, $\ddot{q}_i$, $\boldsymbol{\omega}_{i-1}$, and $\dot{q}_i$, among which $\mathbf{z}_{i-1}$ and $\boldsymbol{\omega}_{i-1}$ are variables of link $(i-1)$, and $\boldsymbol{\omega}_{i-1}$ forms the linear recursive relation. So the required permutation types are regular. The characteristics of the inverse dynamics for computing the NE equations of motion are concluded in Table 4.

2.2. Forward Dynamics Problem

The forward dynamics problem can be stated as: Given an input force/torque vector $\boldsymbol{\tau}(t)$ and a vector of external forces/torques exerted on the last link of the manipulator $\mathbf{k}(t)$, compute the joint acceleration vector $\ddot{\mathbf{q}}(t)$, based on an appropriate manipulator dynamic model, from values of $\boldsymbol{\tau}(t)$, $\mathbf{k}(t)$, the joint position $\mathbf{q}(t)$, and the joint velocity $\dot{\mathbf{q}}(t)$. The resultant $\ddot{\mathbf{q}}(t)$ is then integrated to give new values of $\dot{\mathbf{q}}(t)$ and $\mathbf{q}(t)$, and the process is repeated for the next input force/torque vector. Thus, the forward dynamics problem is essential for the real-time dynamic simulation of robot arm motion.

Among various methods for solving this problem [14-16], the composite rigid-body method [14] (see Table 2), based on the computation of the NE equations of motion, is widely used to develop efficient parallel algorithms [16-18]. The composite rigid-body method is suitable for parallel processing because efficient parallel algorithms for the inverse dynamics computation have been well developed and can be used to speed up the computation time. In fact, most of the forward dynamics parallel algorithms are all based on the composite rigid-body method and are much alike [16-18]. Essentially, they all utilize the recursive doubling technique [19,9] to reformulate the serial algorithm of the composite rigid-body method into the linear recursive equation form to reduce the order of computation. Another advantage of using the composite rigid-body method is that the inertia matrix and the joint acceleration vector can be obtained at the same time from the same computation. This is of great advantage because various model-based control schemes utilize the inertia matrix [20].

The structure of this algorithm in Table 2 is very similar to that of the inverse dynamics. The equations in Table 2 posses uniformity of operations. Equations (2-a) to (2-f) are applied to different joint j. That means the variables

Table 2. The Composite Rigid-Body Method for Forward Dynamics
Computation.

/* Compute the inertia matrix $H(q)$ */

$$M_j = M_{j+1} + m_j \tag{2-a}$$

$$c_j = \frac{1}{M_j}\{m_j(s_j + p_j^*) + M_{j+1}(c_{j+1} + p_j^*)\} \tag{2-b}$$

$$E_j = E_{j+1} + M_j[(c_{j+1} + p_j^* - c_j)^T \cdot (c_{j+1} + p_j^* - c_j)I_{3\times3} \tag{2-c}$$
$$\quad - (c_{j+1} + p_j^* - c_j)(c_{j+1} + p_j^* - c_j)^T] + I_j$$
$$\quad + m_j[(s_j + p_j^* - c_j)^T \cdot (s_j + p_j^* - c_j)I_{3\times3}$$
$$\quad - (s_j + p_j^* - c_j)(s_j + p_j^* - c_j)^T]$$

where $1 \le j \le n - 1$.

$$F_j = z_{j-1} \times (M_j c_j) \lambda_j + M_j z_{j-1} (1 - \lambda_j) \tag{2-d}$$

$$N_j = E_j z_{j-1} \lambda_j \tag{2-e}$$

$$f_{jj} = F_j, \quad n_{jj} = N_j + c_j \times F_j, \quad 1 \le j \le n \tag{2-f}$$

$$\begin{aligned} f_{i,j} &= f_{(i+1),j} \\ n_{i,j} &= n_{i+1,j} + p_i^* \times f_{(i+1),j} \end{aligned} \quad 1 \le i \le j-1, \ 2 \le j \le n \tag{2-g}$$

$$h_{ij} = z_{i-1}^T n_{i,j} \lambda_j + z_{i-1}^T f_{i,j} (1 - \lambda_j) \tag{2-h}$$

where $1 \le i \le j$, $1 \le j \le n$.

/* Compute the bias torque vector **b** */

$$b = C(q, \dot{q})\dot{q}(t) + G(q) \tag{2-i}$$

/* Solve the system of linear equations */

$$H(q)\ddot{q}(t) = \tau(t) - b \tag{2-j}$$

where the initial conditions are: $M_n = m_n$, $c_n = s_n + p_n^*$, and $E_n = I_n$.

with different index j which corresponds to the jth joint, e.g., c_1, c_2, $\cdots$, c_{n-1}, are obtained through identical computation with distinct operands. Equation (2-i) is solved via the inverse dynamics algorithm which has been shown to be highly operational uniform. Equations (2-g) and (2-h) are equations with two indices i and j, so they can be viewed as "two-dimensional" problems. They also posses uniform computation because for each j, $1 \le j \le n$, all n_{ij}'s (or h_{ij}'s), $1 \le i \le j$, are obtained through identical computation. We identify this uniformity because parallel architectures can be designed to make use of this operational uniformity. Finally, equation (2-j) in Table 2 is a system of linear equations which is a very common computational problem and has also shown its regularity [21,22].

As the type of parallelism is concerned, it is natural to perform the job-level parallelism based on the above discussion. Equations (2-a) and (2-c) in Table 2 are in HLR form. Equation (2-b) is in HHLR form. Equation (2-g) can be considered as a set of homogeneous linear recurrence equations (SHLR) [16]. Equation (2-i) is actually an inverse dynamics problem which can be solved by the NE equations in Table 1. Each of these equations is suitable to be implemented on $O(n)$ or $O(n^2)$ array processors [16-18].

The fundamental operations of the forward dynamics problem are basic matrix-vector computations and a number of scalar additions, subtractions, multiplications, and reciprocal especially for the linear system solver. The degree of parallelism is considered at the large grain granularity in most of the past research. However, the forward dynamics equations can also be parallelized at the medium or fine grain granularities like the inverse dynamics problem.

For the data dependency, they are more complex than the inverse dynamics problem due to some special types of data dependency. Besides the HLR type, there are the HHLR type, the set of homogeneous linear recursive equations type, and the linear system solver (LSS) type dependencies. Correspondingly, the communication requirements are also stricter. The major requirement is still the permutation capability. However, in order to implement the linear system solver, the broadcast and one-to-one connections are inevitable. The characteristics of the forward dynamics problem is also summarized in Table 4.

2.3. Forward Kinematics Problem

The forward kinematics problem can be stated as: Given the n measured joint variables (q_1, q_2, $\cdots$, q_n), where $q_i \equiv \theta_i$ (joint angle) for a revolute joint or $q_i \equiv d_i$ (joint displacement) for a prismatic joint, find the position and orientation of the manipulator end-effector in the Cartesian space. Using the Denavit-Hartenberg matrix representation for establishing the link coordinate frames, the solution to this problem is the successive multiplication of the 4×4 homogeneous link transformation matrices for an n-link manipulator

$$\mathbf{T} = \mathbf{A}_0^1 \mathbf{A}_1^2 \mathbf{A}_2^3 \cdots \mathbf{A}_{i-1}^i \cdots \mathbf{A}_{n-1}^n \tag{1}$$

where $\mathbf{A}_{i-1}^i$ is the D-H link transformation matrix which relates the ith coordinate frame to the $(i-1)$th coordinate frame [23,24]. The above successive matrix multiplication equation can be reformulated in a homogeneous linear recursive form

$$\mathbf{T}_0^1 = \mathbf{A}_0^1$$
$$\mathbf{T}_0^i = \mathbf{T}_0^{i-1} \mathbf{A}_{i-1}^i \quad \text{for } i = 2, \cdots, n , \tag{2}$$

from which the configuration of all the coordinate frames can be obtained at the time lower bound [9]. Considering the type of parallelism for the above equations, it is natural to perform the job-level parallelism. That is, apply the recursive doubling technique to n linear array processors to solve the homogeneous linear recursive equations at the time complexity of $O(\lceil \log_2 n \rceil)$. The fundamental operations of this problem are 4×4 matrix multiplication, and more importantly, the sine and cosine functions (trigonometric function calculations). To speed up these fundamental operations, most of the past research on this problem considered finer grain parallelism, including the medium grain granularity [25] and the fine grain granularity [26] by designing a single chip processor with high vector computational speed and efficient trigonometric function generator. The required data dependency is only the HLR type, and the communication requirement is the permutation capability to provide the recursive doubling interconnection. The characteristics of the forward kinematics problem is tabulated in Table 4.

2.4. Forward Jacobian Problem

The Jacobian matrix relates the joint velocities of the manipulator in the joint-variable space to the linear and angular velocities of the manipulator end-effector in the Cartesian space

$$\dot{\mathbf{x}}(t) \triangleq \begin{bmatrix} \mathbf{v}(t) \\ \boldsymbol{\omega}(t) \end{bmatrix} = \mathbf{J}(\mathbf{q})\dot{\mathbf{q}}(t) \tag{3}$$

where $\mathbf{v}(t)$ and $\boldsymbol{\omega}(t)$ are, respectively, the linear and angular velocities of the manipulator end-effector, $\dot{\mathbf{q}}(t)$ is the n-dimensional joint velocity vector of the manipulator, and $\mathbf{J}(\mathbf{q})$ is the 6×n manipulator Jacobian matrix and it is a function of the joint variables. Furthermore, the transpose of the Jacobian matrix relates static contact forces/moments to the set of joint torques as

$$\boldsymbol{\tau}(t) = \mathbf{J}^T(\mathbf{q})\mathbf{F}(t) \tag{4}$$

where $\mathbf{F}(t) \triangleq (F_x, F_y, F_z, M_x, M_y, M_z)^T$ is a 6-dimensional static force/moment vector, $\tau(t)$ is an n-dimensional joint torque vector, and the superscript "T" denotes transpose operation on vectors/matrices.

Existing methods in computing the Jacobian are mostly confined to uniprocessor computers. In particular, Orin/Schrader [27], and Yeung/Lee [28] exploited the linear recurrence characteristics of the Jacobian equations. These methods differed from each other only by a different selection of the reference coordinate frame for computation. The reference coordinate frame is selected such that all the vectors and matrices and the Jacobian computed are referred to that reference coordinate system. They all have the computational order of $O(n)$ for an n-jointed manipulator.

Here we shall adopt the Generalized-k algorithm (see Table 3) developed by Yeung and Lee [28]. Based on the linear recurrence property, this algorithm computes the Jacobian at any given desired reference coordinate frame k in the order of $O(n)$. The equations in Table 3 show high uniformity of operations again. In particular, equations (3-a)-(3-d) are all in homogeneous linear recursive form, among which equations (3-a) and (3-b) are the forward recursive equations, and equations (3-c) and (3-d) are the backward recursive equations. In fact, the forward and backward recursive equations can be processed together. For example, equations (3-a) and (3-c) can be computed simultaneously to produce all the matrices ${}^k\mathbf{R}_i$, $i = 1, \cdots, n$, and so do equations (3-b) and (3-d). Although the forward and backward recursive equations can be processed together, this will require some irregular data dependencies [28]. We call this special type as *forward and backward homogeneous linear recursive* (FBHLR) type data dependency. This requires the communication requirement to include irregular permutation and broadcast. Since the equations to obtain Jacobian matrix are so simple, most of the past research focused on speeding up the computation at finer grain granularities as in the forward kinematics problem. The characteristics of the forward Jacobian problem can be found in Table 4.

2.5. Inverse Jacobian Problem

The inverse Jacobian problem can be stated as: Given the linear and angular velocity vector of the manipulator end-effector in the Cartesian space, find the corresponding joint rates. The most direct method of obtaining the inverse Jacobian is to calculate the Jacobian matrix and then invert it. Unfortunately, the Jacobian matrix becomes singular at the robot's deadpoints and the inverse of the Jacobian is not defined because the Jacobian matrix is not of full rank. Moreover, since the Jacobian is a $6 \times n$ matrix, its inverse does not exist when n is not equal to six. For these unpleasant cases, the concept of generalized inverse has been applied [29]. The inverse Jacobian algorithms for a general manipulator can be

Table 3. Forward Jacobian Computation

Given a desired reference coordinate frame k, $0 \le k \le n$, this algorithm computes the Jacobian matrix with respect to the desired reference coordinate frame k for an n-jointed manipulator.

J1. [Forward Recurrence.]

for $i = k+1$ step 1 to n, do

$$^k\mathbf{R}_i = {}^k\mathbf{R}_{i-1} \, {}^{i-1}\mathbf{R}_i \qquad (3\text{-a})$$

$$^k\mathbf{p}_i = {}^k\mathbf{p}_{i-1} - {}^k\mathbf{R}_i \, {}^i\mathbf{p}_i^* \qquad (3\text{-b})$$

end {for}

J2. [Backward Recurrence.]

for $i = k-1$ step -1 to 1, do

$$^k\mathbf{R}_i = {}^k\mathbf{R}_{i+1} \, {}^i\mathbf{R}_{i+1}^T \qquad (3\text{-c})$$

$$^k\mathbf{p}_i = {}^k\mathbf{p}_{i+1} + {}^k\mathbf{R}_{i+1} \, {}^{i+1}\mathbf{p}_{i+1}^* \qquad (3\text{-d})$$

end {for}

J3. [Jacobian Computation.]

for $i = 1$ step 1 to n, do

$$^k\boldsymbol{\beta}_i = (1 - \lambda_i) \, {}^k\mathbf{R}_i \, \mathbf{z}_0 \qquad (3\text{-e})$$

$$^k\boldsymbol{\mu}_i = (1 - \lambda_i)({}^k\boldsymbol{\beta}_i \times (-{}^k\mathbf{p}_i)) + \lambda_i({}^k\mathbf{R}_i \, \mathbf{z}_0) \qquad (3\text{-f})$$

end {for}

where initially $\mathbf{z}_0 = (0, 0, 1)^T$, $^k\mathbf{R}_k = \mathbf{I}_{3\times3}$, $^k\mathbf{p}_k = \mathbf{0}_{3\times1}$.

Table 4. Characteristics of Basic Robotics Algorithms.

CHARACTERISTICS

Algorithms	Type of Parallelism	Degree of Parallelism	Uniformity of Operations	Fundamental Operations	Data Dependency	Communication Requirement
Inverse Dynamics	Job level	Large grain	Yes	Matrix-Vector	HLR	(Regular) Permutation
Forward Dynamics	Job level	Large grain	Yes	Scalar ops. Reciprocal Matrix-Vector	HLR,IHLR SHLR, PNE System of Linear Eqs.	one-to-one Permutation Broadcast
Forward Kinematics	Job level	Medium or Fine grain	Yes	Matrix Mult. Trigonometric	HLR	(Regular) Permutation
Forward Jacobian	Job level	Medium or Fine grain	Yes	Matrix-Vector	HLR (Forward & Backward)	(Irregular) Permutation Broadcast
Inverse Jacobian (Direct)	Job level	Medium or Fine grain	Yes	Scalar ops. Reciprocal Vector ops.	Global	Permutation Broadcast
Inverse Jacobian (iterative)	Job level	Medium or Fine grain	Yes	Scalar ops. Reciprocal Vector ops.	Local	Permutation Broadcast
Inverse Kinematics (Direct)	Task level	Fine grain	No	Scalar ops. Reciprocal Square root Trigonometric	Global	one-to-one Broadcast
Inverse Kinematics (iterative)	Job level	Medium or Fine grain	Yes	Scalar ops. Reciprocal Matrix-Vector Trigonometric	Local	one-to-one Permutation Broadcast

divided into two categories. One is to calculate the inverse or the generalized inverse Jacobian explicitly [30]. The other is to consider the inverse Jacobian problem as a system of linear equations and solve the joint rate from the Cartesian velocity implicitly [31]. For practical purposes, the latter approach is easier to be parallelized due to the use of some standard techniques to solve a system of linear equations such as the Gaussian elimination method.

The solution of a system of linear equations is the most common computational problem in linear algebra [21,22] and there has been a long tradition of research on parallel algorithms for solving various types of systems of linear equations whether the matrix is square or rectangular. The techniques to solve the system of linear algebraic equations are all rather regular and homogeneous and can map quite naturally onto many types of multiprocessor architecture, especially one- or two-dimensional arrays of processors. Generally speaking, there are two approaches for solving linear systems [21,22]. The first approach is the direct method which usually includes two processes: factorization of the matrix into a triangular matrix and solving the triangular systems of equations. This approach possesses global data dependency because each successive column, row, or submatrix of the matrix factors depends on all the preceding columns, rows, or submatrices, and the broadcast communication capability is often required. The second approach is the iterative method which has the advantage of local data dependency and is also amenable for parallel processing. However, in general, the iterative method cannot be faster than the direct method.

The common fundamental operations include scalar adds, subtracts, multiplication, vector operations (e.g., inner products), and reciprocal. Parallelism can be performed at the fine grain granularity with subtasks of complexity $O(1)$ as on systolic and wavefront processor arrays [30,32], or at the medium grain granularity with subtasks of complex $O(n)$ as on a linear array of processors [21]. The characteristics of the inverse Jacobian problem is summarized in Table 4.

2.6. Inverse Kinematics Problem

The inverse kinematics problem can be stated as: Given a desired position and orientation of the manipulator end-effector and the geometric link parameters with respect to a reference coordinate system, find the joint angles so that the manipulator can reach the desired prescribed manipulator hand position and orientation.

In general, the inverse kinematic position solution can be obtained by various techniques [24], among which the inverse transform [33] and the iterative method [34,35] are widely discussed. The inverse transform technique yields a set of explicit, non-iterative joint angle equations which involve multiplications, additions, square root, and transcendental function operations. The iterative methods can obtain robot independent joint solution, but they usually have some disadvantages: more computations than the closed-form solution, variable computation time and, more important, convergence problem, especially in the singular and degenerate cases. We shall examine the characteristics of the inverse transform technique and the iterative methods.

The equations for closed-form solution appear highly non-uniform [36] since each joint angle is obtained through very distinct computations which contain a large set of elementary operations including scalar addition, multiplication, reciprocal, square root, trigonometric and inverse trigonometric functions (sine, cosine, and arctangent). The data dependencies among these equations are irregular and global, and the required communication capabilities are one-to-one and broadcast. These irregular and hetero-homogeneous equations are obviously not suitable for the job-level parallelism. The only possible way to parallelly process these equations is to perform a functional decomposition at the fine grain granularity. Unfortunately, this decomposition shows only a limited amount of parallelism with a large amount of sequentialism in the flow of computation [36].

To achieve higher parallelism for the inverse kinematics problem, the iterative method provides a better approach, since nearly every presented iterative method contains the computations of forward kinematics, forward Jacobian, and inverse Jacobian [34,35], which have been shown to be highly parallelized. Tsai and Orin [35] presented a parallel iterative algorithm for the inverse kinematics problem (see Fig. 1). The basic algorithm of their method is based on the integration of the joint velocity and can be expressed by the following equations:

$$\dot{\mathbf{q}}(t) = \mathbf{J}^{-1}(\mathbf{q})\dot{\mathbf{x}}(t) \tag{5}$$

$$\mathbf{q}(t) = \int \dot{\mathbf{q}}(\tau)d\tau \tag{6}$$

where the variables are defined as before. The inputs to this algorithm are the values of the desired manipulator end-effector position and velocity $\mathbf{x}_d$ and $\dot{\mathbf{x}}_d$, respectively, and the outputs are the joint variables q_i's, $i = 1, 2, \cdots, n$. Details about this iterative method can be found in [35]. This iterative method is amenable for parallel processing at the job-level parallelism due to its step by step nature, although some steps can only be processed serially; e.g., convergence testing. As the general property of iterative methods, the data dependency tends to be local; however, convergence testing necessarily involves gathering global information and this requires global communication capability. The characteristics of the inverse kinematics problem is tabulated in Table 4.

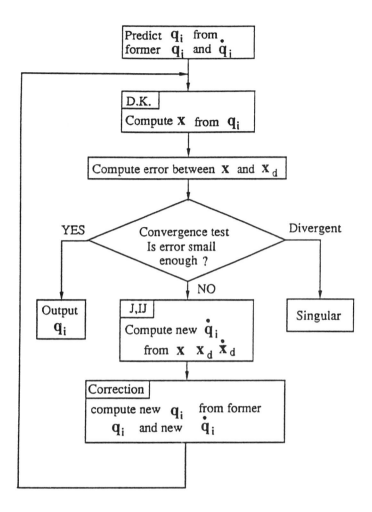

NOTE : x_d $\dot{x}_d$ are the desired end effector position and
velocity respectively. x is the predicted end effector
position. q_i is joint positions. $\dot{q}_i$ is joint rate. DK mean
direct kinematics algorithm. J means forward Jacobian algorithm.
IJ means inverse Jacobian algorithm.

Figure 1. Iterative Method for Inverse Kinematics Solution.

2.7. Common Features Examination and Parallelization of Basic Robotics Algorithms

If we consider these six basic robotics computations as a set of tasks that we need to compute for the control of robot manipulators, then we need to find their common features and characteristics so that a parallel architecture can be designed to efficiently compute these tasks. The characteristics of the six basic robotics algorithms are tabulated in Table 4 and it shows that these algorithms do possess some important common features and characteristics. This is especially true for the inverse dynamics, the forward dynamics, the forward kinematics, and the forward Jacobian computations for the following three reasons. First, they are all suitable to be parallelized at the job-level and the parallelization can be performed at the large, medium, and fine grain granularities simultaneously, although different granularities are emphasized in each individual algorithm. Second, their operations are all uniform for the variables corresponding to each joint, and the most important fundamental operation is the matrix-vector operation. Finally, the strongest common feature is that they are all in homogeneous linear recursive form, for which the recursive doubling technique can be applied to achieve the time lower bound of $O(\lceil \log_2 n \rceil)$. The communication requirement indicates that one-to-one and some regular or irregular permutation capabilities are required for these four computational problems and the broadcast capability is necessary for the forward dynamics and the forward Jacobian algorithms. This indicates that some efficient, versatile network is required in the parallel architecture for their computations.

The inverse Jacobian and the inverse kinematics computations may seem less common to the above four algorithms. However, if less efficient methods to solve these two problems are chosen individually, then these two algorithms may possess some common features to the other four algorithms, and a common parallel architecture can be designed to match all these common characteristics for their computations. From previous discussions, we found that either the direct method or the iterative method for the inverse Jacobian is a proper candidate for parallel processing, while the direct method is more efficient with somewhat complex data dependencies. For the inverse kinematics problem, only the iterative method possesses regular properties similar to the other four computations.

With all the characteristics listed in Table 4, we shall next examine how to reformulate and parallelize these robotics algorithms from their original serial algorithms by complying to their common features. The parallelization process is performed at the job level; that is, we try to express the original algorithms as a sequence of serial steps (jobs). Each individual step is accomplished through the cooperation of all the processors and for each step, the operations of each processor are almost identical by using one of their common features: the uniformity of operations. Hence, each step can be considered to be a single instruction in a serial program. Two different steps (or jobs) are identified after the parallelization process: *single* steps and *macro* steps. The notion of "single instruction" and "subroutine" of a serial program can be used to distinguish between these two steps. A single step corresponds to a single instruction in a serial program, while a macro step corresponds to a subroutine in a serial program. The macro steps require more complex parallel computations for all the processors, for example, the homogeneous linear recursive equation, the hetero-homogeneous linear recursive equation, and the system of linear equations are all macro steps. These macro steps are identified by their completeness and repeatity. The completeness means that the step can be treated as an individual problem. The technique to process these macro steps parallelly needs special consideration and the algorithm to solve these steps is so well-structured that finer decomposition is not helpful or even impossible, for example, the parallel recursive doubling technique for solving the homogeneous linear recursive equation, or the parallel Cholesky factorization technique for solving the system linear equations with a symmetric-positive-definite square matrix. The repeatity means that the problem which can be solved in the step is so important and common that it appears repetitively at many other places; for example, many equations of robotics algorithms are in homogeneous linear recursive form, then the procedure for parallelly solving this problem can be applied to all these places. The method to parallelize each of these macro steps is designed separately. In the following sections, a suitable architecture is designed and analyzed for the computation of these parallel algorithms according to their common features.

3. Design of Reconfigurable DN-SIMD Machine

This section describes an appropriate parallel architecture with the attributes that best match the common features of the six basic robotic parallel algorithms. The important parallel architecture attributes include the type of machine (e.g., SIMD or MIMD mode), number of processors, synchronization requirement, processor capabilities, memory organization, and network requirement. Each of these attributes is affected by one or more features of the six basic robotic algorithms discussed in Section 2. With all these requirements and attributes, the appropriate parallel architecture is a *reconfigurable, dual-network, SIMD* (DN-SIMD) machine for the computation of robotic algorithms.

Because the six basic robotics algorithms are amenable to be parallelized at the job level and all the subtasks possess highly operational uniformity at any degree of granularity, the SIMD mode of operation is a natural choice for our machine. Since the computations are uniform for each manipulator's joint, the number of processors corresponds to the number of manipulator joints, n, arranged in a linear array. Furthermore, since the fundamental operations of these robotic algorithms are mostly matrix-vector operations, we shall design our parallel architecture as a medium-grained machine. Further functional decomposition at the operation level will be performed on each of the single steps of the proposed parallel algorithms and their computations on the proposed parallel machine will be synchronized at each basic operation unit. Finer degree of parallelism can be considered within each basic operation unit to achieve arithmetic-level synchronization. This parallelism requires the detailed design of the processing element (PE), including the synchronization scheme within it. This is beyond the scope of this paper. From Table 4, the required computational capabilities for each PE are scalar addition, multiplication, reciprocal, standard matrix-vector operations, and sine and cosine functions.

Another significant attribute of parallel architectures is the organization of memory. For the job-level parallelism, the data or operands for each processor can be arranged regularly based on distinct manipulator joints, it is reasonable to design n separate memory modules such that one memory module corresponds to one processor. Moreover, since we propose a medium grain machine, the data transfer cycle is based on one basic operation unit. Therefore, the communication frequency is rather high. This requires us to design n global memory modules which are connected to n processors through an interconnection network. Furthermore, due to the neighborhood operands relationship which occurs quite often in the robotics algorithms, another interconnection network which connects a PE to another PE is designed to provide the paths for internal forwarding data (IFD) exchange (i.e., internal direct feedback). This provides the opportunity for the current results to be used immediately without passing through the global memory for data exchange; that is, it avoids unnecessary communication delays.

From the communication requirements of the six robotics algorithms, we need to have reconfigurable interconnection networks which provide a variety of communication patterns, including one-to-one, permutation, and broadcast capabilities. Thus, we require the proposed machine to have two unidirectional networks, one of which connects processors and memories and the other provides the paths for IFD exchange between processors. With all these requirements and attributes, the appropriate parallel architecture is a *reconfigurable, dual-network, SIMD* (DN-SIMD) machine for the computation of robotic algorithms.

The structure of the proposed DN-SIMD machine, as shown in Fig. 2, consists of multiple processing elements, two reconfigurable interconnection networks (RIN1 and RIN2), a set of global data registers (GDRs), three data buffers

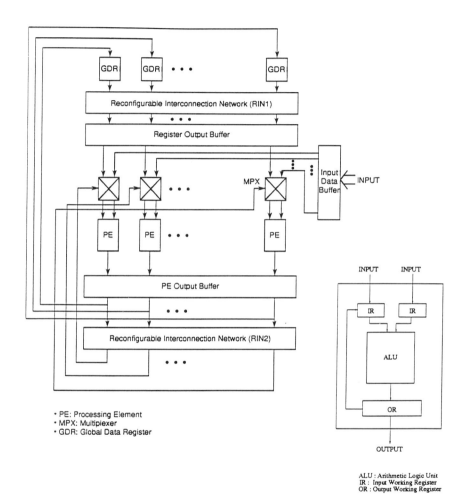

Figure 2. Structure of dual network (DN) SIMD machine.

including register output buffer, PE output buffer and input data buffer (IDB), and a set of multiplexers. All of these are coordinated by a central control unit (CU) which is not shown in Fig. 2. The functions of each element are briefly described here.

1. *Processing Element* (PE). There are n identical PEs. Each PE is essentially an arithmetic logic unit (ALU) with attached working registers (see Fig. 1). All the ALUs perform the same programmable function synchronously in a lock-step fashion under the command of the CU. Some of the PEs can be masked (disabled) for some computation period, while other unmasked or enabled PEs perform computations. Each PE has two input working registers (IWRs) which are used to store two operands for each computation, and one output working register (OWR) which is used to store the current result of each computation. The operands in the IWRs are kept there until they are replaced. Thus, they can be used repetitively if one or two operands are common for a series of continuous computations. An inner loop connection within a PE is designed, which connects the OWR to one of the two IWRs. This provides an immediate inner-PE forwarding path such that the current result can be used as an operand for the next computation immediately.

2. *Global data registers* (GDRs). There are n groups of data registers which correspond to the n global memory modules. In each computation period, the registers with the same relative position in each group can be accessed under the control of the CU. The result of each computation from each PE will be stored in the GDRs only when either the result is the final output or the result will be used in later computations but not the immediate following one, which can make use of the internal forwarding path for data exchange among PEs or inner loop within PEs.

3. *Reconfigurable interconnection networks.* There are two sets of identical interconnection networks: RIN1 and RIN2. The most suitable interconnection network for our special communication requirements will be designed in the next section. The RIN1 connects the GDRs to the PEs. This provides the paths for sending required operands to the appropriate PEs. The RIN2 makes the connection from the outputs of PEs to the inputs of PEs; this provides the direct paths for internal forwarding data exchange among PEs. It should be noted that, if necessary, the output of PE i can be stored into its corresponding memory module i. This is not affected by the RIN2.

4. *Data buffers.* There are three sets of data buffers. The register output buffer allows the "current computation" and the "RIN1 reconfiguration and operand fetch for the next computation" be processed at the same time. The PE output buffer allows the "current computation" and the "RIN2

reconfiguration and output data storing" be processed simultaneously. The input data buffer (IDB) is the buffer for operands directly from external input data.

5. *Multiplexer.* The n multiplexers in advance of the n PEs are used to select proper operands to enter PEs from three possible sources: GDR, IDB, and IFD exchange. They are also under the control of the control unit.

With the functions of these elements described above, the basic mathematical operations performed by each PE of the DN-SIMD machine involve at most two operands,

$$T = A \circ B \tag{7}$$

where A and B are two arbitrary operands and they can be scalar, vector or matrix, and "$\circ$" indicates the operation performed by the PE. When either A or B is null, the computation only involves one operand such as the transpose of a matrix. The operands A or B may come from five different sources. They are GDRs through RIN1, IFD exchange through RIN2, IDB, IWR within PE, and OWR within PE through inner loop connection. The result T may be sent to two possible destinations: GDRs directly, or PEs through RIN2 via IFD exchange. We assume that the time to transfer one operand from the GDR or the IDB to a PE (i.e., operand fetching) is the same as the time to transfer the output result from a PE to the GDR (i.e., result storing) and equals to the computation time of one basic PE operation. This time interval is called a *cycle*. Since operands fetching, computation, and result storing can be performed simultaneously due to the data buffers designed in this system, a three-stage pipelined operation can be performed on our DN-SIMD machine. Since a computation usually needs two operands A and B, and if A and B come from different sources, then they can be transferred to a PE simultaneously in one period. In this case, the three-stage pipelined operation proceeds normally. However, if A and B come from the same source (e.g., GDR or IDB), then it will take 2 cycles to transfer them. This situation is called the *double transmission required* (DTR) computation. In this case, a delay period must be added to the pipeline operation to synchronize the operation. This DTR computation obviously will slow down the system speed. Hence, we need to minimize the number of DTR computations in a computational task.

4. Design of Reconfigurable Interconnection Network

Since a suitable interconnection network is a key issue to achieve high efficiency in the design of the proposed DN-SIMD machine [37], some important design criteria should be considered for the interconnection network. The first is its communication capability. This should be examined based on three basic forms of connection: one-to-one, permutation, and broadcast connections. The other criteria are the cost and the fault tolerance of a network. Finally, we need to develop a proper control scheme for a network, especially for a reconfigurable network. We shall choose a most suitable interconnection network for our special communication requirements. Its properties are discussed and its control scheme is developed. For simplicity of system control, the two networks (RIN1 and RIN2) in our DN-SIMD machine are designed to be the same.

4.1. Choice of Interconnection Network

To design or choose a proper interconnection network for our DN-SIMD machine, the application domain should be well understood. This information can be acquired from the results of mapping parallel algorithms onto the DN-SIMD machine [2]. According to the mapping results, the required interconnection types for the basic robotic computations are concluded in Fig. 3. They include *uniform module shifts, broadcast,* and some *mixed-type* connections.

Definition 1 (Uniform module shift): An input node j is connected to an output node $(j + d) \bmod N$ for all j, $0 \le j < N$, given a positive or negative integer distance d. Here N is the number of input nodes (or output nodes)

It should be noted that the connection types which the robotic computations require are not "real" uniform module shifts. The major difference is that no *end wrap around* is necessary in our application; that is, the dash lines in Fig. 3(a) are unnecessary.

All the three basic forms of network connection including one-to-one, permutation, and broadcast, are required in our application, especially the permutation capability is our major concern. Moreover, through the discussion of the structure of our DN-SIMD machine in Section 3, we hope that each data transfer can be completed in exactly one pass (one cycle) through the network, otherwise extra cycle delays will have to be added into the system, resulting in longer computation time. Here, the cost is less concerned, since in the robotic computation domain, the number of PEs, N, which is equal to the number of manipulator's joints/links, is usually not a large number, says $N \le 12$. Based on the above facts, we found that the Generalized Cube network, which can provide our required permutation and broadcast connections in one pass, is a good candidate [38]. This Generalized Cube network is closely studied in the following subsections.

4.2. Network Properties

The Generalized Cube network is a multistage cube-type network topology as shown in Fig. 4. This network has N input ports and N output ports, where $N = 2^k$, for some positive integer k. The network ports are numbered from 0 to

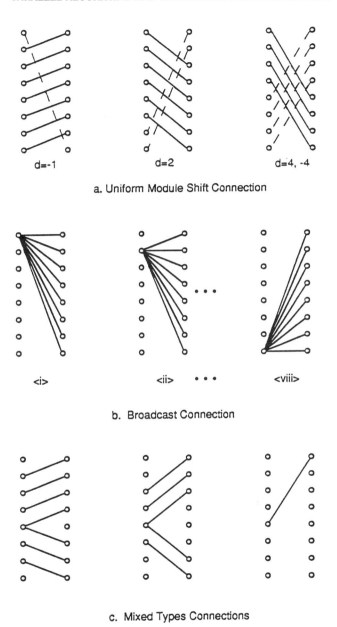

a. Uniform Module Shift Connection

b. Broadcast Connection

c. Mixed Types Connections

Figure 3. Required interconnection types for robotic algorithms.

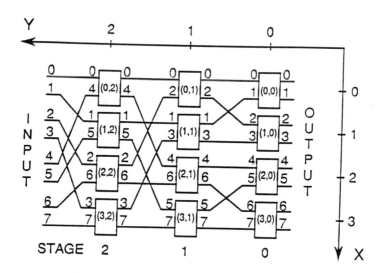

(a) Coordinate Definition of Generalized Cube Network

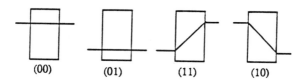

(b) Link Types of Interchange Box

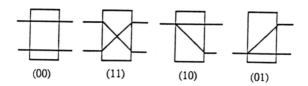

(c) Operating Modes of Interchange Box

(The binary number below each box is the control code defined for it)

Figure 4.

Generalized cube network and link types and operating modes of interchange box.

$N-1$. In Fig. 4(a), N is equal to 8. The Generalized Cube topology has $m = \log_2 N$ stages, where each stage consists of a set of N links connected to $N/2$ interchange boxes. Here, the four-function switch boxes are used. This box provides four connection steps as shown in Fig. 4(b). The labels of the links entering the upper and lower inputs of an interchange box are used as the labels for the upper and lower outputs, respectively (see Fig. 4(a)). The topology connections in the Generalized Cube network can be defined as follows.

Definition 2 (Connections of Generalized Cube): Let $P = P_{m-1} \cdots P_1 P_0$ be the binary representation of an arbitrary link label. Then stage i of the Generalized Cube network topology contains the $cube_i$ interconnection function; that is, it pairs links P and $cube_i(P)$ for an arbitrary P, $0 \le P < N$, where

$$cube_i(P_{m-1} \cdots P_1 P_0) = P_{m-1} \cdots P_{i+1} \overline{P_i} P_{i-1} \cdots P_1 P_0,$$
$0 \le i < m$, $0 \le P < N$.

So at the ith stage, $0 \le i \le m-1$, link P and link $cube_i(P)$ will enter the same interchange box from the upper and lower input ports respectively, where $cube_i(P)$ is a label that differs from the label P in just the ith bit position. The *exchange* setting at some switch box at stage i will implement the $cube_i$ function in mapping the switch box input labels to the box outputs. Thus, data items inputed to that interchange box are transferred as specified by the $cube_i$ interconnection function. When the state of the switch box is set to *straight*, data items inputed are transferred according to the *identity* function, where $identity(P_{m-1} \cdots P_0) = P_{m-1} \cdots P_0$. Since each interchange box is individually controlled, each stage i may perform the $cube_i$ interconnection function on some subset of the data items depending on the settings of the interchange boxes. Stage i is the only stage which can map a source to a destination with an address different from the source address in the ith bit position.

An interconnection network in an SIMD environment can be described as a set of interconnection functions, where each is a permutation (bijection) on the set of PE address, or network input/output port labels. For example, when interconnection function f is applied, network input S is connected to output $f(S) = D$ for all S, $0 \le S < N$, simultaneously. This concept will be used in the following discussion about the communication properties of the Generalized Cube. The Generalized Cube has been proved to possess some important properties about one-to-one and broadcast connections [38].

Property 1 (one-to-one connection): The Generalized Cube network can perform any one-to-one connection, and the connection is unique [38].

Property 2 (Broadcast): The Generalized Cube network provides the connections that any source can broadcast to any set of destinations [38].

The following lemma will be useful in the subsequent proofs.

Lemma 1: In the Generalized Cube network, the stage i input and output links used in the path from S to D is $d_{m-1} \cdots d_{i+1} s_i s_{i-1} \cdots s_0$ and $d_{m-1} \cdots d_i s_{i-1} \cdots s_0$, respectively.

Proof: For an arbitrary source $S = s_{m-1} \cdots s_1 s_0$ and any desired destination $D = d_{m-1} \cdots d_1 d_0$, this Lemma can be proved by showing that the Generalized Cube network can always provide an interconnection function f such that $f(S) = D$. From the definition of the Generalized Cube network, we observe that the two input-output links to an interchange box in stage i differ only in the ith bit position. Therefore, when an interchange box at stage i is set to *exchange*, it is implementing the *cube$_i$* interconnection function. So if the stage i box in the path from S to D is set to *exchange* when $s_i \neq d_i$, then the interconnection function defined by this setting can be expressed as

$$f = f_0 f_1 \cdots f_i \cdots f_{m-1}$$

where
$$f_i = \begin{cases} cube_i , & \text{if } s_i \neq d_i , \quad 0 \leq i \leq m-1 \\ \text{identitiy function} , & \text{if } s_i = d_i . \end{cases}$$

So
$$f(S) = f_0 f_1 \cdots f_i \cdots f_{m-2} (f_{m-1}(s_{m-1} \cdots s_1 s_0))$$

$$= f_0 f_1 \cdots f_i \cdots (f_{m-2}(d_{m-1} s_{m-2} \cdots s_1 s_0))$$

$$= \cdots$$

$$= f_0(d_{m-1} d_{m-2} \cdots d_1 s_0) = d_{m-1} d_{m-2} \cdots d_1 d_0$$

$$= D .$$

By observing the f function, the required results can be obtained. ∎

Next, we need to consider the permutation capability, which is the most important communication capability for our robotic computations. Notice that not all permutations are performable (e.g., shuffle connection [38]). Conflict will occur if we want to use the same link to transfer two different data at the same time. Hence, it is important for us to check if our necessary permutations can be performed on the Generalized Cube network.

Theorem 1 (Uniform module shift): The Generalized Cube network can perform any uniform module shift.

Proof: We want to show that for any two desired sources $S_1 , S_2 , (0 \leq S_1 , S_2 \leq N-1 , S_1 \neq S_2)$, the path from S_1 to $(S_1 + d) \bmod N$ and the path from S_2 to $(S_2 + d) \bmod N$ do not conflict. We shall prove this by contradiction.

Assume $S_1 = s^1_{m-1} s^1_{m-2} \cdots s^1_0$, $S_2 = s^2_{m-1} s^2_{m-2} \cdots s^2_0$, $d = d_{m-1} d_{m-2} \cdots s_0$

$$(S_1 + d) \bmod N = d_{m-1}^1 d_{m-2}^1 \cdots d_0^1, \quad (S_2 + d) \bmod N = d_{m-1}^2 d_{m-2}^2 \cdots d_0^2.$$

If there is a conflict at stage i, then it means that these two paths will use the same output link at stage i; that is, from Lemma 1, $d_{m-1}^1 \cdots d_i^1 s_{i-1}^1 \cdots s_0^1 = d_{m-1}^2 \cdots d_i^2 s_{i-1}^2 \cdots s_0^2$, then $s_{i-1}^1 \cdots s_0^1 = s_{i-1}^2 \cdots s_0^2$ and $d_{m-1}^1 \cdots d_i^1 = d_{m-1}^2 \cdots d_i^2$.
Since

$$d_{m-1}^1 \cdots d_i^1 \cdots d_0^1 = [(s_{m-1}^1 \cdots s_i^1 \cdots s_0^1) + (d_{m-1} \cdots d_i \cdots d_0)] \bmod N$$

and

$$d_{m-1}^2 \cdots d_i^2 \cdots d_0^2 = [(s_{m-1}^2 \cdots s_i^2 \cdots s_0^2) + (d_{m-1} \cdots d_i \cdots d_0)] \bmod N,$$

we have

$$d_{i-1}^1 \cdots d_0^1 = [(s_{i-1}^1 \cdots s_0^1) + (d_{i-1} \cdots d_0)] \bmod 2^i$$

$$= [(s_{i-1}^2 \cdots s_0^2) + (d_{i-1} \cdots d_0)] \bmod 2^i = d_{i-1}^2 \cdots d_0^2.$$

Since $d_{m-1}^1 \cdots d_i^1 = d_{m-1}^2 \cdots d_i^2$ is already known, we have $d_{m-1}^1 \cdots d_0^1 = d_{m-1}^2 \cdots d_0^2$. This results in $(S_1 + d) \bmod N = (S_2 + d) \bmod N$ and $S_1 = S_2$. This contradicts to our assumption that $S_1 \neq S_2$, and this completes our proof. ∎

Up to now, we know that uniform module shifts and broadcast which are necessary in the basic robotic computations can be performed by the Generalized Cube network. It seems that there is no easy way to prove the feasibility of any irregular permutation or any type of combination of broadcast and permutation (e.g., the mixed-type interconnections in Fig. 3(c)) on the Generalized Cube network. In the following subsection, we shall present an easier way to check the feasibility of any type of connection for a small parallel processing system (where N is small).

4.3. Network Switch Control Scheme

There are two kinds of switch control schemes for the MINs: the *centralized* and the *distributed* control schemes [38-40]. In our proposed DN-SIMD machine with a small N and for the efficient pipeline timing consideration, the centralized control scheme for the Generalized Cube network is a better choice [2]. We first define a coordinate (x, y) for each interchange box as shown in Fig. 4(a). Note that, for a system with N PEs, $0 \leq x \leq N/2$ and $0 \leq y \leq \log_2 N$. Based on this coordinate, a *switch control matrix* for each desired communication type is defined. Each entry of this matrix will decide the operating mode (state) of the corresponding interchange box.

Definition 3 (Switch Control Matrix): For a system with N PEs, a $\log_2 N \times \dfrac{N}{2}$ switch control matrix C is used to decide the operating mode of each interchange box for a desired interconnection type. The (i,j)th element of matrix $\mathbf{C}$, $C[i,j]$, decides the operating mode of the interchange box (i,j). The possible values of the element of switch control matrix $\mathbf{C}$ and their corresponding operating modes are defined as follows (see Fig. 4(b)):

$$C[i,j] = \begin{cases} 0, (00)_2, & \text{straight;} \\ 3, (11)_2, & \text{exchange;} \\ 2, (10)_2, & \text{upper–broadcast;} \\ 1, (01)_2, & \text{lower–broadcast;} \\ 4, (100)_2, & \text{no control signal needed.} \end{cases}$$

Notice that the ith column corresponds to stage i interchange boxes. As an example, for a uniform module shift $(d = -1)$, the switch control matrix $\mathbf{C}$ is

$$\mathbf{C} = \begin{bmatrix} 3 & 3 & 3 \\ 0 & 0 & 3 \\ 0 & 3 & 3 \\ 0 & 0 & 3 \end{bmatrix}.$$

Obviously, it is desirable to develop an algorithm which automatically constructs this switch control matrix for a desired interconnection type. Using this algorithm, we can also check if a desired interconnection type is feasible. Let us first consider the one-to-one connection problem. The links through the interchange boxes on the path from a source S to a destination D will only have four possible types as shown in Fig. 4(c), and each type is labeled with a binary number as shown in the figure. The description of the control scheme to build up a path from S to D should include such information as: At each stage, which interchange box should be activated and which state should be set. For example, to build up a path from $S = (110)_2$ to $D = (011)_2$, we need the following control information: interchange box $(2,2)$ is set at state $(1,1)$, interchange box $(0,1)$ is set at state $(0,1)$, and interchange box $(1,0)$ is set at state $(1,0)$. The following theorem indicates how to find the correct interchange box and its proper mode of operation for each stage.

Theorem 2: Assume $S = s_{m-1} \cdots s_1 s_0$ be the source address (input-port number), $D = d_{m-1} \cdots d_1 d_0$ be the destination address (output-port number), and P is the path from S to D. Then at stage i $(i = m-1 \cdots 0)$, P will go through interchange box $(d_{m-1} \cdots d_{i+1} s_{i-1} \cdots s_0, i)$, and the link type for this stage is $(s_i \oplus d_i \mid s_i)$, where "$\mid$" means the "bit concatenation" operation (e.g., $1 \mid 0 = (10)_2$). Note that d_i or s_i is ignored when $i < 0$ or $i > m-1$ (e.g., if $i = m-1$, $d_{m-1} \cdots d_{i+1} s_{i-1} \cdots s_1 s_0 = s_{m-2} \cdots s_1 s_0$).

Proof: From Lemma 1, at stage i, the input link used in the path from S to D is labeled as $d_{m-1} \cdots d_{i+1} s_i s_{i-1} \cdots s_0$. From the topology of the Generalized Cube network and the coordinate that we defined above, we find that the link labeled as $d_{m-1} \cdots d_{i+1} s_i s_{i-1} \cdots s_0$ enters the interchange box with the coordinate (x, i) at stage i, here x is obtained from this label $d_{m-1} \cdots d_{i+1} s_i s_{i-1} \cdots s_0$ by taking the ith bit (s_i) off. For example, at stage $m-1$, the input link label is $s_{m-1} \cdots s_1 s_0$, so $x = s_{m-2} \cdots s_1 s_0$; at stage $m-2$, the input link label is $d_{m-1} s_{m-2} \cdots s_1 s_0$, so $x = d_{m-1} s_{m-3} \cdots s_1 s_0$; at stage 0, the input link label is $d_{m-1} d_{m-2} \cdots d_1 d_0$, so $x = d_{m-1} d_{m-2} \cdots d_1$.

We next need to show that the link type at stage i is $(s_i \oplus d_i \mid s_i)$. Since for the Generalized Cube network, only stage i can change the ith bit of the link label, if S and D differ in the ith bit position; that is, $s_i \oplus d_i = 1$, the link should change direction as the link type (11) or (10), and these two link types are both with bit1=1. Similarly, if S and D agree at the ith bit position; that is, $s_i \oplus d_i = 0$, the link needs not to change direction and goes straightly as the link type (00) or (01). These two link types are both with bit1=0. Another observation shows that if $s_i = 0$, then the link with label $d_{m-1} \cdots d_{i+1} s_i \cdots s_1 s_0$ at stage i enters the upper part of an interchange box as the link type (00) or (10), and these two types are both with bit0=0. Similarly, if $s_i = 1$, the link with label $d_{m-1} \cdots d_{i+1} s_i \cdots s_1 s_0$ will enter the lower part of an interchange box as the link type (01) or (11), and again both of these types are with bit0=1. ■

In the case of permutation, broadcast or any other kinds of multi-source, multi-destination connection, they are considered as the composition of individual one-to-one connection. So we need to consider any combination of the four link types. Two unfeasible interconnections must be noted. First, if the combination includes more than two different link types, then this desired interconnection is unfeasible. Second, if the combination includes two different link types with labels b_1 and b_2, and if $b_1 \oplus b_2 = (11)$, then this desired interconnection is also unfeasible. For example, $b_1 = (00)$ and $b_2 = (11)$ cannot occur at the same interchange box. All the other kinds of combination are feasible.

Before presenting the algorithm for constructing the switch control matrix, the input notation to this algorithm is defined. The desired interconnection type is represented by a set of two-tuple elements, $\{(S_i, D_i) \mid 1 \le i \le N\}$, where N is the number of sources (destinations). Here $S_i = s^i_{m-1} s^i_{m-2} \cdots s^i_0$ is a source, $D_i = d^i_{m-1} d^i_{m-2} \cdots d^i_0$ is a destination, and $m = (\log_2 N)$ is the number of stages. We call it the *set of connection pairs*. For example, the mixed-type connection (*i*) in Fig. 3(c) is represented as $\{(1,0)\ (2,1)\ (3,2)\ (4,3)\ (4,5)\ (5,6)\ (6,7)\}$.

Algorithm C-Building (*Construction of the Switch Control Matrix*).
Input: Set of connection pairs $\{(S_i, D_i)\}$.

Output: A $\log_2 N \times \dfrac{N}{2}$ switch control matrix $\mathbf{C}$.

C0.   [Initialization.] Initialize a $\log_2 N \times \dfrac{N}{2}$ temporary matrix $\mathbf{M}$ with null elements. Note that each element of the matrix $\mathbf{M}$ is a set of binary numbers.

C1.   [Find the control scheme for each one-to-one connection.]
         For each (S_i , D_i) **do**

            For $k = m-1$ **step** -1 **until** 0 **do**
               $M[d^i_{m-1} \cdots d^i_{k+1} s^i_{k-1} \cdots s^i_0 , k]$
                  $= M[d^i_{m-1} \cdots d^i_{k+1} s^i_{k-1} \cdots s^i_0 , k] \cup (s^i_k \oplus d^i_k \mid s^i_k);$
            End {For}
         End {For}

C2   [Feasibility Checking.]
         For each element $M[i,j]$ of $\mathbf{M}$ **do**

            If there are more than two different numbers in $M[i,j]$,
               then STOP (unfeasible);
               else if the two binary numbers $x , y \in M[i,j]$ such that $x \oplus y = (11)_2$,
                  then STOP (unfeasible).
         End {For}

C3   [Construct the switch control matrix $\mathbf{C}$ from $\mathbf{M}$.]
         For each element $M[i,j]$ of $\mathbf{M}$ do

            If $M[i,j] = \varnothing$,
               then $C[i,j] = 4$ (Don't work);
               else if $M[i,j]$ contains only one binary number $(b_1 b_0)$,
                  then $C[i,j] = b_1 b_0$;
               else if $M[i,j]$ contains two binary numbers $(b_1 b_0)$ and $(d_1 d_0)$,
                  then $C[i,j] = c_1 c_0$;
               where $c_1 = b_1 \cdot \bar{b}_0 + d_1 \cdot \bar{d}_0, c_0 = b_0 \cdot d_1 + b_1 \cdot d_0$
               ("·" and "+" indicate the AND and OR operations, respectively).
         End {For}

End. {C-Building}

   Let us illustrate the C-Building algorithm through an example of designing the switch control matrix for a mixed-type interconnection (i) (See Fig. 3(b)). The input to the algorithm is the set of connection pairs for the mixed-type interconnection (i), written as $\{(1,0)\ (2,1)\ (3,2)\ (4,3)\ (4,5)\ (5,6)\ (6,7)\}$.
For Steps C1 and C2:

$(1,0)$: $M[1,2] = (00)$ $M[1,1] = (00)$ $M[0,0] = (11)$

$(2,1)$: $M[2,2] = (00)$ $M[0,1] = (11)$ $M[0,0] = (11),(10)$

$(3,2)$: $M[3,2] = (00)$ $M[1,1] = (01),(00)$ $M[1,0] = (11)$

$(4,3)$: $M[0,2] = (11)$ $M[0,1] = (11),(10)$ $M[1,0] = (10),(11)$

$(4,5)$: $M[0,2] = (11),(01)$ $M[2,1] = (00)$ $M[2,0] = (10)$

$(5,6)$: $M[1,2] = (01),(00)$ $M[3,1] = (10)$ $M[3,0] = (11)$

$(6,7)$: $M[2,2] = (00),(01)$ $M[2,1] = (00),(01)$ $M[3,0] = (11),(10)$

$$\mathbf{M} = \begin{bmatrix} (11),(10) & (11),(10) & (11),(01) \\ (11),(10) & (00),(01) & (00),(01) \\ (10) & (00),(01) & (00),(01) \\ (11),(10) & 10) & (00) \end{bmatrix}$$ and is found feasible.

For Step C3: The switch control matrix $\mathbf{C}$ is

$$\mathbf{C} = \begin{bmatrix} (11) & (11) & (01) \\ (11) & (00) & (00) \\ (10) & (00) & (00) \\ (11) & (10) & (00) \end{bmatrix}.$$

Using the control scheme presented above, the network setting for our required interconnection types can be made [2]. This can also show the feasibility of implementing the required mixed-type connections on the Generalized Cube network.

5. Fault Tolerant Generalized Cube Network for DN-SIMD Machine

A fault tolerant interconnection network can provide reliable communication to some degree even when the system contains a faulty component or components. Among various approaches to achieve fault tolerance [41-43], redundancy which provides several alternate paths for any arbitrary input/output connection is a major skill in fault tolerant design of multistage interconnection networks (MINs). The possible techniques include adding extra links and extra ports [44-50].

In this section, to improve system fault tolerance, a *Fault Tolerant Generalized Cube Network* (FTGCN) is designed, based on the Generalized Cube Network, for our DN-SIMD machine by considering both possibilities of network component failure and PE failure (Fig. 5(a)). The fault tolerance of this design is analyzed under a well-defined fault model, and the revised switch control scheme for the FTGCN is presented.

In the FTGCN, two sets of switch boxes are added to the input and output sides of the Generalized Cube network. They are called *input extra stage* (IES)

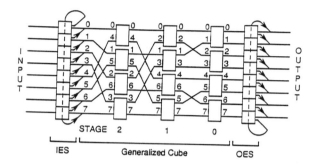

(a) Fault Tolerant Generalized Cube Network

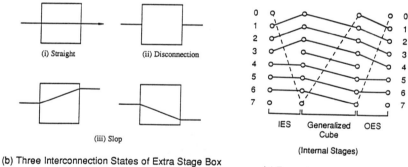

(b) Three Interconnection States of Extra Stage Box

(c) Example of Simplified Graph

Figure 5. Figures about fault-tolerant generalized cube network.

and *output extra stage* (OES). The new added switch box has one input node and two output nodes working at three operating modes: straight, slop, and disconnection (open) as shown in Fig. 5(b). The switch boxes at the input side (IES) make the original input node of the Generalized Cube network connected to its previous nearest neighbor node. Similarly, the switch boxes at the output side (OES) make the original output node of the Generalized Cube network connected to its next nearest neighbor node. The first (i.e., 0th) and the last (i.e., $N-1$th) nodes are seen to be next to each other.

For clarity, in the following discussion, the proposed FTGCN is represented as a simplified graph composed of three stages: IES, Generalized Cube (Internal Stages), and OES. An example is shown in Fig. 5(c).

5.1. Fault Model

The fault model characterizes all the faults that are assumed to occur in the system. The fault model considers both possibilities of network component failure and PE failure.

(i)   Any interchange box or link in the internal stages can fail. The interchange box fault is usually known as a stuck-at-0 or a stuck-at-1 fault. That is, an input terminal is fixed to connect to some output terminal permanently.

(ii)   The fault at the input extra stage or the output extra stage can only be a stuck-at-straight type. That is, at any time the input terminal can connect to the output terminal straightly. This can be achieved by a special by-pass circuit for each switch box at the IES or the OES.

(iii)   The input and output links of the FTGC network can fail. This failure means that some PE will disconnect with the network, so this situation is considered as a PE fault.

(iv)   Any PE can fail; that is, any PE may lose its computation capability.

(v)   Fault occurs independently, and only a single fault occurs at a time.

(vi)   All PEs, interchange boxes, and links have identical reliability.

5.2. General Properties and Switch Control Scheme

Some general properties are shown in the following theorems. These properties are analyzed based on the communication capabilities of one-to-one, permutation, and broadcast.

Theorem 3: Any interconnection type performable by the Generalized Cube is performable by the Fault-Tolerant Generalized Cube network.

Proof: If all the switches at the IES and the OES are set to the straight operating mode, then it is actually a Generalized Cube. ■

Next, we consider the single link fault tolerant capability of the FTGC network. In the following discussion, the switch boxes at the OES are set straight. So the output links at stage 0 of the Generalized Cube (internal stages) are the output links of the FTGC network directly and no fault is assumed at these links.

Theorem 4 (one-to-one connection): In the FTGC network with a single fault, there exists at least one fault-free path between any source and destination.

Proof: If the fault occurs at the IES, a source S can connect to the stage $m-1$ input link S according to our fault model. From Theorem 3, a fault-free path from S to D exists. Consider the case that the fault occurs at the internal link or box. Suppose the connection from a source S to a destination D is required. Here $S = s_{m-1} \cdots s_1 s_0$ and $D = d_{m-1} \cdots d_1 d_0$. The source S can connect to the $(m-1)$th stage input link S or S', where $S' = (S-1) \bmod N = s'_{m-1} \cdots s'_1 s'_0$. It is noticed that the least-order bits of these two input link labels are different. That is, $s'_0 \neq s_0$ or $s'_0 = \bar{s}_0$. Because stage $m-1$ to stage 0 forms a Generalized Cube network, for each input, S or S', there is a path to the destination D. From Lemma 1, the stage i output link used in the path from S to D is $d_{m-1} \cdots d_i s_{i-1} \cdots s_0$ and in the path from S' to D is $d_{m-1} \cdots d_i s'_{i-1} \cdots s'_0$. So these two links must be different for $i = m-1$ to 1. Moreover, the stage i input links used in the paths from S to D and S' to D are $d_{m-1} \cdots d_{i+1} s_i s_{i-1} \cdots s_0$ and $d_{m-1} \cdots d_{i+1} s'_i s'_{i-1} \cdots s'_0$ respectively. So, at least, they are different in the least-order bit. Because no box in stage $m-1$ through 1 has input link labels that differ in the least-order bit position, these two paths enter different interchange boxes at stage $m-1$ through 1. If the interchange box at stage 0 fails, then according to our fault model, at least one input link at stage 0, upper input link or lower input link, will still be functional. Because at stage 0, these two paths use input links $d_{m-1} d_{m-2} \cdots d_1 s_0$ and $d_{m-1} d_{m-2} \cdots d_1 \bar{s}_0$, they enter the same interchange box at stage 0 by the upper input link and the lower input link respectively. And no single fault can disconnect both the paths simultaneously. We have shown that there are two disjoint paths from S to D. If any single fault disconnect one path, the other path must be fault free. ■

Using the similar techniques, we can prove following theorems [2]:

Theorem 5 (Broadcast Connection): In the FTGC network with a single fault, there exists at least one fault-free broadcast path for any broadcast performable by the Generalized Cube.

Theorem 6 (Permutation): In the FTGC network with one fault, all the Generalized Cube performable permutations can be performed in at most two passes.

To develop the switches control scheme for the FTGC network, two more control arrays are defined to describe the proper connection mode for each switch at the IES and the OES. Notice that the switches at the IES and the OES are numbered 0 to $N-1$ from top to bottom.

Definition 4 (Extra Stage Control Array): For a system with N PEs, two N-element extra stage control (ESC) arrays IC and OC are defined to decide the operating mode of each switch of the input extra stage and the output extra stage for some desired interconnection type, respectively. The ith element, $0 \le i \le N-1$, of the ESC array IC (OC) decides the operating mode of switch i at the IES (OES). The possible values of the elements of the ESC array IC or OC and their corresponding operating modes are defined as follows:

$$IC[i] \text{ or } OC[i] = \begin{cases} 0, & \text{disconnect;} \\ 1, & \text{straight;} \\ 2, & \text{slop.} \end{cases}$$

The extra stage control arrays IC and OC and switch control matrix **C** are combined to control the overall FTGC network. Using the C-Building algorithm and from Theorems 3-6, the procedure to generate appropriate control matrix **C** and the ESC arrays, IC and OC, under the single-fault situation can be easily developed. Initially, all the elements of the array IC are set to be zero (0), and because we do not need to control the OES at this time, $OC[i] = 1, i = 0 \cdots N-1$.

(1) Normal case [No fault]:

All the switches of the IES and the OES are set to straight or bypassed, and the FTGC network functions like the Generalized Cube. So $IC[i] = 1, i = 0 \cdots N-1$, and no changes in $C[i,j]$ (meaning that the matrix **C** is the same as that decided by the switch control scheme when the Generalized Cube is considered).

(2) Single fault case:

The fault on the IES is nothing different from the normal case. So we ignore this situation on the following discussion.

(a) One-to-One Connection. Suppose (S, D) is the desired connection pair. Then if (S, D) is unfeasible; that is, the fault lies on the path from S to D,

then $IC[S] = 2, \mathbf{C} = \mathbf{C\text{-}Building}(\{((S-1) \bmod N, D)\});$

else $IC[S] = 1$, $\mathbf{C} = \mathbf{C\text{-}Building}(\{(S, D)\})$.

(b)   Broadcast Connection. Suppose $\{(S,\ D^i)\,|\,0 \leq i \leq N-1\}$ is the desired connection pair set. Then if one of $\{(S, D^i)\}$ is unfeasible, then $IC[S] = 2$, $\mathbf{C} = \mathbf{C\text{-}Building}(\{((S-1) \bmod N, D^i)\})$; else $IC[S] = 1$, $\mathbf{C} = \mathbf{C\text{-}Building}(\{(S, D^i)\})$.

(c)   Permutation. Suppose $\{(S^i, D^i)\,|\,0 \leq i \leq N-1\}$ is the desired connection pair set, then suppose (S^j, D^j) is an unfeasible connection:
Pass1: $IC[k] = 1$, $k = 0 \cdots N-1$, $k \neq S^j$,
$\mathbf{C} = \mathbf{C\text{-}Building}(\{(S^i, D^i)\,|\,0 \leq i \leq N-1,\ i \neq \})$.
Pass2: $IC[S^j] = 2$, $\mathbf{C} = \mathbf{C\text{-}Building}(\{((S-1) \bmod N, D^i)\})$.

5.3. Special Properties on the Application of Robotics Computations and Switch Control Scheme

The FTGC network is used for the system in which the number of PEs, N, is a power of two. While in most cases, the number of manipulator joints is not a power of two. Assume it is $n \neq 2^k$, for any integer k, where N is the smallest integer number which is greater than n and is a number of power 2. In order to discuss the properties of the FTGC network under this situation, it is reasonable to assume that the joint number of the manipulator is $N-1$ in our system. So our system has a spare PE (or port) as a standby device. This is called the *Redundant Model*.

5.3.1. Single Link or Switch Box Fault

Last section reveals that the FTGC network has fault tolerance capability if a single fault is considered. Moreover, it only needs one pass for any one-to-one or broadcast connection under a single fault situation. However, it may need two passes to overcome a single fault if permutation is required. If all the PEs are active, that is, we have N sources and N destinations, then it cannot expect to perform a permutation with less than two passes under a single fault situation because a fault means that some path is disconnected but we need all the N paths simultaneously if one pass is required. This situation is solvable if the redundant model is considered because there is a spare path that can be used.

Suppose $\mathbf{C}$ is the switch control matrix for some uniform module shift permutation in a normal system (no fault) with the Generalized Cube network. If the same permutation is required to be performed in our redundant system with the FTGC network, then the switch control matrix $\mathbf{C}$ remains unchanged, and the extra stage control arrays are set to

$$IC[i] = OC[i] = 1, \quad i = 0 \cdots N-2$$

$$IC[N-1] = OC[N-1] = 0$$

assuming that this permutation is $\{(j, (j+d) \bmod N) \mid 0 \leq j \leq N-1, d \in Z\}$. Now, if some single fault occurs in the redundant system, which results in the connection $(k, (k+d) \bmod N)$ unfeasible for some $k \in [0, N-2]$, the following steps present general rules for modifying the switch control matrix $\mathbf{C}$ and the ESC arrays IC and OC to make the required permutation be performable in one pass. Again, if an extra stage switch is faulty, this system still works as in the normal case and no modification is needed.

Step 1. Let $l = (k+d) \bmod N$.

Step 2. Set $OC[l] = 0$.

Step 3. Set $IC[i] = 2$, for $i = k$ down to 0.

Step 4. Set $OC[i] = 2$, for $i = (l-1) \bmod N$ down to $(l-k-1) \bmod N$.

Step 5. If $OC[N-2] = 2$, then set $OC[N-2] = 0$.

Step 6. The matrix $\mathbf{C}$ remains unchanged.

Let us explain and provide reasons for performing the above steps. For Step 2, because there is a fault on the path from stage $m-1$ input node k to stage 0 output node l, this path is abandoned. So we cut the stage 0 output link l, and the stage $m-1$ input link k will be cut in the following steps. Thus, this faulty path is isolated from the network. After Steps 3 and 4, each source will connect to its desired destination again. Source k will connect to stage $m-1$ input node $(k-1) \bmod N$ through the IES and then it will be connected to stage 0 output node $([(k-1) \bmod N]+d) \bmod N$.   Since $([(k-1) \bmod N]+d) \bmod N = ([(k+d) \bmod N]-1) \bmod N = (l-1) \bmod N$, it will connect to stage 0 output node $([(l-1) \bmod N]+1) \bmod N$ through   the   OES.   Again $([(l-1) \bmod N]+1) \bmod N = l \bmod N = l$. So input node k is connected to output node l as desired. Similar reason can be applied to source $k-1$, $k-2$, $\cdots$ 0, if $k > 0$. The source $S > k$ connects to stage $m-1$ input node S and then connects to stage 0 output node $(S+d) \bmod N$. Because $S > k$, $(S+d) \bmod N$ is not in the range $[(l-1) \bmod N$ down to $(l-k-1) \bmod N]$, the stage 0 output node $(S+d) \bmod N$ will connect to $(S+d) \bmod N$ straightly through the OES. So again, a source $S > k$ connects to its desired output node $(S+d) \bmod N$. In Step 5, setting $OC[N-2] = 2$ means that the stage 0 output node $N-2$ is connected to output node $N-1$ which connects to PE $(N-1)$. But PE $(N-1)$ is not used in the redundant system, this connection is not needed and should be cut. Some examples are shown in Fig. 6(a).

For the mixed-type connection, similar scheme can be used. This is demonstrated with some examples in Fig. 7. In these cases, the switch control matrix $\mathbf{C}$ needs to be changed. This can be accomplished by utilizing the algorithm C-

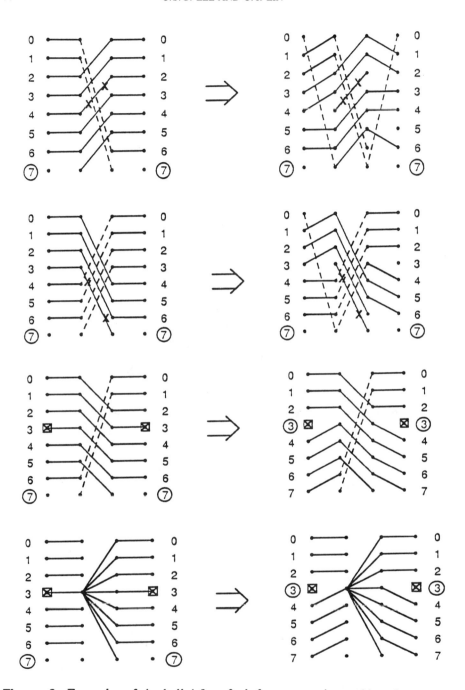

Figure 6. Examples of single link/box fault for permutation and broadcast-type connections.

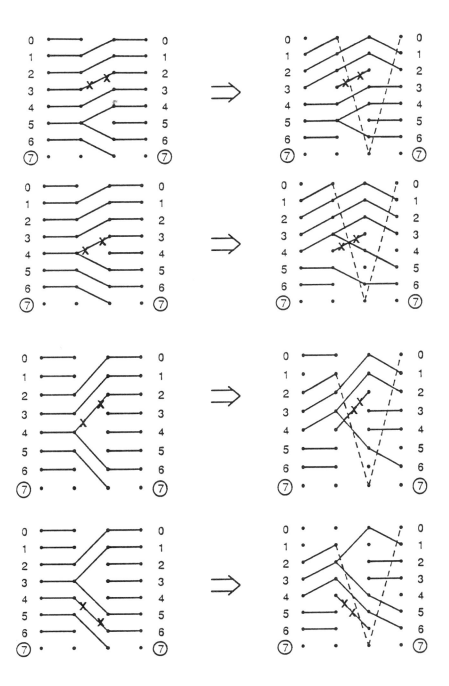

Figure 7. Examples of single link/box fault for mixed-type connection.

Building.

5.3.2. PE Fault

Besides the single link or switch box failure, we could have some PE fails in our DN-SIMD machine. When a PE fails, it means that the PE loses its computation capability temporarily or permanently. At this time, to make the system perform correctly, the task originally assigned to this faulty PE should be transferred to another PE. In the redundant model, it is natural to replace the faulty PE by the spare PE to maintain the same efficiency.

Another similar situation to be considered here is that the PE does not lose its computation capability, but its link to or from the interconnection network fails. Then this PE is isolated from the system and cannot be used either. Under this situation, this PE is considered as a faulty PE and should be replaced by the spare PE also.

The above two faulty situations raise the question on how to replace a faulty PE without changing the required interconnection types of the network and without making some required permutation unfeasible. It is shown that the replacement of the PE can be done in the FTGC network by just changing the switch modes at the IES and the OES, while the connection modes of interchange boxes at internal stages (i.e., the Generalized Cube) remained unchanged. The main idea is to make the active PEs next to each other. This is illustrated and demonstrated by some examples in Fig. 6(b).

If a PE fault occurs in the redundant system, the following steps present general rules for modifying the ESC arrays IC and OC to make the required permutation feasible in one pass.

Step 1.   In the normal situation (no fault), set $IC[i] = OC[i] = 1$, $0 \le i \le (N-1)$, and $IC[N-1] = OC[N-1] = 0$.

Step 2.   If a PE k fails, $0 \le k < N-1$, then execute Steps 3 through 6.

Step 3.   Set $IC[k] = 0$.

Step 4.   Set $IC[i] = 2$ for $i = k+1$ to $N-1$.

Step 5.   Set $OC[i] = 2$ for $i = k+1$ to $N-1$.

Step 6.   The matrix **C** remains unchanged.

The active PEs are PE0, 1, ... , $k-1$, $k+1$, ... , N, if $k \ne 0$, or the active PEs are PE1, 2, ... , N, if $k = 0$.

The correctness of these general rules is obvious. We just replace PE$(k+1)$ as k', PE$(k+2)$ as $(k+1)'$ and so on until PE N as $(N-1)'$. Then, after this new setting of the IES and the OES, the system with PE0, 1,..., $k-1$, k', $(k+1)'$,..., $(N-1)'$

is exactly the same as the original system with active PE0, 1,..., $k-1$, k,..., $N-1$. So the switch control matrix remains unchanged.

6. Mapping of Parallel Robotic Algorithms onto the Dual-Network SIMD Machine

Since our DN-SIMD machine was designed to best match the common characteristics of the six basic robotics parallel algorithms, the scheduling of their computations in our system is more straightforward with less difficulties as compared with other general mapping problems [51,52]. Based on this characteristics matching, a systematic and efficient mapping procedure is developed to map the parallel robotic algorithms onto the proposed medium-grained DN-SIMD machine.

The proposed mapping procedure consists of three stages as shown in Fig. 8. In the first stage, each of the single steps of these parallel robotic algorithms is further decomposed to a set of "subtasks" and each subtask possesses the basic mathematical form of consisting two operands. On the other hand, each of the macro steps in these algorithms is viewed as a subtask and is not decomposed at this stage. The first stage results in a series of parallel subtasks. In the second stage, these subtasks are reordered to reduce the number of DTR operations through a *neighborhood scheduling* algorithm. The reordered subtasks will be mapped onto the DN-SIMD machine directly in the third stage. In the final stage, the actual implementation of the macro steps in the parallel algorithms on the DN-SIMD machine is performed.

6.1. Subtask Assignment

Since the proposed DN-SIMD is a medium-grained machine and is synchronized at each basic mathematical operation, each parallel algorithm must be decomposed into a series of subtasks. Each subtask is either in the basic mathematical form which involves at most two operands or in a well-defined macro step. Although this functional decomposition can be easily performed on the single steps, it is not the case for the macro steps, in which the data dependencies are so complex that the decomposition based on basic computational unit is not obviously feasible. So the macro step will be viewed as a single subtask in this stage. Consider the decomposition of the following equation

$$K = L \times (C + E) + G \times C . \tag{8}$$

Here we use three temporary variables, T_1, T_2, and T_3 to rewrite Eq. (8) into four simple equations in the basic mathematical form:

$$\begin{aligned} T_1 &= C + E, \\ T_2 &= L \times T_1, \\ T_3 &= G \times C, \\ K &= T_2 + T_3. \end{aligned} \tag{9}$$

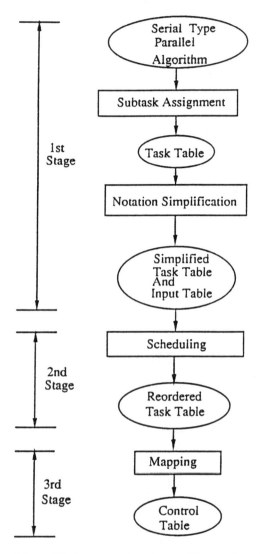

Note : Circle means the output of its previous block
and the input of its next block

Figure 8. A Systematic Mapping Procedure.

This same technique is applied to our decomposition process for single steps. Here, two sets of variables are introduced: T_i's represent the immediate results (temporary variables) or the final outputs. If T_i is a macro subtask, then it is specially denoted as $\overline{T}_i$. I_i's represent the external input variables; that is, the variables that do not come from the outputs of other computations.

To ease the subtask scheduling in the second stage, notation simplification is performed on the above task table to produce a *simplified task table*. In this table, two arrays are defined: TB[i] contains the identification of subtasks T_i's and OP[i] represents the corresponding operation for subtask TB[i]. Each element of OP[i] is either a macro subtask or in the form of $A \circ B$, where A and B may be T_i ($\overline{T}_i$) or I_i. Moreover, the superscript on A or B indicates the difference between the index i of the result, $T_{j_1}[i]$, and the index k or l of its operand $T_{j_2}[k]$ or $T_{j_3}[l]$, where $T_{j_1}[i] = T_{j_2}[k] \circ T_{j_3}[l]$. For example, the subtask $T_1[i] = T_2[i+2] \circ T_3[i-1]$ is denoted as $T_1 = T_2^2 \circ T_3^{-1}$. If their indices are equal, that is, $i = k$ or $i = l$, then the superscript is omitted. For example, the subtask $T_4[i] = T_5[i] \circ T_6[i]$ is denoted simply as $T_4 = T_5 \circ T_6$. The simplified task table is the final result of this stage and will be used as the input for the next stage.

6.2. Subtask Scheduling

To schedule the subtasks for computation, we first observe all the possible operand sources and their combinations for each computation. The operand may be one of the four possible types denoted as S_I, S_{OI}, S_T, and S_{OT} which correspond to four kinds of different sources. S_I denotes the operand from the IDB and it needs one period of transmission time. S_{OI} denotes the operand which is fetched by the previous computation (subtask) from the IDB and is still in the IWR within the PE, so no transmission is required for this operand. S_T denotes the operand from the GDR and this operand requires one cycle of transmission time via the network RIN1. S_{OT} denotes the operand from other sources including the following three possibilities: (i) The operand which is fetched by the previous computation from the GDR and is still in the IWR within the PE, so no transmission time is required; (ii) Current computation result through the inner loop; (iii) Current computation result through the internal forwarding path with data exchange provided by the network RIN2. The transmission time for the last two cases is ignored when compared to the system cycle time. Using these notations, all the possible combinations of operand sources including the situation of only one operand are listed below:

$$(S_I', S_I'') \qquad (S_I, S_I)$$
$$(S_I, S_{OI}) \qquad (S_{OI}, S_{OI})$$
$$(S_I, S_{OT}) \qquad (S_{OT}, S_{OT})$$
$$(S_I, S_T) \qquad (S_T, S_T)$$
$$(S_{OI}, S_{OT}) \qquad (S_I)$$
$$(S_{OI}, S_T) \qquad (S_{OI})$$
$$(S_{OT}, S_T) \qquad (S_T)$$
$$(S_T', S_T'') \qquad (S_{OT})$$

where the prime superscripts are used to distinguish different operands from the same kind of source. Among these situations, the combinations (S_I', S_I'') and (S_T', S_T'') are DTR operations and require two cycles to transmit two operands through the same transmission path. It is possible to eliminate DTR operations, if we reorder the processing sequence without violating the constraint of precedence relation. That is, in these two situations, one operand S_T' (or S_T'') can become the type S_{OT}, or S_I'(or S_I'') can become the type S_{OI}. Then, the DTR operation phenomena can be avoided and the unnecessary transmission can also be avoided for the efficient use of the same data repetitively and instantly.

A neighborhood scheduling algorithm for scheduling and reordering the execution of these subtasks to minimize the total number of DTR operations has been developed and is considered here.

Definition 1. For two subtasks in the kth and lth rows of the simplified task table, $TB[k]$ and $TB[l]$, assume $OP[k] = A \circ B$ and $OP[l] = C \circ D$, where A, B, C, and D are operands, each with one of these possible types: $\{I_j, T_j, T_j^i\}$. Then the subtask $TB[k]$ is called a *neighborhood* of $TB[l]$ if all the following conditions are satisfied:

(i) $k < l$,

(ii) $C = TB[k]$ or $C = TB^i[k]$ or $C = A$ or $C = B$ or
      $D = TB[k]$ or $D = TB^i[k]$ or $D = A$ or $D = B$.

From the above definition, we know that if subtask $TB[l]$ has a previous subtask $TB[k]$ as its neighborhood ($k < l$) and moreover, if these two subtasks are next to each other; i.e., $l = k + 1$, then at least one operand of subtask $TB[l]$ comes directly from the result or operand of subtask $TB[k]$ without accessing the GDR or the IDB. This obviously will save the communication time to access global memories, and the subtask $TB[l]$ will never be a DTR subtask, thus minimizing the number of DTR subtasks.

Definition 2. A subtask in the kth row of the simplified task table $TB[k]$ is called a *double transmission required* (DTR) subtask if the following two conditions are satisfied:

(i) Its operand is one of these types:

$OP[k] = TB[m] \circ TB[n]$ for some m, $n < k$ and $m \neq n$.

$OP[k] = TB^i[m] \circ TB[n]$ for some m, $n < k$ and $m \neq n$.

$OP[k] = TB[m] \circ TB^j[n]$ for some m, $n < k$ and $m \neq n$.

$OP[k] = TB^i[m] \circ TB^j[n]$ for some m, $n < k$ and $m \neq n$.

$OP[k] = I[m] \circ I[n]$ for $m \neq n$.

(ii) $k = 1$ or $TB[k-1]$ is not a neighborhood of $TB[k]$ for $k > 1$.

Notice that for $OP[k] = TB[m] \circ I[n]$ and $OP[k] = TB^i[m] \circ I[n]$, the subtask $OP[k]$ is not a DTR subtask because its two operands can be transmitted simultaneously through two different set of connection lines. Moreover, a subtask involves only one operand is obviously a non-DTR subtask.

From the above definition, whether a subtask is a DTR subtask depends on its "position" in the simplified task table. A DTR subtask can become a non-DTR subtask if it is moved to the place exactly behind its neighborhood. Since it is possible that the movement of a DTR subtask may introduce another new DTR subtask, this reordering process is desirable only when it complies with the precedence constraint of the original algorithm and the number of DTR subtasks in the reordered task table is less than that in the original table. This forms the *scheduling problem*; that is, to reorder the processing sequence of subtasks to reduce the number of DTR subtasks as far as possible without violating the precedence constraint of the original algorithm. This reordering process can be performed by the following efficient neighborhood scheduling algorithm.

Algorithm N-Scheduling (*Neighborhood Scheduling Algorithm*).
Input: Simplified Task Table with n rows (i.e., n subtasks).
Output: Reordered Task Table.

N1. [Main Loop] Check each subtask to see if it is a DTR subtask. If yes, try to change its position.
For $k = 1$ **step** 1 **until** n **do**

N2. [Check DTR]
Check if $TB[k]$ is a DTR subtask according to definition 2? If not, go to step N4.

N3. [Main Body] Try to change the position of a DTR subtask to make it into a non-DTR subtask.
If $OP[k] = (TB[m]$ or $TB^a[m]) \circ (TB[n]$ or $TB^b[n])$,

    then let $i \leftarrow max(m, n)$;
    else let $i \leftarrow 1$; {* $OP[k] = I[m] \circ I[n]$ *}
End {If}

While $i < k-1$ **do**

    If $TB[i]$ is a neighborhood of $TB[k]$, **then**

        If {$TB[i+1]$ is a DTR subtask} or {the insertion of $TB[k]$ between $TB[i]$ and

        $TB[i+1]$ will not make $TB[i+1]$ a DTR subtask},

            then insert $TB[k]$ behind $TB[i]$ to make $TB[k]$ the new $(i+1)$th subtask;

                go to step N4

        End {If}

    End {If}

    Let $i \leftarrow i+1$;

  End {While}

N4.   Continue {main loop}

       End {For}

END. {N-Scheduling}

The N-Scheduling algorithm checks each subtask from the first to the last one to see if there is any DTR subtask. If there is, then it tries to make it a non-DTR subtask by changing its position without increasing the number of DTR subtasks subject to the precedence constraint of the original algorithm. In step N2, a subroutine is called to check if a subtask $TB[k]$ is a DTR subtask. This is done according to definition 2. The third step is the main body of this algorithm. First, the feasible range to which the subtask $TB[k]$ can be moved is defined. This range is called the *feasible insertion range* of subtask $TB[k]$. If the two operands of $TB[k]$ are both from the previous subtasks $TB[m]$ and $TB[n]$, then under the precedence constraint, the subtask $TB[k]$ can only be inserted after $TB[m]$ and $TB[n]$. So the starting position of the feasible insertion range is behind the $\max(m, n)$th subtask. If the operands of subtask $TB[k]$ are both from the input buffer; that is, $TB[k] = I[m] \circ I[n]$, then the subtask $TB[k]$ can be moved to any place before itself, since there is no precedence constraint in this case. So the starting position of its feasible insertion range is 1. The ending position of the feasible insertion range in both cases are the position before the subtask $TB[k]$ itself. After deciding the feasible insertion range for the subtask $TB[k]$, we then consider the possibility of inserting $TB[k]$ behind its neighborhood in this range. If $TB[i]$ is a neighborhood of $TB[k]$ in this range, then $TB[k]$ is inserted between $TB[i]$ and $TB[i+1]$ only when either this insertion will not make $TB[i+1]$ a DTR subtask if it is not a DTR subtask originally, or $TB[i+1]$ is a DTR subtask originally. This constraint assures that a new DTR subtask will not be produced in this range after the movement of the subtask $TB[k]$. If the constraint is satisfied and the subtask $TB[k]$ is moved away its original position, the subtask behind it originally, $TB[k+1]$, may become a DTR subtask depending on its new preceding subtask $TB[k-1]$. Even in this worst case situation, the total number of DTR subtasks remains unchanged. Moreover, the new DTR subtask $TB[k+1]$ can possibly be made a non-DTR subtask on a later process of the N-Scheduling algorithm. If no such $TB[i]$ is found, the subtask $TB[k]$ remains unmoved. This shows that after each iteration, the number of DTR subtasks can only be reduced. Although the reordered subtasks obtained from the N-Scheduling algorithm are possibly not optimal, they perform quite efficiently in the mapping procedure. The time complexity of this reordering process is $O(n^2)$, where n is the number of subtasks in the simplified task table.

6.3. Mapping Procedure

The reordered task table produced by the N-Scheduling algorithm can be mapped onto the proposed DN-SIMD machine in a rather straightforward way because these subtasks are all single-step, simple subtasks. If the subtasks are macro steps, then their mapping requires further consideration. Our mapping procedure at this final stage consists of two phases. In the first phase, the subtasks including single steps and macro steps which are viewed as single steps temporarily are mapped onto the DN-SIMD machine in a row directly. The actual mapping of the macro steps is considered in the second phase. The output of the mapping procedure is a control table. This table consists of ten columns and indicates the exact movement of the central control unit. The first column represents the identification of subtasks appearing in processing order. It also represents the result of the corresponding subtask. The second column indicates the first operand; it may be T_j (T_j^i, $\overline{T}_j$, $\vec{T}_j^i$) or I_j for some i. The third column indicates the source of the first operand, and there are five possibilities: the GDR, the IDB, the IFD, the IWR and the OWR within the PE. The fourth column describes which network is used (RIN1 or RIN2) and the required connection type on it to transmit the first operand if necessary. Columns 5 to 7 contain the same information as the previous three columns, but for the second operand if it exists. Column 8 indicates the operation performed in this subtask. Column 9 indicates the destination of the result; it may be the GDR, the IFD, or both. If the IFD is needed, the connection type of network RIN2 is specified. Column 10 contains some comment on this subtask. For a macro subtask, these columns possess somewhat different meanings. Columns 2-7 indicate the corresponding information for the initial conditions of the macro subtask (similar to the parameters for a subroutine in a serial program). Columns 9-10 indicate the corresponding information for the final result of the macro subtask (similar to the return values of a subroutine in a serial program).

At the end of phase 1 of the mapping procedure, the control table is obtained. Since there may be macro subtasks in the control table, further mapping must be performed in phase 2. The mapping of HLR and HHLR macro subtasks are considered next for demonstration.

6.3.1. HLR Macro Subtask Mapping

The first-order homogeneous linear recurrence equation is defined as: Given $x(0) = a(0) =$ null, and $a(i)$, $1 \leq i \leq n$, find all the $x(i)$ for $1 \leq i \leq n$ from the following recursive equation

$$x(i) = x(i-1) \circ a(i). \tag{10}$$

An efficient technique called the recursive doubling technique has been found to solve this recursive equation efficiently on an SIMD machine [9]. Using this technique, the parallel algorithm to solve Eq. (10) and the mapping diagram of this algorithm onto the proposed DN-SIMD machine are shown in Fig. 9. This diagram possesses the same information as a control table including the sources of operands, destination of result, network used and required connection types for each iteration. It takes an order of $O(\lceil \log_2(n+1) \rceil)$ iterations to produce the final results. Also notice that, in Fig. 9, we assume that the initial conditions $a(i)$'s come from the IDB. In fact, they may also come from the GDR depending on whether $a(i)$'s are external input variables or not. In that case, its mapping diagram is exactly the same except that the $a(i)$'s are from the GDR through the network RIN1 at the beginning. Similarly, the final results $x(i)$'s can be stored in the GDR or directly fedback to PEs depending on the necessity of the next subtask.

6.3.2. HHLR Macro Subtask Mapping

The first-order hetero-homogeneous linear recurrence equation is defined as: Given $a(i)$, $b(i)$, $0 \leq i \leq n$, $a(i) \neq$ identity, $a(i)$, $b(i) \neq$ null, find all the $x(i)$ for $1 \leq i \leq n$ from the following recursive equation

$$x(i) = a(i) \times x(i-1) + b(i), \tag{11}$$

where $\times$ and $+$ are associative operators. The recursive doubling technique can also be used to solve this HHLR equation. The parallel algorithm for the HHLR equation has been presented by Lee and Chang [16], and is shown in Fig. 10 for convenience. This mapping of the parallel algorithm for the HHLR equation consists of two parts. In the first part, the scheme for the HLR macro subtask mapping is used to compute $X^{(k)}(i)$ for $k = 1$ to $s-1$, where $s \equiv \lceil \log_2(n+1) \rceil$, and store each iterative outcome $(X^{(1)}(i), X^{(2)}(i), \cdots, X^{(s-1)}(i))$ into the GDR, since all of these outcomes will be used in the second part of the mapping scheme. The

Algorithm FOHRA *(First-Order Homogeneous Recurrence Algorithm).*

F1. [Initialization] Given the terms a_i, $0 \leq i \leq n$, let $X^{(k)}(i)$ be the ith sequence at the kth splitting and $s = \lceil \log_2(n+1) \rceil$. Set the sequence at the initial step, $X^{(0)}(i) \leftarrow a_i$, $0 \leq i \leq n$.

F2. [Compute x_i parallelly]
for $k \leftarrow 1$ **to** s, **do**

$$X^{(k)}(i) = \begin{cases} X^{(k-1)}(i - 2^{k-1}) * X^{(k-1)}(i) & , \text{ if } 2^{k-1} \leq i \leq n \\ X^{(k-1)}(i) & , \text{ if } 0 \leq i < 2^{k-1} \end{cases}$$

end {for}

Set $x_i \leftarrow X^{(s)}(i)$, $1 \leq i \leq n$.

END FOHRA.

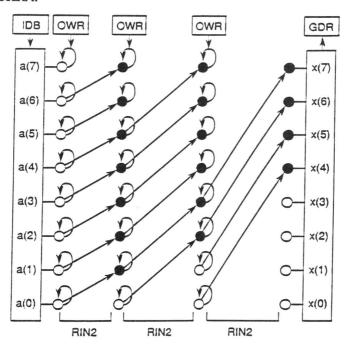

Figure 9.

Mapping Diagram of First-Order Homogeneous Linear Recurrence Equations on DN-SIMD Machine.

mapping diagram for this first part of the mapping is shown in Fig. 10. In the second part, the scheme for the HLR macro subtask mapping is modified to compute $Y^{(k)}(i)$ for $k = 1$ to s. The mapping diagram for this part are shown in Fig. 11. It takes a total of $2\lceil \log_2(n+1)\rceil - 1$ iterations to obtain the final results $Y^{(s)}(1)$, $Y^{(s)}(2)$, $\cdots$, $Y^{(s)}(n)$. Again, in Fig. 10, we assume that $a(i)$ and $b(i)$ are from the IDB. If they come from the GDR, we must then update proper sources and required network at the beginning of the mapping diagram. This change will introduce an extra delay cycle in Fig. 10 because at the beginning, $b(i)$, $X^{(0)}(i)$ are all from the GDR. Similarly, the destination of the final result, $Y(i)$'s can be either the GDR or the PEs according to the necessity of the next subtask.

7. Inverse Dynamics Computation on DN-SIMD Machine

Based on the mapping procedure in the last section, the inverse dynamics computation is used as an example to show the mapping process onto the proposed DN-SIMD machine. We assume that the manipulator has $n = 7$ joints. So, the proposed DN-SIMD machine has $N = 8$ processing elements (or processors).

The recursive NE equations of motion have been known for their computational efficiency and are parallelized on the job level as shown in Table 5. All the processors of the proposed DN-SIMD machine will cooperate to compute each individual equation in order. The equations (1-a),(1-b),(1-c),(1-g),(1-h) are all in homogeneous linear recursive form and are considered to be macro subtasks.

According to the systematic mapping procedure, subtask assignment is first performed on this algorithm. Since the proposed DN-SIMD is a medium-grained machine and is synchronized at each basic mathematical operation, each parallel algorithm must be decomposed into a series of subtasks. Each subtask is either in the basic mathematical form which involves at most two operands or in a well-defined macro subtask, in which the data dependencies are so complex that the decomposition based on basic computational unit is not obviously feasible. The decomposition result and the subtask assignment are shown in Table 6. The input table is also shown in this table. Here, two sets of variables are introduced. T_i's represent the immediate results (temporary variables) or the final outputs. If T_i is a macro subtask, then it is specially denoted as $\overline{T}_i$. I_i's represent the external input variables; that is, the variables that do not come from the outputs of other computations.

In Table 6, we find that there are five DTR subtasks (subtasks 17, 22, 36, 37, and 40). It is possible to eliminate DTR operations, if we reorder the processing sequence without violating the constraint of precedence relation. After the neighborhood scheduling, the reordered task table is produced as shown in Table 7, in

Algorithm FOHHRA *(First-Order Hetero-homogeneous Recurrence Algorithm).*

I1. [Initialization] Given a_i, b_i, $0 \le i \le n$, let $X^{(k)}(i)$, $Y^{(k)}(i)$ be the ith sequences at the kth level, and set $X^{(0)}(i) = a_i$, $Y^{(0)}(i) = b_i$, for $0 \le i \le n$, and $s = \lceil \log_2(n+1) \rceil$.

I2. [Compute x_i parallelly]
for $k \leftarrow 1$ to s, do

$$X^{(k)}(i) = \begin{cases} X^{(k-1)}(i-2^{k-1}) * X^{(k-1)}(i), & \text{if } 2^{k-1} \le i \le n \\ X^{(k-1)}(i), & \text{if } 0 \le i < 2^{k-1} \end{cases}$$

$$Y^{(k)}(i) = \begin{cases} X^{(k-1)}(i) * Y^{(k-1)}(i-2^{k-1}) + Y^{(k-1)}(i), & \text{if } 2^{k-1} \le i \le n \\ Y^{(k-1)}(i), & \text{if } 0 \le i < 2^{k-1} . \end{cases}$$

end {for}

Set $x_i \leftarrow Y^{(s)}(i)$, $1 \le i \le n$.

End FOHHRA.

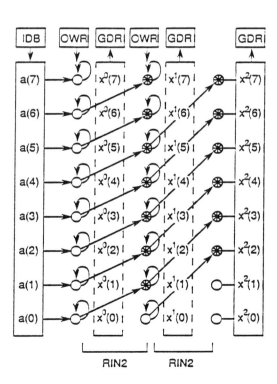

Figure 10. Mapping Diagram of the First Part Parallel Computation of First-Order Hetero- Homogeneous Linear Recurrence Equations.

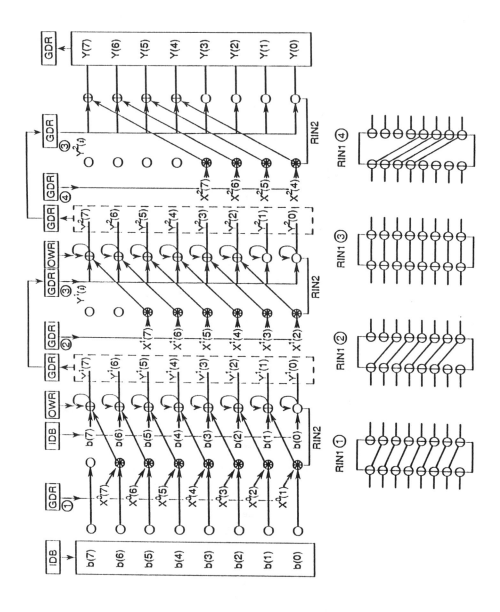

Figure 11.   Mapping Diagram of the Second Part Parallel Computation of
First-Order Hetero-Homogeneous Linear Recurrence Equations.

Table 5. Algorithm: Parallel Recursive Newton-Euler Equations of Motion.

N1. For $i = 1$ to n parallel Do

$$^0\mathbf{R}_i = {}^0\mathbf{R}_{i-1} \, {}^{i-1}\mathbf{R}_i$$

N2. For $i = 1$ to n parallel Do

$$\mathbf{z}_i = {}^0\mathbf{R}_i \, \mathbf{z}_0 \quad , \quad \mathbf{z}_0 = [0, 0, 1]^T$$

$$\mathbf{p}_i^* = {}^0\mathbf{R}_i \, {}^i\mathbf{p}_i^*$$

$$\mathbf{s}_i = {}^0\mathbf{R}_i \, {}^i\mathbf{s}_i$$

N3. For $i = 1$ to n parallel Do

$$b_i = \mathbf{z}_{i-1} \dot{q}_i \, (1 - \lambda_i)$$

For $i = 1$ to n parallel Do

$$\boldsymbol{\omega}_i = \boldsymbol{\omega}_{i-1} + b_i$$

N4. For $i = 1$ to n parallel Do

$$b_i = (\mathbf{z}_{i-1} \, \ddot{q}_i + \boldsymbol{\omega}_{i-1} \times \mathbf{z}_{i-1} \, \dot{q}_i) \, (1 - \lambda_i)$$

For $i = 1$ to n parallel Do

$$\dot{\boldsymbol{\omega}}_i = \dot{\boldsymbol{\omega}}_{i-1} + b_i$$

N5. For $i = 1$ to n parallel Do

$$b_i = \dot{\boldsymbol{\omega}}_i \times \mathbf{p}_i^* + \boldsymbol{\omega}_i \times (\boldsymbol{\omega}_i \times \mathbf{p}_i^*) +$$

$$(\mathbf{z}_{i-1} \, \ddot{q}_i + 2 \, \boldsymbol{\omega}_i \times (\mathbf{z}_{i-1} \, \dot{q}_i)) \, \lambda_i$$

For $i = 1$ to n parallel Do

$$\dot{\mathbf{v}}_i = \dot{\mathbf{v}}_{i-1} + b_i$$

N6. For $i = 1$ to n parallel Do

$$\mathbf{a}_i = \dot{\boldsymbol{\omega}}_i \times \mathbf{s}_i + \boldsymbol{\omega}_i \times (\boldsymbol{\omega}_i \times \mathbf{s}_i) + \dot{\mathbf{v}}_i$$

N7. For $i = 1$ to n parallel Do

$$\mathbf{F}_i = m_i \mathbf{a}_i$$

N8. For $i = 1$ to n parallel Do

$$^i\boldsymbol{\omega}_i = {}^i\mathbf{R}_0 \, \boldsymbol{\omega}_i = ({}^0\mathbf{R}_i)^T \, \boldsymbol{\omega}_i$$

$$^i\dot{\boldsymbol{\omega}}_i = {}^i\mathbf{R}_0\,\dot{\boldsymbol{\omega}}_i = ({}^0\mathbf{R}_i)^T\,\dot{\boldsymbol{\omega}}_i$$

$$^i\mathbf{N}_i = {}^i\mathbf{I}_i\,{}^i\dot{\boldsymbol{\omega}}_i + {}^i\boldsymbol{\omega}_i \times ({}^i\mathbf{I}_i\,{}^i\boldsymbol{\omega}_i)$$

$$\mathbf{N}_i = {}^0\mathbf{R}_i\,{}^i\mathbf{N}_i$$

N9.   For $i = 1$ to n parallel Do

$$\mathbf{f}_i = \mathbf{f}_{i+1} + \mathbf{F}_i$$

N10. For $i = 1$ to n parallel Do

$$b_i = \mathbf{N}_i + (\mathbf{p}_i^* + \mathbf{s}_i) \times \mathbf{F}_i + \mathbf{p}_i^* \times \mathbf{f}_{i+1}$$

For $i = 1$ to n parallel Do

$$\mathbf{n}_i = \mathbf{n}_{i+1} + b_i$$

N11. For $i = 1$ to n parallel Do

$$\tau_i = \begin{cases} (\mathbf{n}_i)^T\,\mathbf{z}_{i-1} &, \quad \text{if } \lambda_i = 0 \\ (\mathbf{f}_i)^T\,\mathbf{z}_{i-1} &, \quad \text{if } \lambda_i = 1 \end{cases}$$

where

$^{i-1}\mathbf{R}_i$:   3×3 rotation matrix indicate the orientation of link i coordinates referenced to link $(i-1)$ coordinates.

$^i\mathbf{p}_i^*$:   the origin of the ith coordinate frame with respect to the $(i-1)$th coordinate system, expressed with respect to link i coordinates. (a constant)

$^i\mathbf{s}_i$:   position of the center of mass of link i from the origin of the ith coordinate system, expressed with respect to link i coordinates. (a constant)

$^i\mathbf{I}_i$:   inertia matrix of link i about its center of mass, expressed with respect to link i coordinates. (a constant)

$^i\boldsymbol{\omega}_i$:   angular velocity of link i with respect to the ith coordinate frame.

$^i\mathbf{N}_i$:   total external moment exerted on link i at the center of mass, expressed with respect to link i coordinates.

λ_i equal to 0 if link i is rotational; equal to 1 if link i is translational.

Table 6.

Table of Subtask Assignment and Input Table for Inverse Dynamics Algorithm.

Subtask Assignment				Input Variable	
$\bar{T}_1[i]$	$^0\mathbf{R}_i$	$T_{26}[i]$	$\mathbf{a}_i$	$I_1[i]$	$^{i-1}\mathbf{R}_i$
$T_2[i]$	$\mathbf{s}_i$	$T_{27}[i]$	$\mathbf{F}_i$	$I_2[i]$	$^i\mathbf{s}_i$
$T_3[i]$	$\mathbf{p}_i^*$	$T_{28}[i]$	$^i\mathbf{R}_0$	$I_3[i]$	$^i\mathbf{p}_i^*$
$T_4[i]$	$\mathbf{z}_i$	$T_{29}[i]$	$^i\boldsymbol{\omega}_i$	$I_4[i]$	$\mathbf{z}_0$
$T_5[i]$	$\mathbf{z}_{i-1}\dot{q}_i$	$T_{30}[i]$	$^i\dot{\boldsymbol{\omega}}_i$	$I_5[i]$	$\dot{q}_i$
$T_6[i]$	$\mathbf{z}_{i-1}\dot{q}_i(1-\lambda_i)$	$T_{31}[i]$	$^i\mathbf{I}_i\,{}^i\dot{\boldsymbol{\omega}}_i$	$I_6[i]$	$1-\lambda_i$
$\bar{T}_7[i]$	$\boldsymbol{\omega}_i$	$T_{32}[i]$	$^i\mathbf{I}_i\,{}^i\boldsymbol{\omega}_i$	$I_7[i]$	$\ddot{q}_i$
$T_8[i]$	$\boldsymbol{\omega}_{i-1} \times (\mathbf{z}_{i-1}\dot{q}_i)$	$T_{33}[i]$	$^i\boldsymbol{\omega}_i \times (^i\mathbf{I}_i\,{}^i\boldsymbol{\omega}_i)$	$I_8[i]$	λ_i
$T_9[i]$	$\mathbf{z}_{i-1}\ddot{q}_i$	$T_{34}[i]$	$^i\mathbf{N}_i$	$I_9[i]$	m_i
$T_{10}[i]$	$T_8[i]+T_9[i]$	$T_{35}[i]$	$\mathbf{N}_i$	$I_{10}[i]$	$^i\mathbf{I}_i$
$T_{11}[i]$	$T_{10}[i](1-\lambda_i)$	$\bar{T}_{36}[i]$	$\mathbf{f}_i$		
$\bar{T}_{12}[i]$	$\dot{\boldsymbol{\omega}}_i$	$T_{37}[i]$	$\mathbf{p}_i^* + \mathbf{s}_i$		
$T_{13}[i]$	$\dot{\boldsymbol{\omega}}_i \times \mathbf{p}_i^*$	$T_{38}[i]$	$(\mathbf{p}_i^* + \mathbf{s}_i) \times \mathbf{F}_i$		
$T_{14}[i]$	$\boldsymbol{\omega}_i \times \mathbf{p}_i^*$	$T_{39}[i]$	$\mathbf{N}_i + T_{38}[i]$		
$T_{15}[i]$	$\boldsymbol{\omega}_i \times (\boldsymbol{\omega}_i \times \mathbf{p}_i^*)$	$T_{40}[i]$	$\mathbf{p}_i^* \times \mathbf{f}_{i+1}$		
$T_{16}[i]$	$T_{13}[i]+T_{15}[i]$	$T_{41}[i]$	$T_{39}[i] + T_{40}[i]$		
$T_{17}[i]$	$2\boldsymbol{\omega}_i \times (\mathbf{z}_{i-1}\dot{q}_i)$	$\bar{T}_{42}[i]$	$\mathbf{n}_i$		
$T_{18}[i]$	$\mathbf{z}_{i-1}\ddot{q}_i + T_{17}[i]$	$T_{43}[i]$	$(\mathbf{n}_i)^T$		
$T_{19}[i]$	$T_{18}[i]\,\lambda_i$	$T_{44}[i]$	$(\mathbf{n}_i)^T\mathbf{z}_{i-1}$		
$T_{20}[i]$	$T_{16}[i]+T_{19}[i]$	$T_{45}[i]$	$(\mathbf{f}_i)^T$		
$\bar{T}_{21}[i]$	$\dot{\mathbf{v}}_i$	$T_{46}[i]$	$(\mathbf{f}_i)^T\mathbf{z}_{i-1}$		
$T_{22}[i]$	$\dot{\boldsymbol{\omega}}_i \times \mathbf{s}_i$	$T_{47}[i]$	$T_{46}[i](1-\lambda_i)$		
$T_{23}[i]$	$\boldsymbol{\omega}_i \times \mathbf{s}_i$	$T_{48}[i]$	$T_{44}[i]\lambda_i$		
$T_{24}[i]$	$\boldsymbol{\omega}_i \times (\boldsymbol{\omega}_i \times \mathbf{s}_i)$	$T_{49}[i]$	τ_i		
$T_{25}[i]$	$T_{22}[i]+T_{24}[i]$				

Table 7. Reordered Subtask Table for Inverse Dynamics Algorithm.

ROW	TB[ROW]	OP[ROW]	ROW	TB[ROW]	OP[ROW]
1	$\bar{T}_1$	$\bar{T}_1^{-1} I_1$	26	T_{25} †	$T_{22} + T_{24}$
2	T_2	$\bar{T}_1 I_2$	27	T_{26}	$T_{25} + \bar{T}_{21}$
3	T_3	$\bar{T}_1 I_3$	28	T_{27}	$T_{26} I_9$
4	T_{37}	$T_3 + T_2$	29	T_{38}	$T_{37} \times T_{27}$
5	T_4	$\bar{T}_1 I_4$	30	$\bar{T}_{36}$	$\bar{T}_{36}^{+1} + T_{27}$
6	T_5	$T_4^{-1} I_5$	31	T_{40}	$T_3 \times \bar{T}_{36}^{+1}$
7	T_6	$T_5 I_6$	32	T_{28}	$(T_1)^T$
8	$\bar{T}_7$	$\bar{T}_7^{-1} + T_6$	33	T_{29}	$T_{28} \bar{T}_7$
9	T_{23}	$\bar{T}_7 \times T_2$	34	T_{30}	$T_{28} \bar{T}_{12}$
10	T_{24}	$\bar{T}_7 \times T_{23}$	35	T_{31}	$I_{10} T_{30}$
11	T_{17}	$\bar{T}_7 \times T_5$	36	T_{32}	$I_{10} T_{29}$
12	T_8	$\bar{T}_7^{-1} \times T_5$	37	T_{33}	$T_{29} \times T_{32}$
13	T_9	$T_4^{-1} I_7$	38	T_{34}	$T_{33} + T_{31}$
14	T_{18}	$T_{17} + T_9$	39	T_{35}	$\bar{T}_1 T_{34}$
15	T_{10}	$T_8 + T_9$	40	T_{39}	$T_{35} + T_{38}$
16	T_{11}	$T_{10} I_6$	41	T_{41}	$T_{39} + T_{40}$
17	$\bar{T}_{12}$	$\bar{T}_{12}^{-1} + T_{11}$	42	$\bar{T}_{42}$	$\bar{T}_{42}^{+1} + T_{41}$
18	T_{22}	$\bar{T}_{12} \times T_2$	43	T_{43}	$(\bar{T}_{42})^T$
19	T_{13}	$\bar{T}_{12} \times T_3$	44	T_{44}	$T_{43} T_4^{-1}$
20	T_{14}	$\bar{T}_7 \times T_3$	45	T_{45}	$(\bar{T}_{36})^T$
21	T_{15}	$\bar{T}_7 \times T_{14}$	46	T_{46}	$T_{45} T_4^{-1}$
22	T_{16}	$T_{13} + T_{15}$	47	T_{47}	$T_{46} I_6$
23	T_{19}	$T_{18} I_8$	48	T_{48}	$T_{44} I_8$
24	T_{20}	$T_{19} + T_{16}$	49	T_{49}	$T_{47} + T_{48}$
25	$\bar{T}_{21}$	$\bar{T}_{21}^{-1} + T_{20}$			

† This is a DTR task.

$T_1, T_7, T_{12}, T_{21}, T_{36}$, and T_{42} are "homogeneous linear recursive equations."

which only one DTR subtask remains (subtask 25). Notice that, in Table 7, notation simplification is used.

The reordered task table produced in the above stage can be mapped onto the proposed DN-SIMD machine in a rather straightforward way like a serial program. The final stage of our mapping procedure consists of two phases. In the first phase, the subtasks including simple subtasks and macro subtasks which are viewed as simple subtasks temporarily are mapped onto the DN-SIMD machine in a row directly. The actual mapping of the macro subtask is considered in the second phase. To show the mapping process in the first phase, an arbitrary simple subtask, say T_8 (TB[12]), is considered as demonstration. First, assume there is no faulty element in the system. Since the number of joints is 7 and we have 8 PEs (ports), the first 7 PEs (ports) are used and let PE 7 idle as a standby device. In this case, we just keep the IES and OES of RIN1(RIN2) be at "straight" state. Since $T_8 = \overline{T_7}^{-1} \times T_5$ and by observing the previous subtask, T_{17} (see Table 6), we can decide that the first operand, $\overline{T_7}^{-1}$, of T_8 comes from GDR through RIN1. Moreover, the connection type of RIN1 should be uniform module shift with $d=1$, since the superscript of $\overline{T_7}$ is -1. The second operand of T_8, T_5, can be gotten from IWR, since it has been used in the previous subtask, T_{17}. The above is shown in Fig. 12(a). Similar concept can be used to other subtasks to complete the first phase mapping procedure. The output of the mapping procedure is a control table as shown in Table 8, which indicates the sources of operands, destinations of results, network used and required connection types for each (simple) subtask. The central control unit will use this control table to control the system.

Since there are six macro subtasks, further mapping must be performed in phase 2. These macro subtasks are all HLR equations. An efficient technique called the recursive doubling can be used to solve this recursive equation efficiently on an SIMD machine as shown in Fig. 12(b). This diagram possesses the same information as a control table.

Now, assume link 4 between stage 2 and stage 1 in Fig. 4(a) is disconnected due to some faulty element. In this case the reconfiguration of the networks to isolate this fault is shown in Fig. 13(a) and Fig. 13(b). The ability to reconfigure the networks under a single fault demonstrates the high reliability of the proposed DN-SIMD machine for robotic computations.

8. Conclusions

To design a global architecture for a set of parallel robotic algorithms, the characteristics of these algorithms are identified according to six fundamental features: degree of parallelism, uniformity of operations, fundamental operations, data dependency, and communication requirements. Considering the

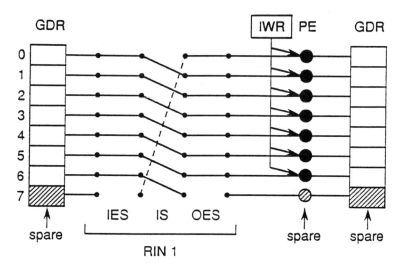

(a) Simple subtask mapping.

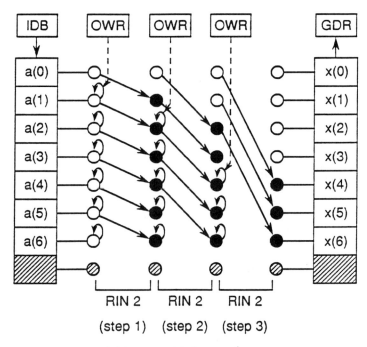

(b) Macro subtak mapping.

Figure 12.
Demonstration of mapping result under normal situation (no faulty element).

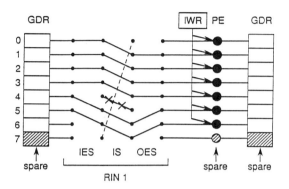

(a) Simple subtask mapping.

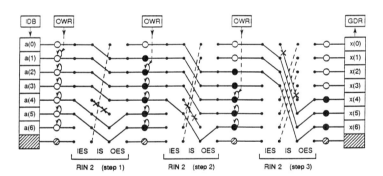

(b) Macro subtask mapping.

Figure 13. Demonstration of mapping result under faulty situation.

Table 8. Control Table for Inverse Dynamics Algorithm.

1	2	3	4	5	6	7	8	9			10
	Operand	Source	Network	Operand	Source	Network	Operation	Output Destination			Comment
T_i	1	1	Type	2	2	Type		GDR	IFD	RIN2	
$\bar{T}_1$	$\bar{T}_1^{-1}$	IFD	*	I_1	IDB	-	MM	×		-	* HLR Eqn.
T_2	$\bar{T}_1$	OWR	-	I_2	IDB	-	MV	×		-	
T_3	$\bar{T}_1$	IWR	-	I_3	IDB	-	MV	×		-	
T_{37}	T_3	OWR	-	T_2	GDR	RIN1-1	VA	×		-	
T_4	$\bar{T}_1$	GDR	RIN1-1	I_4	IDB	-	MV	×	×	2	
T_5	$\bar{T}_4^{-1}$	IFD	RIN2-2	I_5	IDB	-	SV	×		-	
T_6	T_5	OWR	-	I_6	IDB	-	SV	×		-	
$\bar{T}_7$	$\bar{T}_7^{-1}$	IFD	*	T_6	OWR	-	VA	×		-	*HLR Eqn.
T_{23}	$\bar{T}_7$	OWR	-	T_2	GDR	RIN1-1	VC	×		-	
T_{24}	$\bar{T}_7$	IWR	-	T_{23}	OWR	-	VC	×		-	
T_{17}	$\bar{T}_7$	IWR	-	T_5	GDR	RIN1-1	VC	×		-	
T_8	$\bar{T}_7^{-1}$	GDR	RIN1-2	T_5	IWR	-	VC	×		-	
T_9	$\bar{T}_4^{-1}$	GDR	RIN1-2	I_7	IDB	-	SV	×		-	
T_{18}	T_9	OWR	-	T_{17}	GDR	RIN1-1	VA	×		-	
T_{10}	T_9	IWR	-	T_8	GDR	RIN1-1	VA	×		-	
T_{11}	T_{10}	OWR	-	I_6	IDB	-	SV	×		-	
$\bar{T}_{12}$	$\bar{T}_{12}^{-1}$	IFD	*	T_{11}	OWR	-	VA	×		-	*HLR Eqn.
T_{22}	$\bar{T}_{12}$	OWR	-	T_2	GDR	RIN1-1	VC	×		-	

Table 8. Control Table for Inverse Dynamics Algorithm (continued).

T_i	Operand 1	Source 1	Network Type	Operand 2	Source 2	Network Type	Operation	Output Destination			Comment
								GDR	IFD	RIN2	
T_{13}	$\bar{T}_{12}$	IWR	-	T_3	GDR	RIN1-1	VC	×		-	
T_{14}	$\bar{T}_7$	GDR	RIN1-1	T_3	IWR	-	VC			-	
T_{15}	$\bar{T}_7$	IWR	-	T_{14}	OWR	-	VC	×		-	
T_{16}	T_{15}	OWR	-	T_{13}	GDR	RIN1-1	VA			-	
T_{19}	T_{18}	GDR	RIN1-1	I_8	IDB	-	SV			-	
T_{20}	T_{19}	OWR	-	T_{16}	GDR	RIN1-1	VA	×		-	
$\bar{T}_{21}$	$\bar{T}_{21}^{-1}$	IFD	*	T_{20}	OWR	-	VA			-	*HLR Eqn.
T_{25}	T_{22}	GDR	RIN1-1	T_{24}	GDR	RIN1-1	VA			-	
T_{26}	T_{25}	OWR	-	$\bar{T}_{21}$	GDR	RIN1-1	VA			-	
T_{27}	T_{26}	OWR	-	I_9	IDB	-	SV			-	
T_{38}	T_{27}	OWR	-	T_{37}	GDR	RIN1-1	VC	×		-	
$\bar{T}_{36}$	$\bar{T}_{36}^{-1}$	IFD	*	T_{27}	IWR	-	VA	×	×	3	*HLR Eqn.
T_{40}	$\bar{T}_{36}^{-1}$	IFD	RIN2-3	T_3	GDR	RIN1-1	VC	×		-	
T_{28}	T_1	GDR	RIN1-1	-	-	-	VT			-	
T_{29}	T_{28}	OWR	-	$\bar{T}_7$	GDR	RIN1-1	MV	×		-	
T_{30}	T_{28}	IWR	-	$\bar{T}_{12}$	GDR	RIN1-1	MV			-	
T_{31}	I_{10}	IDB	-	T_{30}	OWR	-	MV	×		-	
T_{32}	I_{10}	IWR	-	T_{29}	GDR	RIN1-1	MV			-	

Table 8. Control Table for Inverse Dynamics Algorithm (continued).

	2	3	4	5	6	7	8	9			10
1	Operand	Source	Network	Operand	Source	Network	Operation	Output Destination			Comment
T_i	1	1	Type	2	2	Type		GDR	IFD	RIN2	
T_{33}	T_{32}	OWR	-	T_{29}	IWR	-	VC			-	
T_{34}	T_{33}	OWR	-	T_{31}	GDR	RIN1-1	VA			-	
T_{35}	T_{34}	OWR	-	$\bar{T}_1$	GDR	RIN1-1	MV			-	
T_{39}	T_{35}	OWR	-	T_{38}	GDR	RIN1-1	VA			-	
T_{41}	T_{39}	OWR	-	T_{40}	GDR	RIN1-1	VA			-	
$\bar{T}_{42}$	$\bar{T}_{42}^{-1}$	IFD	*	T_{41}	OWR	-	VA			-	*HLR Eqn.
T_{43}	$\bar{T}_{42}$	OWR	-	-	-	-	VT			-	
T_{44}	T_{43}	OWR	-	T_4^{-1}	GDR	RIN1-2	VI	×		-	
T_{45}	$\bar{T}_{36}$	GDR	RIN1-1	-	-	-	VT			-	
T_{46}	T_{45}	OWR	-	T_4^{-1}	GDR	RIN1-2	VI			-	
T_{47}	T_{46}	OWR	-	I_6	IDB	-	SP	×		-	
T_{48}	T_{44}	GDR	RIN1-1	I_8	IDB	-	SP			-	
T_{49}	T_{48}	OWR	-	T_{47}	GDR	RIN1-1	SA	×		-	Result

Connection type 1 : Straight connection.
Connection type 2 : Uniform Module connection ($d = 1$).
Connection type 3 : Uniform Module connection ($d = -1$).

characteristics matching between the common features of the robotic algorithms and the architecture features, a medium-grained, DN-SIMD machine is designed. It consists of two sets of reconfigurable interconnection networks. One provides the communication between the PEs and the GDRs. The other provides the internal direct feedback paths among PEs to avoid unnecessary data storing and routing time. This machine performs three-stage pipelined operations and is synchronized at each basic mathematical calculation. Based on the requirements of communication capability, fault tolerance, and cost, the Generalized Cube network is chosen as an efficient interconnection network in the DN-SIMD machine. Its properties are investigated and a centralized switch control scheme is also developed to set the switching boxes to proper states for any required connection types. To improve the system fault tolerance, two extra stages are added to the original Generalized Cube network to achieve the Fault Tolerant Generalized Cube network. This FTGC network provides reliable communication even if some single fault occurs to the component of the interconnection network, including the box fault or the link fault. In the redundant model, in which there is at least one spare PE, the FTGC network provides an easy solution to maintaining satisfactory system performance under single fault situation.

With the parallel robotics algorithms and the proposed DN-SIMD parallel machine, a systematic mapping procedure to schedule the subtasks of the parallel algorithms onto the parallel architecture is developed. This mapping procedure consists of three stages. At the first stage, mathematical decomposition is performed on the parallel algorithms to achieve a series of subtasks and each subtask is either in the basic mathematical form which involves at most two operands, or a well-structured macro subtask such as the linear recurrence equations. At the second stage, to shorten the communication time, the processing sequence of subtasks is reordered to minimize the total number of DTR subtasks using the Neighborhood Scheduling algorithm. At the final stage, the reordered subtasks are mapped onto the DN-SIMD machine. In this process, the single-step subtasks can be mapped directly, while the macro-step subtasks need further design and special technique such as the recursive doubling technique for solving the linear recurrence equations. Finally, the algorithm of inverse dynamics computation was used as an example to illustrate the mapping procedure under normal and faculty situations.

9. References

[1]   C.S.G. Lee and C.T. Lin, "Characterization of Robotics Parallel Algorithms and Mapping onto a Reconfigurable SIMD Machine," *3rd Annual Conference on Aerospace Computational Control,* JPL, Oxnard, CA, pp.

460-476, 1989.

[2]   C. T. Lin, "Parallel Algorithms and Reconfigurable Architecture for Robotics Computations," MSEE Thesis, School of Electrical Engineering, Purdue University, West Lafayette, IN, August 1989.

[3]   L. H. Jamieson, "Characterizing Parallel Algorithms," in *The Characteristics of Parallel Algorithms,* L. H. Jamieson et al. (Eds.), The MIT Press, 1987.

[4]   J. Y. S. Luh, M. W. Walker, and R. P. C. Paul, "On-line Computational Scheme for Mechanical Manipulator," *Trans. ASME J. Dynam. Syst., Meas. Contr.,* Vol. 102, pp. 69-76, June 1980.

[5]   C. S. G. Lee, T. N. Mudge, and J. L. Turney, "Hierarchical Control Structure Using Special Purpose Processor for the Control of Robot Arm," *Proc. 1982 Conf. Patt. Recog. and Image Processing,* Las Vegas, Nevada, pp. 634-640, June 14-17, 1982.

[6]   R. Nigam, C. S. G. Lee, "A Multiprocessor-Based Controller for the Control of Mechanical Manipulators," *IEEE J. of Robotics and Automation,* Vol. RA-1, No. 4, pp. 173-182, Dec. 1985.

[7]   T. Kanade, P. K. Khosla, and N. Tanaka, "Real-Time Control of the CMU Direct Arm II Using Customized Inverse Dynamics," *Proc. of IEEE Conf. on Decision and Contr.,* pp. 1345-1352, Dec. 1984.

[8]   L. Lathrop, "Parallelism in Manipulator Dynamics," *Int'l J. of Robotics Res.,* Vol. 4, No. 2, pp. 80-102, Summer 1985.

[9]   C. S. G. Lee and P. R. Chang, "Efficient Parallel Algorithm for Robot Inverse Dynamics Computation," *IEEE Trans. on Syst. Man. Cybern.,* Vol. SMC-16, No. 4, pp. 532-542, July/Aug. 1986.

[10]   J. Y. S. Luh and C. S. Lin, "Scheduling of Parallel Computation for a Computer-controlled Mechanical Manipulator," *IEEE Trans. on Syst. Man. Cybern.,* Vol. SMC-12, No. 2, pp. 214-234, March 1982.

[11]   H. Kasahara, and S. Narita, "Parallel Processing of Robot-arm Control Computation on a Multimicroprocessor System," *IEEE J. of Robotics and Automation,* Vol. RA-1, No. 2, pp. 104-113, June 1985.

[12]   C. L. Chen, C. S. G. Lee, and E. S. H. Hou, "Efficient Scheduling Algorithms for Robot Inverse Dynamics Computation on a Multiprocessor System," *IEEE Trans. on Syst. Man. Cybern.,* Vol. SMC-18, No. 5, pp. 729-743, September/October 1988.

[13]   C. S. G. Lee and C. L. Chen, "Efficient Mapping Algorithms for Scheduling Robot Inverse Dynamics Computation on a Multiprocessor System,"

to appear in *IEEE Trans. on Syst. Man. Cybern.*

[14]   M. W. Walker and D. E. Orin, "Efficient Dynamic Computer Simulation of Robot Mechanisms," *Trans. ASME J. Dynam. Syst. Meas. and Contr.,* Vol. 104, pp. 205-211, Sept. 1982.

[15]   R. Featherstone, "The Calculation of Robot Dynamics Using Articulated-body Inertia," *Int. J. Robotics Res.,* Vol. 2, No. 1, pp. 13-30, Spring 1983.

[16]   C. S. G. Lee and P. R. Chang, "Efficient Parallel Algorithms for Robot Forward Dynamics Computation," *IEEE Trans. on Syst. Man. Cybern.,* Vol. SMC-18, No. 2, pp. 238-251, Mar./Apr. 1988.

[17]   M. Amin-Javaheri and D. E. Orin, "A Systolic Architecture for Computation of the Manipulator Inertia Matrix," *IEEE Trans. on Syst. Man. Cybern.,* Vol. SMC-18, No. 6, pp. 939-951, Nov/Dec 1988.

[18]   A. Fijany and A. K. Bejczy, "A Class of Parallel Algorithms for Computation of the Manipulator Inertia Matrix," *IEEE Trans. on Robotics and Automation,* Vol. RA-5, No. 5, pp. 600-615, October 1989.

[19]   P. M. Kogge and H. S. Stone, "A Parallel Algorithm for the Efficient Solution of a General Class of Recurrence Equations," *IEEE Trans. on Computer,* Vol. c-22, pp. 789-793, Aug. 1973.

[20]   A. K. Bejczy, T. J. Tarn, and X. Yun, "Robust Robot Arm Control with Nonlinear Feedback," *Proc. IFAC Symp. on Robot Control,* Barcelona, 1985.

[21]   A. H. Sameh, "Solving the Linear Least Squares Problem on a Linear Array of Processors," *Proc. Purdue Workshop on Algorithmically-Specialized Computer Organizations, 1982.*

[22]   G. A. Geist, M. T. Heath and E. Ng, "Parallel Algorithms for Matrix Computations," in *The Characteristics of Parallel Algorithms,* L. H. Jamieson et al. (Eds.), The MIT press, 1987.

[23]   J. Denavit and R. B. Hartenberg, "A Kinematic Notation for Lower-pair Mechanisms based on Matrices," *ASME J. Appl. Mech.,* Vol. 23, 1955.

[24]   K. S. Fu, R. C. Gonzalez, and C. S. G. Lee, *Robotics: Control, Sensing, Vision, and Intelligence,* New York: McGraw-Hill, 1987.

[25]   A. Fijany and J. G. Pontnau, "Parallel Computation of the Jacobian for Robot Manipulators," *Proc. IASTED,* Santa Barbara, May 1987.

[26]   S. S. Leung and M. A. Shanblatt, "Real-Time DKS on a Single Chip," *IEEE J. Robotics and Automation,* Vol. RA-3, No. 4, Aug. 1987.

[27]   D. E. Orin and W. W. Schrader, "Efficient Computation of the Jacobian for Robot Manipulators," *the Int. J. of Robotics Research,* Vol. 3, No. 4, pp. 66-75, Winter 1984.

[28]   T. B. Yeung and C. S. G. Lee, "Efficient Parallel Algorithms and VLSI Architectures for Manipulator Jacobian Computation," *IEEE Trans. on Syst. Man. Cybern.,* Vol. SMC-19, No. 5, September/October 1989.

[29]   C. R. Rao and S. K. Mitra, *Generalized Inverse: Theory and Applications,* Wiley, NY, 1974.

[30]   P. R. Chang and C. S. G. Lee, "Residue Arithmetic VLSI Array Architecture for Manipulator Pseudo-Inverse Jacobian Computation," *IEEE Trans. on Robotics and Automation,* Vol. RA-5, No. 5, pp. 569-582, October 1989.

[31]   R. Featherstone, "Position and Velocity Transformations Between Robot End-effector Coordinates and Joint Angles," *the Int. J. of Robotics Research,* Vol. 2, No. 2, Summer 1983, pp. 35-45.

[32]   S. Y. Kung, *VLSI Array Processors,* Prentice-Hall International Inc. 1988

[33]   R. P. Paul, B. E. Shimano, and G. Mayer, "Kinematic Control Equations for Simple Manipulators," *IEEE Trans. on Syst. Man. Cybern.,* Vol. SMC-11, pp. 494-455, 1981.

[34]   L. W. Tsai and A. P. Morgan, "Solving the Kinematics of the Most General Six- and Five-DOF Manipulators by Continuation Methods," *Trans. ASME, J. Mechanism, Transmission, Automation in Design,* Vol. 107, pp. 189-200, June 1985.

[35]   Y. T. Tsai and D. E. Orin, "A Strictly Convergent Real-time Solution for Inverse Kinematics of Robot Manipulators," *J. of Robotic Systems,* Vol. 4, No. 4, pp. 477-501, 1987.

[36]   C. S. G. Lee and P. R. Chang, "A Maximum Pipelined CORDIC Architecture for Robot Inverse Kinematic Position Computation," *IEEE J. Robotics and Automation,* Vol. RA-3, No. 5, pp. 445-458, Oct. 1987.

[37]   K. Hwang and F. A. Briggs., *Computer Architecture and Parallel Processing,* New York:McGraw Hill, 1984.

[38]   R. J. McMillen, *A Study of Multistage Interconnection Networks: Design, Distributed Control, Fault Tolerance, and Performance,* Ph. D. Thesis, Purdue U. Dec. 1982.

[39]   T. Lang and H. S. Stone, "A Shuffle-exchange Network with Simplified Control," *IEEE Trans. on Computers,* Vol. C-25, Jan. 1976, pp. 55-65.

[40] C. L. Wu and T. Y. Feng, "Routing Techniques for a Class of Multistage Interconnection Networks," *1978 Int'l Conf. Parallel Processing,* Aug. 1978, pp. 197-205.

[41] C. L. Wu and T. Y. Feng, *Tutorial: Interconnection Networks for Parallel Distributed Processing,* IEEE Computer Society Press, 1984.

[42] J. Hayes, "A Graph Model for Fault-tolerant Computing Systems," *IEEE Trans. on Computers,* Vol. C-25, Sept. 1976, pp. 875-883.

[43] J. Fortes and C. Raghavendra, "Graceful Degradable Processor Arrays," *IEEE Trans. on Computers,* Vol. C-34, Nov. 1985, pp. 1033-1044.

[44] K. Padmanabhan and D. H. Lawrie, "A Class of Redundant Path Multistage Interconnection Networks," *IEEE Trans. on Computers,* Dec. 1983, pp. 1099-1108.

[45] L. Ciminiera and A. Serra, "A Connecting Network with Fault Tolerance Capabilities," *IEEE Trans. on Computers,* June 1986, pp. 578-580.

[46] K. Yoon and W. Hegazy, "The Extra Stage Gamma Network," *13th Symp. Computer Architecture,* Computer Society Press, Silver Spring, Md., 1986, pp. 175-182.

[47] C. Raghavendra and A. Varma, "Fault-tolerant Multiprocessors with Redundant-path Interconnection Networks," *IEEE Trans. on Computers,* Vol. C-35, Mar. 1986, pp. 307-316.

[48] V. Kumar and S. Reddy, "Augmented Shuffle-exchange Multistage Interconnection Networks," *IEEE Computer,* Vol. 20, June 1987, pp. 30-40.

[49] W. Lin and C. L. Wu, "A Fault-tolerant Mapping Scheme for a Configurable Multiprocessor System," *IEEE Trans. on Computers,* Vol. C-38, No. 2, Feb. 1989, pp. 227-237.

[50] G. B. Adam III, *A Fault-Tolerant Interconnection Network and Image Processing Application for the PASM Parallel Processing System,* Ph. D. Thesis, Purdue U. Dec. 1984.

[51] R. H. Kuhn, "Efficient Mapping of Algorithms to Single-stage Interconnection," *7th Annual Symp. on Computer Architecture,* May 1980, pp. 182-189.

[52] S. H. Bokhari, "On the Mapping Problem," *IEEE Trans. on Computers,* Vol. C-30, March 1981, pp. 207-214.

Trajectory Planning for Robot Control: A Control Systems Perspective

Sunil K. Singh
Thayer School of Engineering, Dartmouth College
Hanover, NH 03755

1   Introduction

The development of autonomous controllers in general, and 'intelligent' robots in particular, has led to active research in 'motion planning'. The planning problem has been interpreted and solved in various ways by different researchers. Most planners completely ignore the dynamics of the system. This has led to the current trend of dividing the problem into smaller sub-problems, and solving each one separately. The three typical sub-problems may be identified as:

1. **Problem P1: path planning.** *Given a manipulator and a description of its environment, plan a path between two specified positions where the path avoids collision with obstacles in the environment and is optimal with regard to a geometric performance index, e.g. shortest path.* With reference to Figure 1, this problem would entail generating the spatial sequence $\{x_i\}$ which moves the object from the start to the goal position without colliding with the obstacles. This problem has been widely researched both as a purely geometric problem and at an algorithmic level. A good review may be found in [1]. It should be noted that **P1** does not account for the dynamics of the system. The attendant *low-level* problem of dynamics is left for the second stage.

2. **Problem P2: trajectory planning.** *Given a path to be followed by the end-effector, the differential equations that describe the manipulator dynamics and the constraints on the torques/forces available from the actuators, find the time history of the positions along the path so that a given performance index is minimized.* This is the input to the next stage.

CONTROL AND DYNAMIC SYSTEMS, VOL. 40

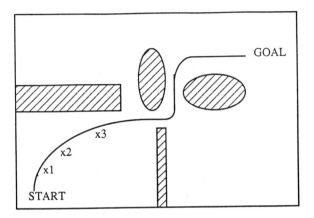

Figure 1: The path and trajectory planning problem.

3. **Problem P3: trajectory tracking.** *Given a reference trajectory, and a description of the manipulator as a non-linear dynamical system, design a feedback controller which tracks the given trajectory accurately.* The solution to **P3** is complicated by the nature of the system to be controlled. The controller must ensure accurate tracking in the face of parameter uncertainty and be insensitive to disturbances. Several control schemes have been developed for accurate tracking, ranging from simple linear controllers to those employing nonlinear compensation (see, for example, [2]).

2 Trajectory Planning and Feedback Control

In order to obtain a clear understanding of the role of trajectory planners, it is helpful to first establish their relationship to the feedback controller. Figures 2a and 2b show two possible methods of moving the object in Figure 1 through the sequence of points generated by the path planner. In 2a the input to the feedback controller is the initial configuration x_i and the next desired configuration x_{i+1}. If, for example, the feedback controller is based on a second order system, it will guide the object along the trajectory of the form

$$\ddot{x} + 2\zeta\omega_n\dot{x} + \omega_n^2 x = 0 \tag{1}$$

The velocity profile generated for one such sequence is shown in Figure 3. Alternatively, if the feedback controller was chosen to be a minimum-time

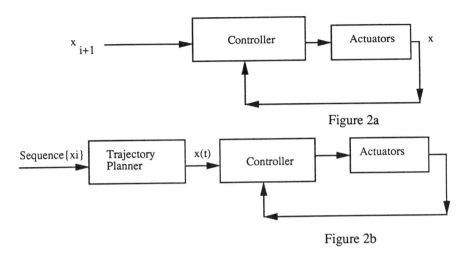

Figure 2a

Figure 2b

Figure 2: Alternative implementations of planners and controllers in robotic systems.

controller with the same maximum available control as the second-order system, the velocity profile that would be generated would look like the one shown in Figure 3. The resulting motion would start and end at each of the set-points x_i with zero velocity, which is clearly undesirable for efficient task executions. In 2b, a trajectory planner is placed between the path planner and the controller (later on we shall see how the path and trajectory planner can also be integrated). The role of the trajectory planner is to accept the sequence $\{x_i\}$ from the path planner and convert it to a time parameterized signal $\{x_i(t)\}$, preferably in some *optimal* manner. The feedback controller is now charged with the task of merely tracking the signal provided by the trajectory planner.

2.1 Trajectory Planning for Specified Paths

Many robot applications require the end effector to track a prescribed path. In such situations, the planner must be able to prescribe the timing and velocity along the path. Since the path may be described in Cartesian coordinates, $\bar{z}$, but the dynamics are more conveniently expressed in joint coordinates, $\bar{q}_p$, an immediate concern is to decide which coordinate system is more appropriate for planning the trajectory. The manipulator is characterized by the

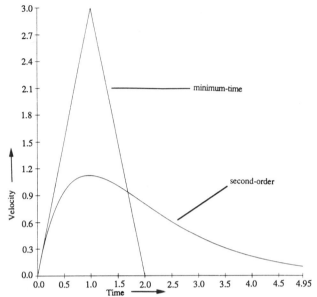

Figure 3: Velocity profiles for a second order and time optimal controller to move from START to GOAL.

following kinematics relation

$$\bar{z} = \Psi(\bar{q}_p), \quad \Psi : R^n \to R^n \tag{2}$$

The manipulator has n degrees of freedom and $\bar{z} \in R^n$ denotes the position and orientation of the end-effector, while $\bar{q}_p \in R^n$ represents the vector of generalized coordinates. In [4] it was demonstrated that the description of the path in task space can be converted to the descripton in joint space at a sufficient number of discrete points using a recursive algorithm. For most industrial robots, the control is done at the joint level. This demands the transformation from Cartesian to joint positions and velocities, using the inverse of the kinematic map and the Jacobian respectively:

$$\bar{q}_p = \Psi^{-1}(\bar{z}) \tag{3}$$

$$\dot{\bar{q}}_p = J^{-1}(\bar{q}_p)\dot{\bar{z}} \tag{4}$$

This is computationally burdensome. Therefore, trajectory planning at the joint level is more attractive. Finkel [5] discussed the specification of the trajectory in great detail. Given the desired spatial sequence of points through which the trajectory must pass, the trajectory may be specified as a polynomial in time. This polynomial may be chosen to satisfy the desired boundary conditions. The motion may also be made as smooth as desired by choosing a polynomial of sufficiently high degree. However, if a single polynomial were employed to pass through the entire trajectory, then the degree of the polynomial would be extremely high. To avoid using very high degree polynomials to describe the entire trajectory, which could have undesirable properties, the use of lower order polynomials splined together at knot points has been suggested. Since the velocity must be continuous at all times, a second degree polynomial appears to be the minimum requirement. Cubic polynomials have been shown to have desirable features. Trajectory planning has been dealt with at two levels in literature. Originally, it was as a constraint satisfaction problem, i.e., the coefficients specifying the polynomial were chosen such that the trajectory would pass through the prescribed points and satisfy the continuity requirements on the velocity and acceleration. This imposes the restriction that the number of parameters required to describe the trajectory should be exactly equal to the number of constraints. These ideas can be found in [5,6]. Both employed piecewise cubic splines to express individual joint trajectories. Since each spline must pass through the knot point two position constraints are imposed; further, if the velocity and acceleration are to be continuous, then two additional constraints need to be imposed. A cubic trajectory can exactly satisfy this problem, since it has only four unknown parameters. However, at the beginning and the end of the motion, each knot point introduces three constraints, giving rise to a total of five in the first and last stages. Paul [7] accounted for this by introducing a fourth degree polynomial in the first and last stages. Lin, Chang and Luh [6] and Finkel [5] introduced two additional knot points to solve the problem. The resultant problem can then be shown to be a solution to a set of equations, involving a tridiagonal matrix, which is uniquely solvable by the L-U decomposition scheme. More recently, the emphasis has shifted to the optimal planning of manipulator trajectories. In other words, the trajectory must not only pass through the knot points and have continuous first (and second, if required) derivatives, but must also be optimal with respect to a specified performance index

$$J = \int_0^{t_f} L(\bar{q}, \dot{\bar{q}}, \bar{u}) dt$$

An additional benefit from this is that the trajectory can be planned taking into account the complete non-linear dynamics of the manipulator and the state-dependent constraints on the available torques/forces. Its implementation in real-time is rendered difficult by the complicated non-linear and coupled nature of the differential equations describing the manipulator dynamics. An alternative to real-time optimal control is to divide the problem into two stages; off-line optimal trajectory planning, followed by on-line tracking. With this objective in mind, off-line trajectory planning has received considerable attention lately [10-14].

2.2   Formulation of the Trajectory Planning Problem

Let the path be described by a set of $N + 1$ knot points in the task coordinate system, $\bar{z}$. The inverse kinematics solution from Eq. (3) can be used to come up with the corresponding set of joint vectors, i.e. for joint i, the following vector can be obtained

$$(\hat{q}_{i1p}, \hat{q}_{i2p}, \ldots, \hat{q}_{iN+1p})$$

The dynamics of the manipulator may be described as:

$$\dot{\bar{q}}_p = \bar{q}_v \tag{5}$$

$$\dot{\bar{q}}_v = [M(\bar{q}_p)]^{-1}(-h(\bar{q}_p, \bar{q}_v) - g(\bar{q}_p) + \bar{u}) \tag{6}$$

where $M[\bar{q}_p]$ is the symmetric inertia matrix, $h(\bar{q}_p, \bar{q}_v)$ is the Coriolis and centripetal torque or force vector, $g(\bar{q}_p)$ is the gravity torque and force vector and $\bar{u}$ is the actuating torque and/or force vector.

The problem can now be reduced to the determination of a time function which will pass through the $N + 1$ points and satisfy the requirement that the derivatives be continuous up to order $m - 1$ for an m-spline trajectory. In addition it must satisfy several inequality constraints along the path. Typically, these inequality constraints describe the limitations on the actuation torques/forces, and on the joint velocities, and may be expressed as

$$\bar{v}(\bar{q}_p, \bar{q}_v) \leq \bar{u} \leq \bar{w}(\bar{q}_p, \bar{q}_v) \tag{7}$$

$$\bar{r}(\bar{q}_p) \leq \bar{q}_v \leq \bar{s}(\bar{q}_p) \tag{8}$$

It must be noted that the constraints on the torques/forces have been represented as functions of joint velocities and displacements for the purpose of generality, and additional constraints have been imposed on the joint velocities independently, because the motion of the manipulator can become unstable at high speeds. Sometimes, the torque constraints cannot be incorporated directly as in the above equation. Then it must be converted into bounds on the velocities, accelerations and the jerk along the path. A general expression of the inequality constraint is

$$F(\bar{q}_p, \bar{q}_v, \bar{u}) \leq 0 \tag{9}$$

The performance index to be minimized may be expressed as

$$J = \int_0^{t_f} L(\bar{q}_p, \bar{q}_v, \bar{u}) dt \tag{10}$$

For time minimization cases, $L = 1$.

The problem as posed falls under the category of state constrained optimal control problems which are extremely difficult to solve [8,9]. However, by expressing the trajectory as piecewise polynomials it can be simplified, as shall be seen now. In stage k, let the displacement of joint i be represented by the following polynomial of degree m

$$q_{ik} = a_0 + a_1 t + \frac{a_2}{2} t^2 + \ldots + \frac{a_m}{m!} t^m \tag{11}$$

Let q_{ik}^r denote the r-th derivative of joint i in stage k. The following statements can be easily verified:

1. In stage k, q_{ik}^m is a constant.

2. $a_r = \hat{q}_{ik}^r$, where $\hat{q}_{ik}^r$ is the value of q_{ik}^r at the beginning of stage k

Let h_k represent the duration of stage k. The following recurrence relation follows immediately:

$$q_{ik+1}^r = q_{ik}^r + q_{ik}^{r+1} h_k + \frac{q_{ik}^{r+2}}{2!} h_k^2 + \ldots + \frac{q_{ik}^m}{m-r!} h_k^{(m-r)} \tag{12}$$

where $r = 1, 2, \ldots, m - 1$. For our purposes, it will be assumed that the values of the first $m - 1$ derivatives have been specified at the initial and final points. This is reasonable because we shall be dealing with quadratic and cubic splines, where the first derivative represents the velocity of the joint, and the second derivative the acceleration. We are now in a position to state the dimensionality of the problem to be solved. The unknown variables are the q_{ik}^m, for n joints and N stages, giving rise to Nn variables. In addition, the time intervals h_k in each stage are as yet undetermined, which gives rise to N additional unknowns. There are therefore $Nn + N$ unknowns. The number of equality constraints to be satisfied are Nn position constraints and $(m - 1)n$ additional constraints for the $(m - 1)$ derivatives to be satisfied at the final point for each joint. This means that the number of free variables at our disposal is $Nn + N - Nn - (m - 1)n = N - (m - 1)n$. It follows that:

1. If $N < (m - 1)n$, then the problem has no solution.

2. If $N = (m - 1)n$, then it will have a unique solution. This solution may be obtained by solving the system of algebraic equations using any one of several well-known techniques.

3. If $N > (m-1)n$, then an optimization scheme may be used to determine the extra variables.

When the trajectory is composed of second order splines, a more elegant method for trajectory planning is to use dynamic programming. Given the $N + 1$ knot points, the search for the optimal solution can be shown to be equivalent to the search over the range of possible velocity values of any one moving manipulator joint at the discrete knot points (i.e. in the phase space of any one manipulator link). The computational efficiency of the algorithm can be greatly enhanced by using a recursive refinement scheme. Details of the results may be found in [11,12].

3   Path and Trajectory Planners Embedded in the Feedback Control Loop

This approach was motivated by two observations. Firstly, the inability of existing motion planners to handle dynamics seemed an unnecessary limitation. Secondly, since the robotic system is essentially a dynamic system, the evolution of its states (positions and velocities) could be completely specified in time by prescribing the time history of the control inputs to the system. The traditional path planning problem is therefore merely a subset of the more general path-and-trajectory planning problem which could be interpreted as determining the most 'suitable' control inputs to execute the desired task subject to both geometric and dynamic constraints- in other words, an optimal control problem. This has the additional advantage that dynamic constraints such as torque-speed characteristics of the actuators and the dynamic behavior of the system would be included automatically in the general formulation. Predictably, the first few approaches incorporating the planner embedded within the controller dealt with obstacle avoidance. Therefore these studies combined path and trajectory planning (sub-problems **P1** and **P2**). In this case, the output of the motion planner is not merely a sequence of configurations in space, but a time history of the joint positions. Correspondingly, the input to the planner is a description of the geometry of the manipulator and its environment, the manipulator dynamics, and the initial and final states. The criterion for planning is defined in terms of a *performance index*. One such planner was designed in [20]. The objective was to design a general algorithm capable of handling varying performance indices and generating optimal motions as part of a CAD system. The main features of the planner are shown in Figure 4.

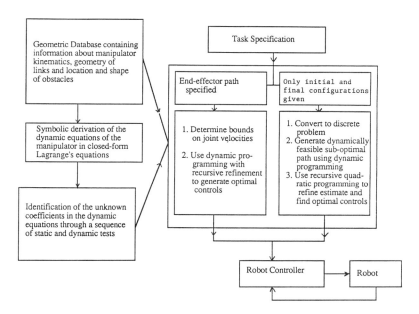

Figure 4: Block diagram showing the principal features of the path and trajectory planning CAD system.

3.1   Problem Formulation

The overall optimal control problem may be expressed as

$$\text{minimize}\quad J = h_0(\bar{x}(0)) + \int_0^T f_0(\bar{x}, \bar{u}, t)dt + l_0(\bar{x}(T)) \qquad (13)$$

subject to

$$\dot{\bar{x}} = f(\bar{x}(t), \bar{u}(t)) \qquad (14)$$
$$D(\bar{x}(0)) = 0 \qquad (15)$$
$$E(\bar{x}(T)) = 0 \qquad (16)$$
$$F(\bar{x}(t), \bar{u}(t)) \leq 0 \qquad (17)$$
$$G(\bar{x}(t)) \leq 0 \qquad (18)$$

where $\bar{x} \in R^{2n}, \bar{u} \in R^n$ and

$$
\begin{aligned}
h_0 &: R^{2n} \to R \\
l_0 &: R^{2n} \to R \\
f_0 &: R^{2n} \times R^n \to R
\end{aligned}
$$

$$f \; : \; R^{2n} \times R^n \to R^{2n}$$
$$F \; : \; R^{2n} \times R^n \to R^j$$
$$G \; : \; R^{2n} \to R^r$$

There are j mixed inequality constraints (involving both state and control variables), and r pure state inequality constraints. All the above functions are assumed to be at least C^1 (C^k implies k times differentiable). The formulation above is in the form of a general optimal control problem, and it includes the framework for all manipulator motion planning problems. The final time T may be fixed or free. We will now translate each of the above conditions to the case of manipulator motion planning. The control vector $\bar{u}$ represents the torque/force applied at each joint and is of dimension n for an n degree-of-freedom manipulator. The state vector $\bar{x}$ corresponds to $(\bar{q}_p, \bar{q}_v)$ where $\bar{q}_p$ denotes the generalized coordinate vector representing the configuration of the manipulator. Thus $\bar{x}$ corresponds to the vector of joint positions and velocities and is of dimension $2n$. The state equations describing the dynamics of the system are given by Eqs. (5) and (6). The mixed constraints involving both control and state vectors arise from the restrictions on the torques/forces available from the actuators as expressed by Eq. (9) and have been separated from pure state dependent constraints which do not involve the control variables. For reasons explained in §1, control, and therefore planning, is more attractive in joint space. Therefore it is desirable to express the obstacle avoidance conditions in joint space. The most obvious solution is the use of the configuration space obstacles [18]. The generation of the *C-functions* to represent the obstacle in configuration space is not a trivial problem . This is because not only must the *C-functions* be used to generate the *C-Space* obstacle but their domain of applicabilty must also be determined [19].

In [22] the obstacle avoidance conditions were obtained through the use of the distance function between the obstacle K_i and the moving link K_j as:

$$d_{ij}(\bar{x}) = \min\{|z_i - z_j| : z_i \in K_i(\bar{x}), z_j \in K_j(\bar{x})\} \qquad (19)$$

and enforcing the condition that $d_{ij} \geq d_{\min}$. This required the computation of the minimum distance between two objects which turns out to be computationally the most intensive part of their procedure. A further complication arose from the fact that any optimizing algorithm (in their case, the interior penalty function method) requires the computation of the gradient of the constraints and the cost function, which was not guaranteed to exist for the distance function except for strictly convex objects.

An attempt to introduce time optimal path-planning with obstacle avoidance was made in [21]. In this work, the algorithm in [13] was used as the main tool to compute the minimum time over a specified path, $t_p(\bar{a})$, where the path was represented by the parameter vector $\bar{a}$. Different representations were chosen for the path, such as straight line segments joined by circular arcs and splines. For a given representation of the path the minimum time algorithm was used, but a penalty function was added to the cost. The constrained minimization that was performed was of the form:

$$J = t_p(\bar{a}) + \sum_{i=1}^{q} \frac{w_i}{d_i^2} + \sum_{i=1}^{n} \frac{w_{\theta_i}}{(\theta_i - \theta_{imin})^2} + \frac{w_{\theta_i}}{(\theta_i - \theta_{imax})^2} \qquad (20)$$

where w_i represents a weighting factor, d_i is the shortest distance from the links to the i-th obstacle and q is the number of obstacles. This penalty function penalized the planner for trajectories which came too close to the obstacles. This, too, required computing distances between objects. This was done by using a table of various general geometric shapes for computing minimum distance to the obstacles. Since this work concentrated on time-minimization only, a 'pruning technique' was used to weed out paths which were not globally optimal.

In [20] the use of the distance function was eliminated by pointwise mapping of the link boundary into task space and expressing the collision avoidance conditions as inequality constraints. The obstacle boundary may be expressed in the task space as

$$p(\bar{z}) \leq 0 \quad \bar{z} \in R^3, \quad p : R^3 \to R \qquad (21)$$

It is assumed that p is smooth for which the first derivatives exist. In the coordinate system fixed to link j (the local coordinate system), the shape of the link may be approximated by the set S_j which contains the coordinates of the nodal points, referenced to the local coordinate system, of a discrete mesh covering the link boundary. For a given configuration of the manipulator, let $^0T_j(\bar{q}_p, \bar{a})$ represent the mapping that maps an element from the local coordinate system attached to link j, to the task coordinate system. The vector $\bar{q}_p$ denotes the generalized coordinates for the manipulator and $\bar{a}$ denotes the vector of link parameters that describe the geometry of the manipulator. The obstacle avoidance condition may now be expressed as follows:

Given any point b, such that $b \in S_j$, the collision avoidance condition can be represented by the inequality

$$h_b(\bar{q}_p) > 0 \qquad (22)$$

where $h_b(\bar{q}_p) = p \circ t$.

$$t(\bar{q}_p) =^0 T_j(\bar{q}_p, \bar{a})b;$$

h_b is continuously differentiable in $\bar{q}_p$.

A discussion of some practical issues arising from discretization and measures to improve the computational efficiency may be found in [20].

3.2   Discrete Approximation of the Continuous Problem Formulation

We shall discretize the continuous system described by Eqs. (13-18) in §3.1 in order to apply numerical methods. The time interval $[0, T]$ is decomposed into $N - 1$ intervals by the points $0 = t_1 < t_2 < \ldots < t_N = T$. The discrete system is then described by the following recurrence relations

$$\bar{x}_{k+1} = \bar{x}_k + \int_{t_k}^{t_{k+1}} f(\bar{x}_k, \bar{u}_k)dt \qquad k = 1, \ldots N - 1 \tag{23}$$

$$\bar{x}_1 = \bar{x}(0) \tag{24}$$

The functional to be minimized assumes the form

$$J = h_0(\bar{x}_1) + \sum_{k=1}^{N-1} \int_{t_k}^{t_{k+1}} f_0(\bar{x}_k, \bar{u}_k)dt + l_0(\bar{x}_N) \tag{25}$$

The inequality constraints from Eqs. (17) and (18) are enforced at the discrete points. We thus obtain the following conditions corresponding to Eqs. (15-18).

$$D(\bar{x}_1) = 0 \tag{26}$$

$$E(\bar{x}_N) = 0 \tag{27}$$

$$F(\bar{x}_k, \bar{u}_k) \leq 0, \qquad k = 2, \ldots, N - 1 \tag{28}$$

$$G(\bar{x}_k) \leq 0, \qquad k = 2, \ldots, N - 1 \tag{29}$$

In the continuous system, $\bar{u}$ was a vector function. In the discrete approximation, the control is a finite dimensional vector, $\bar{w}$, given by:

$$\bar{w} = [\bar{u}_1, \bar{u}_2, \ldots, \bar{u}_{N-1}] \in R^{n(N-1)} \tag{30}$$

Given the complete control vector $\bar{w}$ and the initial state vector $\bar{x}_1$ we can determine the complete state vector from Eq. (23) as:

$$\bar{x} = [\bar{x}_1, \bar{x}_2, \ldots, \bar{x}_N] \in R^{2nN}$$

Substituting this into Eq. (25) we can obtain the value of the functional J. Thus the only independent variables which can be varied in conformity with the constraints are the components of $\bar{w}$. We denote this dependency as $\bar{x} = \bar{x}(\bar{w})$ and $J = J(\bar{x}(\bar{w}), \bar{w})$. Moreover, we are concerned only with the case where the initial and final state vectors are specified. In such cases, Eqs. (26) and (27) may be written as

$$\bar{x}_1 = \bar{x}(0) \tag{31}$$

$$\bar{x}_N = \bar{x}(T) \tag{32}$$

3.2.1 The Resultant Discrete Optimal Control Problem

The overall finite dimensional problem we are dealing with has now been converted into a discrete optimal control problem with state constraints, as described below:

$$\text{minimize} \quad J = h_0(\bar{x}_1) + \sum_{k=1}^{N-1} \hat{f}_{0k}(\bar{w}) + l_0(\bar{x}_N) \tag{33}$$

$$\text{where} \quad \hat{f}_{0k}(\bar{w}) = \int_{t_k}^{t_{k+1}} f(\bar{x}_k, \bar{u}_k) dt \tag{34}$$

and $\bar{x}_k$ is defined by the recurrence relation from Eq. (23) with initial condition given by Eq. (24) and $\bar{w}$ is defined as in Eq. (30). The constraints from Eqs. (28) and (29) may be written

$$F(\bar{x}_k(\bar{w})) \leq 0 \qquad k = 2, \ldots, N-1 \tag{35}$$

$$G(\bar{x}_k(\bar{w})) \leq 0 \qquad k = 2, \ldots, N-1 \tag{36}$$

The equation describing the final state is

$$\bar{x}_N = \bar{x}_1 + \sum_{k=1}^{N-1} \int_{t_k}^{t_{k+1}} f(\bar{x}_k, \bar{u}_k) dt \tag{37}$$

For variable-time problems the vector $\bar{w}$ is

$$\bar{w} = [u_{11}, u_{21}, \ldots, u_{n1}, u_{12}, \ldots, u_{n-1N-1}, u_{nN-1}, \delta t_1, \delta t_2, \ldots \delta t_{N-1}]$$

where u_{ij} refers to the i component of the vector $\bar{u}$ in the j stage. The overall problem now has $n(N-1) + (N-1)$ variables, with $2n$ equality constraints furnished by Eq. (37) and a total of $(N-2)(j+r)$ inequality constraints from Eqs. (35) and (36). It is now in the form of a non-linear programming problem and any suitable algorithm may be used to solve it.

This approach was implemented in [20]. The particular algorithm chosen to solve the optimization problem in this case is from the software IDESIGN

[23]. It is a Recursive Quadratic Programming algorithm which exploits the *Active Set Strategy*. The main aspect of this technique that makes it attractive for this case is that the quadratic sub-problem solved at each iteration employs the linearized constraints from the *Active Set* only [25], as compared to, for example, that of Han [24], which incorporates all constraints at each iteration. *This means that the set of constraints used for computing the next improved estimate will contain only those which are in violation by more than a specified tolerance.* This is specially meaningful for obstacle avoidance because in a particular configuration only a small percentage of the points on the boundaries of the links would violate the constraints and would therefore become candidates for generating the inequality constraints. For an initially estimated path the dynamic programming algorithm was employed to provide a dynamically feasible estimate which was also sub-optimal in the sense that the velocity profile was optimized *for the given geometric path*. The main planner then used the recursive quadratic programming algorithm, which incorporates obstacle avoidance conditions, to improve the original estimate and at the same time ensure geometric and dynamic feasibility.

The above approach was implemented on a two-link manipulator moving from (90 deg, 0 deg) to (30 deg, −120 deg) while attempting to avoid the disk obstacle shown in Figure 5. The maximum torque was restricted to 3 N-m for link 1 and 2 N-m for link 2. The initial estimate provided was of first folding the outer link completely from 0 deg to −120 deg while holding link 1 stationary and then moving link 1 from 90 deg to 30 deg while holding the outer link stationary. For 16 stages the total number of variables were 48 (32 control and 16 time steps). The number of nodal points on link 2 were 20 so that theoretically there would have been 300 inequality constraints from the obstacle avoidance and another 16 from the requirement that the time intervals lie within a certain range. However, because of our use of the active set strategy, the maximum number of inequality constraints used at any time during the search was 36.

Figure 5a shows the motion of the two-link manipulator and how it avoids the disk obstacle. Notice that the strategy is to initially move the inner link away from the obstacle and fold the outer link down. Figure 5b shows the variation of the control torques for the two joints. Each control switches twice and is extremal (or *bang-bang*) in both cases.

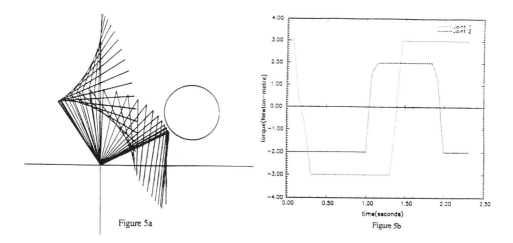

Figure 5a   Figure 5b

Figure 5: The motion of the two-link manipulator to avoid the disk obstacle and the corresponding joint torques.

4   Real-Time Planners: Implementations and Issues

The above discussion has indicated that there has been considerable emphasis on designing trajectory planners *separate* from the feedback control loop. The task of the feedback controller in such cases is to simply guide the joints along the trajectory generated by the planner. With increasing emphasis on *intelligent* controllers, this separation between the trajectory planner and feedback controller was recognized to be a significant drawback, mainly because the trajectory was planned off-line. This limits the capability of the system to react to a dynamic environment or make decisions in real-time. This prompted several researchers to investigate the possibility of a planner embedded inside the feedback control loop. The problem of obstacle avoidance in the context of real-time control was considered in [15]. The obstacle boundary was used to create a potential field around the obstacle which repelled the manipulator while the goal attracted it. Non-linear dynamics were taken into consideration by first casting the dynamics equations in 'operational space coordinates' in the form

$$\Lambda(x)\ddot{x} + \mu(x,\dot{x}) + p(x) = F \tag{38}$$

and decoupling them by explicit non-linear feedback

$$F = \Lambda(x)F^\star + \mu(x, \dot{x}) + p(x) \tag{39}$$

The operational space coordinates are related to the joint space coordinates by $x = J^T(q_p)q_p$. Here $F^\star$ represents the command vector of the force on the decoupled end-effector. The command vector was determined as the gradient of a potential field which was the sum of an attractive potential $U_{x_d}(x)$ attracting the end-effector to a goal x_d and a repelling field $U_o(x)$ created by the obstacles which repelled the end-effector. The command vector was therefore

$$F^\star = -\nabla[U_{x_d}(x)] - \nabla[U_o(x)] \tag{40}$$

For real-time implementation the non-linear decoupling control structure was updated only over several sampling time instants. No consideration was given to moving towards the goal in the most optimal manner. The computation of the potential function required evaluation of the minimum distance between the links and obstacles. A real-time obstacle avoidance scheme for the end-effector of a two-link planar manipulator, moving amidst circular or elliptical obstacles, was implemented in [16]. This scheme required the evaluation of the distance between the end-effector and the obstacle and modified the trajectory using a pointwise optimization strategy.

Motivated by the desire to accomplish real-time trajectory planning incorporating some level of decision-making, a *decentralized tracking and centralized planning structure* has been proposed [26]. The approach adopted is an indirect one: rather than accounting for the complete non-linearities of the system under control, the decisions can be made using simplified linear reference models which attempt to capture the behavior *as much as possible*; this would allow the planning of fairly complex strategies (such as optimal motion) and the execution phase for the controller can then be defined as forcing the actual system to track the trajectories planned by the linear reference model *as closely as possible*. Since planning and execution are coupled, (the tracking capability of the controller is of interest to the planning phase since it must compensate for the errors) the characteristics of the tracking controller are of immense importance to the planner. These include:

1. How 'good' is the tracking capability of the controller? This may be answered by characterizing the region of ultimate boundedness of the system. Of course, this presupposes the convergence and stability of the closed-loop system.

2. What is the transient behavior of the system before it settles into this region?

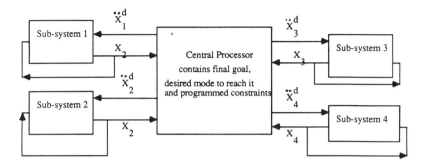

Figure 6: The decentralized sub-system controllers and central planner.

3. What is the magnitude of the control effort required during tracking?

4. How do various system parameters affect the ability to track (and therefore influence the planner which must compensate for expected deviations)?

The block diagram in Figure 6 shows the organization of the decentralized controllers for each sub-system, tracking a trajectory $\ddot{\bar{x}}_i$ which has been issued by the central processor which communicates with each sub-system. Each sub-system is equipped with position, velocity and force sensors which measure the current state of the system, $\bar{x}_i$, and send this information to the central processor. The centralized processor then uses the knowledge of the current states and the desired goal, *programmed by the designer*, together with the constraints under which the task must be accomplished, also programmed by the designer, to generate the command $\ddot{\bar{x}}_i$.

5 Design of Decentralized Tracking Controllers With Constrained Control

Several works have dealt with tracking controllers in robotic systems. Wen and Bayard [38] displayed exponential stability of the closed-loop through carefully chosen Lyapunov functions. Paden and Panja [36] described a globally stable tracking controller which employed an outer dynamic compensation loop to feed forward the nominal joint torques and other non-linear terms. Variable structure controllers have been developed by Young [29] and Slotine and Sastry [30]. They offer a computationally simple implementation compared to the above ones requiring dynamic compensation. In [30] and a sequel to that work [31] the use of variable structure control for tracking was discussed in considerable detail. The tracking control of a system with non-linearities can be reduced to a problem of robust stabilization of an uncertain system [37]. Several works have dealt with this approach using discontinuous control [34] and continuous approximations of the discontinuous control [32]. In a related work [35] the authors investigated the stability domains for a class of linear time-invariant systems subject to uncertainties with bounded control and the requirement of "sliding mode". From the planning perspective, however, it is desirable to have a 'simple' and decentralized controller. In particular, for non-linear systems where the state vectors represent the generalized position and velocity variables, the system equations may be represented as in Eqs. (5) and (6). The problem of designing real-time trajectory planners may then be stated as follows:

> *If simple linear models are used to facilitate planning for the above nonlinear system, then (1) how can the nonlinearities be accomodated without explicitly compensating for them during the planning stage and (2) how can it be guaranteed that the linear models used for predicting future states and computing controls during planning, will prove accurate despite the nonlinear inter-actions?*

The decentralized approach is specially attractive because the actual tracking controller can be implemented on a microprocessor which uses only sensory data from the sub-system it is controlling.

For purposes of investigating a decentralized control scheme, each position and velocity is viewed as a subsystem of the overall system. Defining the state variables as $\bar{x}_i = (x_{pi}, x_{vi})$, the reference model may be represented as

$$\dot{\bar{x}}_i = A_i \bar{x}_i + B_i u_i + Z_i, \quad i = 1, 2, \ldots n. \tag{41}$$

where

$$A_i = \begin{bmatrix} 0 & 1 \\ 0 & 0 \end{bmatrix}$$

$$B_i = \begin{bmatrix} 0 \\ \beta_i \end{bmatrix}$$

$Z_i = i$-th row of the vector $\bar{f}$ where

$$\bar{f} = -[M(\bar{x}_p, \bar{x}_v)]^{-1}\bar{h}(\bar{x}_p, \bar{x}_v) + ([M(\bar{x}_p, \bar{x}_v)]^{-1} - diag(\beta_1 \ldots \beta_n))\bar{u}$$

Since M^{-1} is positive definite it follows that

$$M^{-1} = (\lambda_{\max}(M^{-1}))(I + E)$$

where $\lambda_{\max}(M^{-1}(\bar{x}))$ is the maximum eigenvalue of $M^{-1}(\bar{x})$ and

$$\|E\| \leq |1 - \frac{\lambda_{\min}}{\lambda_{\max}}| \leq 1$$

Hence β_i can be chosen as $\lambda_{max}(M^{-1}(\bar{x}))$. The objective of the controller is to find a control u_i such that the state $\bar{x}$ of the system tracks a reference trajectory $\bar{x}_d$, which is generated by the reference model

$$\dot{\bar{x}}_i^d = A_i x_i^d + B_i u_i^d, \quad i = 1, 2, \ldots n \tag{42}$$

Employing linear feedback in the error between the reference and actual models, $\bar{e}_i = \bar{x}_i - \bar{x}_i^d$, we obtain

$$u_i = u_i^d + K_i \bar{e}_i + p_i(\bar{e}_i, t)$$

The objective of the uncertain system design is to determine $p_i(\bar{e}_i, t)$. The use of this control results in the following equation describing the error dynamics

$$\dot{\bar{e}}_i = (A_i + B_i K_i)\bar{e}_i + B_i p_i(\bar{e}_i, t) + B_i H_i(\bar{x}, \bar{r}) \tag{43}$$

5.1 Representation of Nonlinear Interactions

The model we use for the plant is

$$\dot{\bar{x}}_i = A_i \bar{x}_i + B_i u_i + B_i H_i(t, \bar{x}), \quad i = 1, 2, \ldots, N. \tag{44}$$

where $H_i(t, \bar{x})$ describes the strength of interactions from other subsystems. The parameters A_i and B_i may be chosen by the designer as explained above. It should be noted that in this formulation, the nonlinear interactions are assumed to satisfy the matching property. There are several assumptions we can make about the structure and behavior of the interactions. We list these below. It must be remembered that the assumption made here would affect our controller structure:

A1. The magnitude of the interactions is bounded, i.e.,

$$\|H_i(t,\bar{x})\| \leq \rho \qquad (45)$$

The advantage of this approach is that only an upper bound on the magnitude of the uncompensated non-linearities/uncertainties is required; no prior knowledge of their *structure* is necessary.

A2. In several other cases, it is assumed that the interactions, rather than obeying a strict upper bound as above, are bounded linearly in states, i.e., there exist non-negative numbers, ξ_{ij} , such that

$$\|H_i(t,\bar{x})\| \leq \sum_{j=1}^{N} \xi_{ij}\|\bar{x}_j\| \qquad (46)$$

A3. A third assumption often employed by control engineers is the "slowly time-varying" property of the interactions, i.e.,

$$\|\dot{H}_i(t,\bar{x})\| \approx 0 \qquad (47)$$

Depending on the type of plant, each of these assumptions may or may not be valid. In the decentralized model representation, all non-linearities are represented as uncertainties to make the problem very general. In actual implementation, estimates of the model non-linearities would be available, which can be fed forward to cancel some of these non-linearities. Then the uncertainty vector would represent only the *deviation* from the estimated model which would be smaller in magnitude. This has significant influence on the tracking ability.

In the next section we give the details of one special decentralized controller that has been implemented for trajectory tracking [39]. Some useful discussions on decentralized controllers may also be found in [41,42,43].

5.2 Decentralized Variable Structure Control: A Case Study

Variable structure control, also called sliding mode control, has been discussed in several works [27,28,29]. Here only the issues relevant to the decentralized sub-systems outlined above will be presented. Let us first examine the error dynamics of each sub-system given by Eq. (43) in the absence of coupling nonlinearities or any uncertainty. In that case, well-known results from the theory of variable structure control can be used to demonstrate the asymptotic stability of both the surface and the origin. These results are then extended for the case of uncertain systems. The essence of designing a variable structure control is to select a sliding surface in the error state space of the form

$$\sigma = G_i^T \bar{e}_i = c_i e_{ip} + e_{iv}, \quad c_i > 0 \tag{48}$$

for the i-th sub-system. The objective of the controller design is to force the system on to this surface, which can be accomplished by satisfying

$$\sigma \dot{\sigma} < 0 \tag{49}$$

Once the system has reached the sliding surface it remains on it and satisfies the following:

$$c_i e_{ip} + e_{iv} = 0 \tag{50}$$

The control required to force the system to remain on the surface can be determined as u_{eq} from $\dot{\sigma} = 0$. This gives

$$u_{eq} = -(G_i^T B_i)^{-1} G_i^T \hat{A} e_i \tag{51}$$

where $\hat{A} = A + BK$. Therefore in the sliding mode the system satisfies

$$\dot{e}_i = [I - B_i (G_i^T B_i)^{-1} G_i^T] \hat{A} e_i \tag{52}$$

which, for the subsystem outlined in Eq. 41 means

$$\dot{e}_{ip} = e_{iv} \tag{53}$$

$$\dot{e}_{iv} = -c_i e_{iv} \tag{54}$$

Therefore the system behaves like a first order system with

$$e_{ip} = e_{ip}(\tau) \exp(-c(t - \tau))$$

$$e_{iv} = e_{iv}(\tau) \exp(-c(t - \tau))$$

Since the control actually employed to maintain the system on the sliding surface is a high frequency switching control (u_{eq} may be treated as the dc component of this high frequency chattering control), it has been proposed that these discontinuous control schemes may be approximated by smooth controls continuous across a thin boundary layer parallel to the sliding surface [30].

5.2.1   Variable Structure Control in Uncertain Systems

The modification of the controller for an uncertain system borrows ideas from [33]. The uncertainties are assumed to satisfy assumption A1, i.e. $\|H_i\| \leq \rho$. The controller employs the uncertainty compensation $u_N(e_i, t)$ given by

$$
u_N(e_i, t) = \begin{cases} -\dfrac{B_i \sigma \frac{\partial \sigma}{\partial e_i}}{\|B_i \sigma \frac{\partial \sigma}{\partial e_i}\|}(\alpha + \rho) & \text{if } \|\sigma\| \geq \epsilon \\[4mm] -\dfrac{B_i \sigma \frac{\partial \sigma}{\partial e_i}}{\|B_i \frac{\partial \sigma}{\partial e_i}\|\epsilon}(\alpha + \rho) & \text{if } \|\sigma\| < \epsilon \end{cases}
\tag{55}
$$

where $\alpha > 0$ is a measure of the available control for compensation of uncertainty and $\epsilon > 0$ is a thin boundary layer parallel to the sliding surface. It must be noted that the control is continuous. The characteristics of this controller that are of primary interest to the planner are the tracking accuracy and the bound on the acceleration of the command trajectory. These are summarized in the following lemmas [39].

Lemma 1. Trajectories of the system (43) under the control

$$
p_i(\bar{e}_i, t) = u_{eq} + u_N(e_i, t)
\tag{56}
$$

approach the origin and are ultimately bounded with respect to the set S_0 where

$$
S_0 = S_\delta \cap S_f
$$
$$
S_f = \{e_{iv} : \|e_{iv}\| \leq \gamma\}
$$
$$
\gamma = \frac{\epsilon}{1 + \frac{\alpha}{\rho} + \frac{c_i \epsilon}{\beta_i \rho}}
\tag{57}
$$

Figure 7 shows the geometric relationship between the various parameters and the relevant sets. It must be remembered that the Lyapunov function approach usually gives a conservative estimate and we would expect the actual region of ultimate boundedness to be smaller. Notice that the 'tightness'of the tracking which is indicated by how close the error remains to the origin is indicated by the magnitude of γ. Better tracking can be achieved with larger available control for compensation of the uncertainty (large α), smaller magnitudes of the uncertainties (small ρ) and larger sub-system inertia (small β_i). Larger values of c_i reduce the tracking error.

For the reference model in the form of a double integrator in Eq. (41) it can be easily shown that u_{eq} in Eq. (56) turns out to be $-ce_{iv}$. The total control effort is therefore

$$
U_i = u_{eq} + u_N(\bar{e}_i, t) + u_i^d
$$

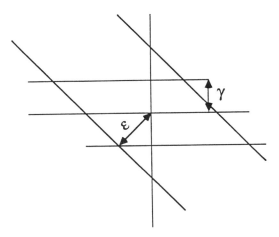

Figure 7: The region of ultimate boundedness of the tracking error.

By construction $\|u_N\| \leq (\alpha + \rho)$. Therefore the control effort required is

$$\|U_i\| \leq c\|e_{iv}\| + (\alpha + \rho) + \|u_i^d\|$$

During the reaching phase, the control effort will be

$$\|U_i\| \leq c\|e_{ivmax}\| + (\alpha + \rho) + \|u_i^d\| \tag{58}$$

Once the solution enters the region S_0 (with respect to which it is ultimately bounded) the control effort reduces to

$$\|U_i\| \leq c\gamma + \rho + \|u_i^d\| \tag{59}$$

Notice that the above equation tells us that the planner must plan trajectories so that it utilizes only u_i^d of the total available control. The remaining control effort is used to compensate for the nonlinear interactions and ensure bounded tracking error. These results are of particular significance since they restrict the maximum acceleration of the reference model trajectory $(\ddot{x}_i^d = \beta_i u_i^d)$ that can be tracked. For a planner which generates trajectories on-line this information can be used during planning.

Transient Behavior of the Tracking Controller

Lemma 1 characterizes the behavior of the system in steady state, but provides no information about system behavior before it settles into the region of ultimate boundedness. This behavior is summarized in the following lemmas:

Lemma 3. For the error dynamics of the decentralized system described by Eq. (41), with the control in Eq. (56) the set

$$S_\delta = \{\sigma : \|\sigma\| < \delta\}$$

is attractive everywhere where

$$\delta = \frac{\epsilon}{1 + \frac{\alpha}{\rho}}$$

As expected, the error system trajectories are confined closer to the surface with increasing α or decreasing For $\rho = 0$ (no uncertainty) the surface becomes asymptotically stable ($\delta = 0$). For very large uncertainties ($\rho \to \infty$) trajectories would still approach to within ϵ of the surface (provided the control ($\alpha + \rho$) is available).

Corollary. Let $\kappa \geq \delta$. The trajectories of the system (41) under feedback control (56) reach the set S_κ in time $T(\bar{e}_i(t_0))$ where

$$T(\bar{e}_i(t_0)) \leq T_1(\bar{e}_i(t_0), \epsilon) + T_2(\epsilon, \kappa) \tag{60}$$

where

$$T_1(\bar{e}_i(t_0), \epsilon) \leq \frac{\|\sigma(t_0)\| - \epsilon}{\alpha \beta_i}, \quad \|\sigma\| \geq \epsilon \tag{61}$$

and

$$T_2(\epsilon, \kappa) \leq \frac{\epsilon - \kappa}{c_0}, \quad \kappa \leq \|\sigma\| < \epsilon \tag{62}$$

where

$$c_0 = \min\{(\alpha + \rho)\frac{\|\sigma\|}{\epsilon} - \rho\} \leq \{\frac{\alpha}{\epsilon} - \rho(1 - \frac{\kappa}{\epsilon})\} \tag{63}$$

Lemma 3 establishes that the set S_δ is asymptotically attractive and all initial states outside the set S_δ will enter it in finite time and remain inside it. The corollary establishes an upper bound on this time. It does not, however, provide any information about the boundedness of the solution during the *reaching* phase, i.e when $\|\sigma\| > \epsilon$, (when the control employed is different from the one employed inside S_ϵ). Lemma 4 below establishes the boundedness of the solution for all times.

Lemma 4. Let the initial state $\bar{e}_{i0} = (e_{ip0}, e_{iv0})$ at time $t = 0$ of the decentralized system (43) under the feedback control (56) be such that $\bar{e}_{i0} \notin S_\epsilon$. Then the solutions of the system during the *reaching* phase ($\|\sigma\| > \epsilon$) are bounded as

$$e_{ivm1} \leq e_{iv}(t) \leq e_{ivm2}$$

$$e_{ipm1} \leq e_{ip}(t) \leq e_{ipm2}$$

for $\sigma_0 < 0$ and

$$e_{ivp1} \geq e_{iv}(t) \geq e_{ivp2}$$

$$e_{ipp1} \geq e_{ip}(t) \geq e_{ipp2}$$

for $\sigma_0 > 0$.

where

$$e_{ivm1} = e_{iv0}\exp(-ct) + \frac{\alpha\beta_i}{c}\{\exp(-ct) - 1\}$$

$$e_{ivm2} = e_{iv0}\exp(-ct) + \frac{\beta_i(\alpha + 2\rho)}{c}\{\exp(-ct) - 1\}$$

$$e_{ivp1} = e_{iv0}\exp(-ct) - \frac{\beta_i\alpha}{c}\{\exp(-ct) - 1\}$$

$$e_{ivp2} = e_{iv0}\exp(-ct) - \frac{\beta_i(\alpha + 2\rho)}{c}\{\exp(-ct) - 1\}$$

$$e_{ipm1} = e_{ip0} + \frac{\alpha\beta_i}{c}t + \frac{e_{iv0}}{c}\{1 - \exp(-ct)\} + \frac{\alpha\beta_i}{c^2}\{\exp(-ct) - 1\}$$

$$e_{ipm2} = e_{ip0} + \frac{\beta_i(\alpha + 2\rho)}{c}t + \frac{e_{iv0}}{c}\{1 - \exp(-ct)\}$$
$$+ \frac{\beta_i(\alpha + 2\rho)}{c^2}\{\exp(-ct) - 1\}$$

$$e_{ipp1} = e_{ip0} - \frac{\alpha\beta_i}{c}t + \frac{e_{iv0}}{c}\{1 - \exp(-ct)\} + \frac{\alpha\beta_i}{c^2}\{1 - \exp(-ct)\}$$

$$e_{ipp2} = e_{ip0} - \frac{\beta_i(\alpha + 2\rho)}{c}t + \frac{e_{iv0}}{c}\{1 - \exp(-ct)\}$$
$$+ \frac{\beta_i(\alpha + 2\rho)}{c^2}\{1 - \exp(-ct)\}$$

This enables us to construct a rectangular region in phase space at any instant within which the solution trajectory is constrained to lie (Figure 8). This is of particular significance for planning since it provides us with estimates of the maximum possible deviations before the solution *settles* down.

In the above formulation, we have expressed the dynamics in the joint-space. However, we can express the dynamics of the end-effector with respect to the Cartesian vector $\bar{X}$ as [15]

$$\bar{F} = M_x(\bar{\theta})\ddot{\bar{x}} + V_x(\bar{\theta}, \dot{\bar{\theta}}) + G_x(\bar{\theta}) \tag{64}$$

where $\bar{F}$ is a force torque vector acting on the end-effector. Using this as our original non-linear system, we can once again convert it into a decentralized

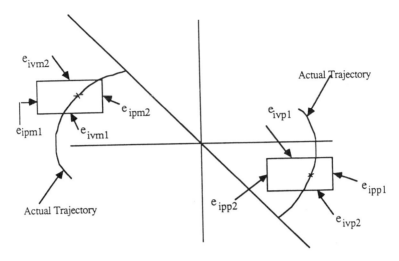

Figure 8: The rectangle bounding the transient behavior of the tracking controller.

system with linear reference models formulated in Cartesian variables. Techniques for efficient determination of the upper bound ρ_x and the limitation on the magnitude of the force vector, $\|F\|$, in this case are open problems.

It is important to understand the reason for designing a controller that tracks a simple linear reference model as compared to designing one which tracks a prescribed trajectory. With the advent of faster processors and accurate estimation of parameters in the dynamics Eqs. (5-6), one can compute the required torque for a given acceleration, velocity and configuration of the system. However, the distinction is that *the simpler reference models are needed to solve the inverse problem*, i.e. determine the desired acceleration given the current position and velocity and the programmed constraints. Also since in all decentralized models, the *tightness* of the tracking is determined by the magnitude of the nonlinear interactions between the various sub-systems, the tracking ability is enhanced if these interactions are small. One way of decreasing these estimates is to estimate the nonlinear interactions and feed them forward to each sub-system. Since these interactions depend on other states, this task must be accomplished by the central processor which has access to all other states. We have thus defined another major task of the centralized processor.

6 The Central Processor:Planning, Prediction and Modification

The central planner outputs the actual desired accelerations of the trajectory that must be tracked by each sub-system; further it is in constant communication with each sub-system so that the desired trajectory is generated based on the *desired goal* and the current state of each sub- system. In this section we shall examine the features of one such system that has been implemented [26]. The implementation of this environment has been done so that there are two levels at which the planner operates:

1. Automatic inclusion of constraints imposed by the *structure* of the decentralized controllers. Since the planner is aware of the limitations of the tracking controller and the controls, it accounts for these automatically while planning trajectories.

2. Transformation of user-defined constraints into executable trajectories.

An example of the first level is the automatic compensation employed by the planner to account for the tracking error. From the previous section we know that at any point in the position sub-space of the system, we can only guarantee the presence of the system inside a certain volume. Thus if a trajectory is planned we can only guarantee that the actual trajectory would be inside a *tube* of known radius as shown in Figure 9; we will call this the *bounding tube*. The cross section of the tube is the compact set characterizing the region of ultimate boundedness in the error phase space

$$\Omega_f = \{(\bar{e}_i) \in R^2 : V(\bar{e}_i) \leq V_f\}$$

where V_f is the final value of V which is a Lyapunov function for the decentralized sub-system.

Another strategy that the planner employs is the *control-switching* strategy. Since the tracking controller exhibits poor performance as a precision controller, the planner uses a two layered controller structure. During the *goal reaching* stage, when the trajectory must be planned on-line and uncertainties in the environment must be accounted for, the tracking controller is employed. Once the object has entered the the region of ultimate boundedness around the goal state, (the goal state is characterized by zero velocity), the control is switched to a pure position controller (we have implemented a constant gain PID controller) which ensures zero position error.

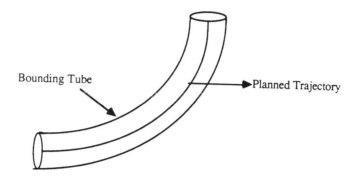

Figure 9: The bounding tube

We now turn our attention to the environment that the designer must program to execute tasks. There are essentially three major functions of the planner:

- ability to make future predictions about system states based on linear reference models.

- ability to coordinate motions between various sub-systems.

- ability to modify trajectories if new constraints are added to the system.

6.1   The Desired Goal and Mode

The first information that the designer must program is the *ultimate objective*. This is specified in the form of the *goal state* denoted by the state $\bar{x}_f$. The designer must then prescribe the *desired* mode for accomplishing this task, which tells the planner of *how it should try to reach the goal state*. Some of the possibilities are:

Accomplish the motion in minimum time.

Since the reference model that each decentralized sub-system attempts to follow is characterized as a linear second-order system, any optimal control strategy may be derived using the theory of linear systems. In particular, we consider the minimum time solution for transferring the system from the current state $(\bar{x}_c, \dot{\bar{x}}_c)$ to the desired final state $(\bar{x}_f, \dot{\bar{x}}_f)$. Since each sub-system is second order, we know that the optimal solution would involve the use of saturated control at all times with at most one switching of the control variable (in this case the acceleration of each sub-system) which is restricted as

$$- \ddot{x}_{imax} \leq \ddot{x}_i \leq \ddot{x}_{imax} \tag{65}$$

This optimal motion could be used as the primary motion control scheme for moving the object from one point to another (initial and final velocities zero) or as an intermediate control strategy for transitions between two different trajectories (which necessitates matching non-zero velocities at the initial and terminal points). For each subsystem, the minimum-time solution may be computed assuming that the control switches at time t_1 and also assuming, without loss of generality, that the control is $+\ddot{x}_{imax}$ is t_1 and $-\ddot{x}_{imax}$ for time t_2.

$$T_i = t_1 + t_2 \tag{66}$$

For the case of motion from rest to zero terminal velocity this reduces to

$$t_1 = t_2 = \frac{\sqrt{x_{if} - xic}}{\ddot{x}_{imax}} \tag{67}$$

Since this yields a different minimum-time for each subsystem, the actual strategy is to choose the transition time as the maximum value of T_i (the one for the slowest sub-system)

$$T = \max\{T_i\} \tag{68}$$

The remaining subsystems are then forced to use the same time during transition by scaling down the maximum allowable acceleration as

$$\ddot{\hat{x}}_{jmax} = \frac{T_j}{T_i}\ddot{x}_{jmax} \tag{69}$$

User defined parameterization for each sub-system trajectory.

In such cases, the user must actually prescribe the desired mode; for example, the system may be commanded to *move from current state $\bar{x}_{ci}$ to final state $\bar{x}_{fi}$ as a second order system given by*

$$\ddot{\bar{x}}_i = K_p(\bar{x}_{fi} - \bar{x}_{ci}) - K_v\dot{\bar{x}}_i; \tag{70}$$

These strategies are executed by simple commands like **MOVE**$(\bar{x}_c, \bar{x}_f, \text{minimum-time})$.

6.2   The Constraint Set

The constraint on available control provides us with an inequality constraint as given by Eq. (59). There could also be other constraints that must be satisfied during task execution. These must be specifically programmed by the designer. The task of the planner is to transform these into state variable inequality constraints which must be satisfied during trajectory generation. Examples of such constraints include:

- **maintain-safe-distance**(d_p); This tells the planner to move so that at any time its state is such that it can come to a halt within a prescribed distance d_p.

- **maintain-distance-plane**($\bar{n}_1, > d$); This command tells the planner to move so that the distance (in the Euclidean norm sense) between the system states and the plane denoted by the normal n_1 is greater than d.

- **maintain-distance-point**($\bar{a}, > d$); This command tells the planner that tasks must be executed so that the Euclidean norm of the distance between the system states and the point $\bar{a}$ is greater than d.

- **maintain-ratio**($\bar{x}_1, \bar{x}_2, \ldots, \bar{x}_n, \lambda_1, \lambda_2, \ldots, \lambda_n$); This command tells the planner to execute motions so that the ratio of the movement by the sub-systems is proportional to $\lambda_1, \lambda_2, \ldots, \lambda_n$. In some sense, this may be looked upon as an *electronic gearing* where the user can define the gear-ratios.

We will now examine how the planner translates these commands to yield *state variable constraints which are linear in the acceleration of the desired trajectories.*

Prediction of Permissible States

We examine the case where the trajectory must be planned so that the object can be brought to rest within a predetermined distance of travel which is the prescribed *safe distance* d_p. Therefore the state $(\bar{x}_c, \dot{\bar{x}}_c)$ is permissible if and only if

$$\|\bar{s}\| \leq d_p \tag{71}$$

$$s_i = \dot{x}_{ic} T_i + \frac{1}{2} \ddot{x}_{ic} T_i^2 + \|e_{ip}\| \tag{72}$$

$$\ddot{x}_{ic} = -sgn(\dot{x}_{ic})\ddot{x}_{icmax} \tag{73}$$

$$T_i = \frac{-\dot{x}_{ic}}{\ddot{x}_{ic}} \tag{74}$$

where $\ddot{x}_{icmax}$ is determined from the maximum permissible acceleration of the desired trajectory that can be tracked and e_{ip} is the position tracking error.

Trajectory Planning When Steering Away From Given Plane

We now look at the planner which attempts to *steer* the object away from a direction. We characterize this direction by the unit vector $\bar{n}_1$. Determination of $\bar{n}_1$ must be accomplished depending on the objective. Two possibilities are suggested:

1. The object is attempting to steer away from a plane. In this case the plane is characterized by the constant normal $\bar{n}_1$.

2. The object is steering from a given point. In this case the normal n_1 is characterized by

$$n_1 = \frac{\bar{x}_c - \bar{a}}{\|\bar{x}_c - \bar{a}\|} \tag{75}$$

where $\bar{a}$ is the point position. Case 1 is relatively simple since only the component along n_1 must be reduced to zero and $\bar{n}_1$ is a constant vector. The velocity component along this direction is

$$v_{n1} = \dot{\bar{x}}_c \cdot \bar{n}_1 \tag{76}$$

The acceleration along this direction is

$$a_{n1} = \ddot{\bar{x}}_c \cdot \bar{n}_1 \tag{77}$$

As before, for constant $\bar{n}_1$, the component a_{n1} is constant permitting us to define the permissible state using the concept of the *safe distance* d_p as

$$\|s\| \leq d_p \tag{78}$$

$$s = v_{n1}T + \frac{1}{2}a_{n1}T^2 + \sum_{i=1}^{N}\|n_i e_{ip}\| \tag{79}$$

$$T = -\frac{v_{n1}}{a_{n1}} \tag{80}$$

It must be remembered that this gives us an additional constraint that must be satisfied when generating the trajectory to accomplish the desired goal. It does not actually alter the desired goal which must be done by a higher level planner. For Case 2 the normal vector $\bar{n}_1$ varies with the position of the

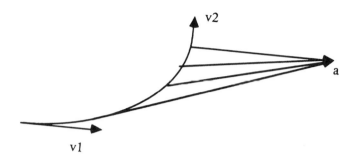

Figure 10: Steering away from a point: the case of variable n_1.

object and must be recomputed each sampling instant, as shown in Figure 10. We now look at this case in more detail.

Steering Away From a Given Point.

Let the position of the point be $\bar{a}$ and the current state of the object be $(\bar{x}_c, \dot{\bar{x}}_c)$. Then the *approach velocity* is given by

$$v_a = \dot{\bar{x}}_c \cdot \frac{\bar{x}_c - \bar{a}}{\|\bar{x}_c - \bar{a}\|}$$

Assuming that this rate is negative (the object is moving towards this point), we are interested in determining the controls that would steer the object away from this point. Suppose we could apply a *constant* component of the acceleration $a_{\max}$ which would reduce the approach velocity v_a to zero. A permissible state could once again be defined by ensuring $s \leq d_p$. The difficulty with this is that a constant value of $a_{\max}$ implies position dependent accelerations of each sub-system, the relationship between the acceleration of the sub-systems and the deceleration of the approach velocity a_a being

$$a_a = \sum_{i=1}^{N} \ddot{x}_i \frac{x_i - a_i}{\|\bar{x} - \bar{a}\|} \tag{81}$$

The permissible commanded accelerations are therefore defined as those for which the commanded accelerations of the sub-systems $\ddot{x}_i$ satisfy the inequality

$$a_a \geq a_{\max}$$

whenever $v_a < 0$.

Note that in all cases the final inequality constraints $s \leq d_p$ and $a_a \geq a_{\mathrm{max}}$ are linear in the variables $\ddot{\bar{x}}_i$ and may be converted to general constraints of the form

$$B\ddot{\bar{x}} - \bar{c} \leq 0 \tag{82}$$

6.3 Generating Desired Trajectories From Knowledge of Goal, Mode and User Defined Constraints

In the previous sections we described how the designer could prescribe the *goal, the desired mode of reaching the goal* and *the constraints that must be satisfied.* We then saw how these constraints could be interpreted as linear inequality constraints by the planner. The next step is to solve for the desired trajectories. The most general solution to these problems is by solving the Quadratic Programming Problem

$$\min(\ddot{\bar{x}} - \ddot{\bar{x}}^d)^T Q(\ddot{\bar{x}} - \ddot{\bar{x}}^d) \tag{83}$$

subject to

$$\|\ddot{x}_i\| \leq \ddot{x}_{imax}, \quad i = 1, 2, \ldots, n \tag{84}$$

which arise from the control limitation in Eq. (59) and

$$B\ddot{\bar{x}} - \bar{c} \leq 0 \tag{85}$$

which arise from the user defined constraints described above.

The trajectory planning problem may now be solved as a Quadratic Programming problem with linear inequality constraints which may be easily implemented in real-time [3]. In our case, the problem is often simplified by the fact that because we are dealing with decentralized systems, each of which is describable as a linear system, the QP problem is tackeld directly in the acceleration subspace, $\ddot{\bar{x}}$, instead of the control subspace. Further, this reduces the problem to a much simpler one since nonlinear dynamics do not need to be accounted for. Although complicated minimizations may be tackled by this QP solver, it is possible to use simpler direct methods for many problems. Some of these are discussed in the next section.

7   Some Examples of Planning Strategies

We now illustrate how the planner generates the desired accelerations from commands programmed by the user. Suppose it is desired to accomplish a straight-line motion between the initial and final points in configuration space in minimum-time. This can be done as follows:

MOVE$(\bar{x}_0, \bar{x}_f,$ minimum-time) and **maintain-ratio**$(\bar{x}_1, \bar{x}_2, \ldots, \bar{x}_n, (\bar{x}_{f1} - \bar{x}_{01}), (\bar{x}_{f2} - \bar{x}_{02}), \ldots, (\bar{x}_{fn} - \bar{x}_{0n}))$;

We have previously discussed (§6.1) how the command **MOVE**$(\bar{x}_c, \bar{x}_f,$ minimum-time) may be used to produce the desired acceleration for each joint by scaling the accelerations of the faster joints so that they are synchronized with the 'slowest' joint. Now we will see how the additional requirement of straight-line motion generates constraints which must also be satisfied. The straight line in configuration space may be described as

$$\bar{x}(t) = \bar{x}_0 + \lambda(\bar{x}_f - \bar{x}_0), \quad 0 \leq \lambda \leq 1 \tag{86}$$

Let the current state of the system at time $t = 0$, be $(\bar{x}_c, \dot{\bar{x}}_c)$. Let the commanded accelerations during the sampling interval T be $\ddot{\bar{x}}$. Then the accelerations of the various sub-systems must satisfy:

$$\frac{x_i(T) - x_{ci}}{x_j(T) - x_{cj}} = \frac{\int_0^T \dot{x}_{ci} d\tau + \int_0^T \int_0^T \ddot{x}_i d\zeta d\tau}{\int_0^T \dot{x}_{cj} d\tau + \int_0^T \int_0^T \ddot{x}_j d\zeta d\tau} = \frac{x_{if} - x_{0i}}{x_{fj} - x_{0j}} \tag{87}$$

For n sub-systems, these equations give us a set of $n - 1$ equality constraints of the form

$$A\ddot{\bar{x}} - \bar{b} = 0 \tag{88}$$

which must be satisfied. Similarly, if it is desired to move in minimum time with the added restriction that the trajectory must be safe with respect to distance d_p, the program is changed to:

MOVE$(\bar{x}_c, \bar{x}_f,$ minimum-time) and **maintain-ratio**$(\bar{x}_1, \bar{x}_2, \ldots, \bar{x}_n, (\bar{x}_{f1} - \bar{x}_{c1}), \ldots, (\bar{x}_{fn} - \bar{x}_{cn}))$ and **maintain-safe-distance**(d_p).

Trajectory Planning for Obstacle Avoidance.

Similarly, let us examine another problem. The object shown in Figure 11 is attempting to reach the *target position* $\bar{x}_f$ from the current position x_c. We now make the following assumptions about the detection of the obstacle:

1. The boundary of the obstacle can be smoothly approximated by $f(\bar{x}) = 0$; and the boundary of the *grown* obstacle by $f(\bar{x}) = \|e_{ip}\| = \hat{f}(\bar{x})$. This accounts for the position uncertainty. We can determine the plane:

$$\nabla \hat{f} \cdot (\bar{x}_f - \bar{x}) = 0 \tag{89}$$

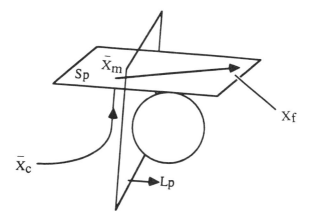

Figure 11: The collision avoidance motion

This plane is *tangent* to the grown obstacle boundary and passes through the final position. We will call this plane the *sliding plane*, S_p.

2. We also determine the *limiting plane*, L_p, perpendicular to S_p and located such that it is a supporting plane for the obstacle.

This entire sequence may be programmed by the following program:

MOVE$(\bar{x}_c, \bar{x}_f,$ minimum-time) and **maintain-safe-distance**(d_p) until obstacle is detected. When obstacle is detected detrmine $\bar{x}_m$ which lies on S_p ahead of the limiting plane L_p (a simple way to do this is to determine point of intersection of normal to L_p from $\bar{x}_c$). Now the following two-phase movement is programmed.

MOVE$(\bar{x}_c, \bar{x}_m,$ minimum-time) and **maintain - distance - plane**$(L_p, > 0)$; then **MOVE**$(\bar{x}_m, \bar{x}_f,$ minimum-time) and **maintain-ratio**$(\bar{x}_1, \bar{x}_2, \ldots, \bar{x}_n, (\bar{x}_{f1} - \bar{x}_{m1}), \ldots, (\bar{x}_{fn} - \bar{x}_{mn}))$;

Figure 11 shows the trajectory followed. The object begins to move in a straight line until it detects the presence of an obstacle in the path. The object is then steered away from L_p and begins to move towards S_p and comes to rest when it intersects S_p. The final motion is accomplished by *sliding* on S_p. A two-dimensional version of the same strategy shows more clearly how the controls are selected. Figure 12a shows the relevant configuration and Figure 12b shows the rectangle describing the acceleration constraints. The

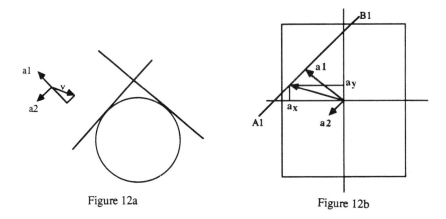

Figure 12a   Figure 12b

Figure 12: The two-dimensional collision avoidance geometry and the acceleration rectangle.

acceleration component a_1 decelerates the movement of the obstacle towards L_p (a line in this case) and the magnitude is determined to be greater than $a_{\max}$. Figure 12a shows this component. The component a_2 controls the movement towards S_p and must therefore lie along the line A_1B_1 in Figure 12b which is within the constraint rectangle. For example, let the desired acceleration direction be as shown by a_2 in Figure 12a with a magnitude $\|a_2\|$. Then the accelerations are chosen as shown in Figure 12b.

Planning Trajectories in Spaces Different from the Controlled Configuration Space

   In many situations, the control may be executed in a different space (the configuration space) from the one in which the task is prescribed (the Task space) and must be planned. Of particular importance in this case is the knowledge of the relationship between the tracking error in the configuration space and that in the task space. From Eq. (3) we know that:

$$\bar{z} = \Psi(\bar{\theta}) \tag{90}$$

Let the bounds on the error in tracking in the control subspace be given as

$$\|\delta\theta_i\| \le \|e_{ip}\| \tag{91}$$

Then we can write the error in task space for small values of $\delta\theta_i$ as

$$\delta\bar{x} = \Psi_{\bar{\theta}}\delta\bar{\theta} \tag{92}$$

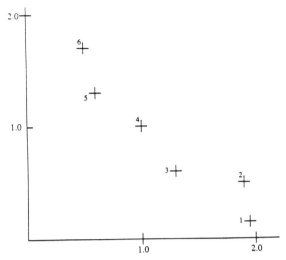

Figure 13: The desired positions in Cartesian Space

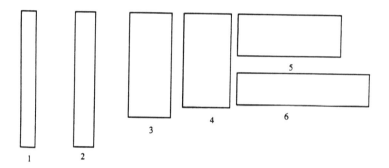

Figure 14: The relative sizes of the uncertainties in Cartesian space

The error in the i-th coordinate direction is therefore

$$\delta x_i = \sum_{j=1}^{N} J_{ij}\delta\theta_j \tag{93}$$

where

$$J = \Psi_{\bar{\theta}} \tag{94}$$

denotes the Jacobian of the transformation evaluated at the current configuration $\bar{\theta}_c$. From Eq. (93) it follows that

$$\|\delta x_i\| \leq \sum_{j=1}^{N} \|J_{ij}e_{jp}\| \tag{95}$$

The primary importance of these results lies in the fact that they indicate that the *magnitudes* of the uncertainties in positions are not constant but

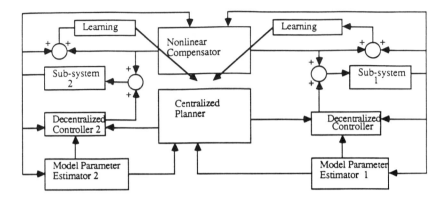

Figure 15: The proposed architecture

configuration and therefore time-dependent. Figure 13 shows 6 different positions that a planar revolute manipulator attempts to reach by moving the two joints. If the tracking error in each joint is 1 degree, the corresponding errors in the Cartesian space are shown by means of the rectangular boxes 1 to 6 in Figure 14. In each case, the desired Cartesian position lies at the geometric center of the box, but because of the uncertainty in position due to tracking errors, we are only guaranteed that the end-effector position lies somewhere within the box. These must be considered when planning a precision task such as an assembly operation.

8   Future Directions and Open Problems

This article has attempted to outline the issues in trajectory planning in automated dynamic systems. The emphasis has been on understanding the complexities as well as the limitations involved in formulating this as a controller design problem. It would be erroneous to conclude, however, that such a trajectory planner could completely replace a *high level* planner, capable of making complicated logical decisions.

Figure 15 shows the proposed architecture for implementing a complete dynamic trajectory planner. The *model parameter estimator* estimates the most appropriate linear model of each sub-system. The central planner issues the command trajectory to be tracked and the *decentralized controller* issues the signal for tracking. The *nonlinear compensator* feeds forward estimates of the nonlinear interactions. The *learning mechanism* updates the estimates of the nonlinear compensator. The importance and the effectiveness of these trajectory planners can be greatly enhanced with more advances in the following:

- Rapid processing of data from multiple sensors. Any trajectory planner must plan the trajectory based on information about its environment. For a robotic system these include position, velocity, force and vision sensors. Obstacle avoidance schemes rely on a knowledge of the location and the shape of the obstacles. The utility of such schemes for real-time operation will be limited unless techniques can be developed that can provide this information rapidly and accurately to the planner.

- Efficient techniques, implementable in real-time, for parameter identification, estimation and learning of system parameters. This will directly influence the accuracy of the models used for planning and prediction of future states.

- Better nonlinear controllers, capable of superior tracking of the planned trajectory, despite deviations from the model assumed for planning. Ideally, we would like to ensure that these tracking controllers ultimately track with zero error. Further, this property of global asymptotic stability should be guaranteed in the face of constrained control. This result can be expressed as a bound on the rate of change of the planned trajectory as a function of the control limitation and the system dynamics.

Finally, there is little doubt that the advent of superior processors can enhance the capability to plan more complex tasks without making assumptions or neglecting the nonlinear dynamic terms inherent in robotic systems.

References

[1] J. T. SCHWARTZ and C. K. YAP, "Advances in Robotics : Vol.1," Lawrence Earlbaum Associates, New York, 1986.

[2] J. Y. S. LUH, M. W. WALKER and R. P. PAUL, "Resolved Acceleration Control of Mechanical Manipulators," *IEEE Transactions on Automatic Control*, Vol. AC-25, No. 3, pp. 468-474 (1980).

[3] M. W. SPONG, J. S. THORP and J. M. KLEINWAKS, "The Control of Robot Manipulators with Bounded Input," *IEEE Transactions on Automatic Control*, Vol. AC-31, No. 6, pp. 483-490 (1986).

[4] R. H. TAYLOR, Planning and Execution of Straight Line Manipulator Trajectories, *IBM Journal of Research and Development*, Vol. 23, pp. 424-436 (1979).

[5] R. A. FINKEL, "Constructing and Debugging Manipulator Programs," Memo 284, Stanford Artificial
     Intelligence Laboratory, Stanford University, CA (1976).

[6] C. S. LIN, P. R. CHANG and J. Y. S. LUH, "Formulation and Optimization of Cubic Polynomial Joint
     Trajectories for Industrial Robots," *IEEE Transactions on Automatic Control*, Vol. AC-28, No. 12, pp
     1066-1073 (1983).

[7] R. P. C. PAUL, "Modeling, Trajectory Calculation and Servoing of a Computer Controlled Arm," AI
     Memo 177, Stanford AI Laboratory, Stanford University, CA (1972).

[8] L. S. PONTRYAGIN, V. G. BOLTYANSKI, R. V. GAMKRELIDZE and E. F. MISCHENKO, "The
     Mathematical Theory of Optimal Processes," New York: Wiley (Interscience Publishers), 1962.

[9] A. MIELE and A. K. WU, "Sequential Gradient Restoration Algorithm for Optimal Control Problems
     With Non-Differential and General Boundary Conditions: Parts 1 and 2," *Optimal Control Application
     and Methods*, Vol. 1, No. 1, pp. 69-88 and pp. 119-130 (1980).

[10] H. H. TAN and R. B. POTTS, "Minimum Time Trajectory Planner for the Discrete Dynamic Robot
     Model with Dynamic Constraints," *IEEE Journal of Robotics and Automation*, Vol. 4, No. 2, pp.
     174-185 (1988).

[11] K. G. SHIN and N. D. McKAY, "A Dynamic Programming Approach to Trajectory Planning of Robotic
     Manipulators," *IEEE Transactions on Automatic Control*, Vol. AC-31, No. 6, pp. 491-500 (1986).

[12] S. K. SINGH and M. C. LEU, "Optimal Trajectory Generation for Robotic Manipulators using Dynamic
     Programming," *ASME Journal of Dynamic Systems, Measurement and Control*, Vol. 109 No.2, pp.
     88-96 (1987).

[13] J. E. BOBROW, S. DUBOWSKY and J. S. GIBSON, "Time Optimal Control of Robotic Manipulators
     Along Specified Paths," *International Journal of Robotics Research*, Vol. 4, No. 3, pp. 3-17 (1985).

[14] M. E. KAHN and B. E. ROTH, "The Near-Minimum Time Control of Open Loop Articulated Kine-
     matic Chains," *ASME Journal of Dynamic Systems, Measurement and Control*, Vol. 93, No.3, pp.
     164-172 (1971).

[15] O. KHATIB, "Real-Time Obstacle Avoidance for Manipulators and Mobile Robots," *The International
     Journal of Robotics Research*, Vol. 5, No. 1, pp. 90-98 (1983).

[16] S. K. KHERADPIR and J. S. THORP, "Robust Real Time Control of Robot Manipulators in the
     Presence of Obstacles," *Proceedings of the IEEE Conference on Robotics and Automation*, Raleigh,
     NC, pp. 1146-1151 (1987).

[17] W. S. NEWMAN, "Automatic Obstacle Avoidance at High Speeds via Reflex Control," *Proceedings of
     the 1989 IEEE Conference on Robotics and Automation*, Scottsdale, Arizona, pp. 1104-1109 (1989).

[18] T. LOZANO-PEREZ, "Spatial Planning : A Configuration Space Approach," *IEEE Transactions on
     Computers*, Vol. C-32, No.2, pp. 108-120 (1983).

[19] B. R. DONALD, "Motion Planning with Six Degrees of Freedom," Technical Report AI-TR-791, Ar-
     tificial Intelligence Laboratory, Massachussets Institute of Technology (1984).

[20] S. K. SINGH and M. C. LEU, "Manipulator Motion Planning in the Presence of Obstacles and Dynamic
     Constraints," (in press), *The International Journal of Robotics Research* (1990).

[21] Z. SHILLER and S. DUBOWSKY, "Global Time Optimal Motions of Robotic Manipulators in The
     Presence of Obstacles," *Proceedings of the IEEE Conference on Robotics and Automation*, Philadel-
     phia, PA, pp. 370-375 (1988).

[22] E. G. GILBERT and D. W. JOHNSON, "Distance Functions and their Applications to Robot Path
     Planning in the Presence of Obstacles," *IEEE Journal of Robotics and Automation*, Vol. RA-1, No. 1,
     pp. 21-30 (1985).

[23] J. S. ARORA "Theoretical Manual for IDESIGN," Technical Report ODL-85.9, Department of Me-
     chanical Engineering, University of Iowa (1985).

[24] S. P. HAN, "A Globally Convergent Method for Non-Linear Programming," *Journal of Optimization Theory and Applications*, Vol. 22, No. 3, pp. 297-309 (1977).

[25] B. N. PSCHENICHNY and Y. M. DANILIN, "Numerical Methods in Extremal Problems," Mir Publishers, Moscow 1978.

[26] S. K. SINGH, "The ConSERTD Effort: A Control System Environment for Real-Time Decision-Making," *Proceedings of the IEEE Conference on Robotics and Automation*, Cincinnati, Ohio (1990).

[27] R. A. DeCARLO, S. H. ZAK and G. P. MATTHEWS, "Variable Structure Control of Nonlinear Multivariable Systems: A Tutorial," *Proceedings of the IEEE*, Vol. 76, No. 3, pp. 212-232 (1988).

[28] V. I. UTKIN, "Variable Structure Systems with Sliding Modes: A Survey," *IEEE Transactions on Automatic Control*, Vol. 22, pp. 212-222 (1977).

[29] K. K. D. YOUNG, "Design of Variable Structure Model Following Systems ," *IEEE Transactions on Automatic Control*, Vol. AC-23, No. 6, pp. 1079-1085 (1977).

[30] J. J. E. SLOTINE and S. SASTRY, "Tracking Control of Non-linear Systems Using Sliding Surfaces with Applications to Robot Manipulators," *International Journal of Control*, Vol. 38, No. 2, pp.465-492 (1983).

[31] J. J. E. SLOTINE, "Sliding Controller Design for Non-linear Systems," *International Journal of Control*, Vol. 40, No. 2, pp. 421-434 (1985).

[32] M. J. CORLESS and G. LEITMANN, "Continuous State Feedback Guaranteeing Uniform Ultimate Boundedness for Uncertain Dynamic Systems," *IEEE Transactions on Automatic Control*, Vol. AC-26, No.5, pp. 1135-1144 (1981).

[33] G. LEITMANN, "On the Efficacy of Nonlinear Control in Uncertain Linear Systems," *ASME Journal of Dynamic Systems, Measurement and Control*, Vol. 102, pp. 95-102 (1981).

[34] S. GUTMAN and Z. PALMOR, "Properties of Min-Max Controllers in Uncertain Dynamical Systems," *SIAM Journal of Control and Optimization*, Vol. 20, No. 6, pp. 850-861 (1982).

[35] M. HACHED, S. M. MADANI and S. H. ZAK, "Stabilization of Uncertain Systems Subject to Hard Bounds on Control with Application to a Robot Manipulator," *IEEE Journal of Robotics and Automation*, Vol. 4, No. 3, pp. 310-323 (1988).

[36] B. PADEN and R. PANJA, "Globally Asymptotically Stable 'PD+' Controller for Robot Manipulators," *International Journal of Control*, Vol. 47, No. 6, pp. 1697-1712 (1988).

[37] M. SPONG and H. RAMIREZ, "Robust Control Design Techniques for a Class of Non-linear Systems," *Proceedings of the American Control Conference*, pp. 1515-1522 (1986).

[38] J. T. WEN and D. S. BAYARD, "New Class of Control Laws for Robotic Manipulators- Part 1. Non-adaptive Case and Part 2. Adaptive Case," *International Journal of Control*, Vol. 47, No. 5, pp. 1361-1406 (1988).

[39] S. K. SINGH, "Decentralized Variable Structure Control for Tracking in Nonlinear Systems," *(in press)* *The International Journal of Control*. Also Technical Report CAR-89-01, Thayer School of Engineering at Dartmouth, Hanover, NH (1989).

[40] S. K. SINGH, "Trajectory Planning for Manipulators Under Uncertainty," *Proceedings of the Fourth IEEE International Symposium on Intelligent Control*, Albany, NY, pp. 152-158 (1989).

[41] L. SHI and S. K. SINGH, "On Decentralized Adaptive Controllers Based on The Direct Method of Lyapunov," Technical Report CAR-90-01, Thayer School of Engineering at Dartmouth, Hanover, NH (1990).

[42] P. A. IOANNOU, "Decentralized Adaptive Control of Interconnected Systems," *IEEE Transactions on Automatic Control*, Vol. AC-31, pp. 291-298 (1986).

[43] D. T. GAVEL and D. D. SILJAK, "Decentralized Adaptive Control: Structural Conditions for Stability," *IEEE Transactions on Automatic Control*, Vol. AC-34, No. 4, pp. 413-426 (1989).

SIMPLIFIED TECHNIQUES FOR ADAPTIVE CONTROL
OF
ROBOTIC SYSTEMS

Izhak Bar-Kana and Allon Guez

Department of Electrical and Computer Engineering

Drexel University

Philadelphia, PA 19104

TABLE OF CONTENTS

CONTROL AND DYNAMIC SYSTEMS, VOL. 40

I. INTRODUCTION

Most adaptive control techniques (Landau, 1979; Åström, 1983) assume the prior knowledge of, or on an upper bound, on the order of the unknown controlled plant. Prior knowledge on the pole-excess in the plant was also needed. This knowledge is then used to implement observer-based controllers of the same order as the plant. Since these assumptions may not be satisfied in realistic large systems, a simplified adaptive control approach was developed (Sobel, Kaufman, and Mabius 1979, 1982; Bar-Kana and Kaufman, 1982a, b). The developers of this approach were very much influenced by the progresses of adaptive control methods of the last years of the seventies, especially by the model reference and stability analysis approach of Landau (1979), Monopoli (1974), and Narendra (1979). However, they were also interested in the simple adaptive control methods of the sixties (Whitaker, 1959; Osburn et al., 1961; Donaldson and Leondes, 1963; Parks, 1966). As from the very beginning they had to deal with real-world large scale systems (Kaufman and colleagues, 1981) the developers of this approach were forced to start by taking into consideration the real world constraints, which only lately have received consideration in the adaptive and general control design under the area of "robustness":

1) The order of real world plants is large and probably unknown, and the models we are able to use for the control design are only low-order approximations of the real plant. Even if the order is known, it is normally too large to allow using controllers of the same (or of larger) order as the plant.

2) The same for the pole-excess, or in the general multivariable case, for the MacMillan degree of the plant.

3) Although the algebraic conditions for the minimum order of stabilizing controllers are an outstanding unsolved problem, most real-world systems can be

easily stabilized with simple controllers. Many plants are even stable to start with. However, stability is only the first and necessary but not sufficient condition, because performance is what the control designer is interested in. Fixed controlers may not be able to maintained the desired performance under various operational conditions.

The problem that arises from the above constraints and assumptions is: can the basic stabilizability of the controlled plants, combined with specific nonlinear adaptive control techniques, fit the right gains to the right operational situation such that performance is obtained while stability is guaranteed? The answer is not necessarily positive, whenever nonstationary or nonlinear controllers are used. Yet, starting with modest results (Sobel, Kaufman and Mabius, 1982), which only presented the first conditions needed for the stable behavior of a basic simplified adaptive control algorithm, this approach was developed and extended to result in a simple, robust, and well-performing algorithm (Bar - Kana, 1989d) whose implementation is feasible for most real-world applications.

The first motivation and the representative examples of this approach are large flexible structures. Whether we treat them as finite systems or as infinite dimensional systems, their order is very large and unknown. Still, they may be stabilizable via low-order controllers. Fixed low-order controllers may not be able, however, to guarantee the desired performance, and this is the job we leave for the nonlinear adaptive controllers. Although one can find examples of plants that cannot be stabilized by controllers of lesser order than the plant, these are not the common examples, if processes, planes, missiles, manipulators, etc... are concerned, rather then the "general model," and as Root-Locus or Bode analysis show when the information on the plant permits such an analysis. The high order of the plant and the "unmodeled dynamics" do not necessarily justify the problems that some adaptive control algorithms met with the well-known now Rohrs examples, (Rohrs and colleagues, 1982, 1985) which are both stable and minimum-phase, and could have been controlled even by a controller of the zero order (constant).

Recently, it was shown for stationary and nonstationary multivariable linear systems (Bar-Kana and Kaufman, 1985; Bar-Kana, 1986a, 1987a, 1989b, 1990) that stable simple adaptive controllers (**SAC**) can be implemented using some vague prior knowledge or assumptions about the basic stabilizability properties of the plant to be controlled. In fact, the simple result states that if a plant can be stabilized by some simple configuration, the adaptive controller can adaptively compute the desired controller gains, without using any real initial knowledge. Stability of the nonlinear adaptive control system is guaranteed if the inverse of some stabilizing configuration is used as parallel feedforward to augment the plant to be controlled (Bar-Kana, 1987a, b, 1989c). The (multivariable) augmented controlled system is now called "almost strictly passive (**ASP**)," because it was shown that there exists some positive definite static (unknown) output feedback, such that the fictitious resulting closed-loop system is strictly passive (also strictly positive real, in linear time-invariant systems), and strict passivity is characterized by some relations which guarantee the robust stability of nonlinear controllers.

In other words, the passivity (or positive realness in stationary linear systems) conditions which are required to guarantee stability of the nonlinear adaptive control system are in fact a different and equivalent formulation of the basic stabilizability properties of the plant to be controlled. The result is a simple and robust direct Model Reference Adaptive Control algorithm, and its applications now include nonminimum-phase and unstable, nonstationary or non-linear examples of: missiles, flexible structures, motors, helicopters, manipulators, and other servo systems. In fact, the applications are now trying to collect the most difficult examples that can be found. The simplicity of implementation may make it attractive for engineers with practical applications on their mind, although theoretically one can ask: how do you know that the system is stabilizable? This question is not so serious when one tries to control his process, plane or manipulator, rather than the general control model. The adaptive control designer must take into consideration that most control designs and even robust control designs are not adaptive. The problem of the control designer is performance rather than stability of the control system. However, without guaranteeing stability, performance can not be even discused. When called to select the fixed gains of the controller, the designer must trade-off among his various choices. High gains imply high maneuverability, but they also imply high cost of control, high noise amplification and "nervous" response even if not necessarily needed. Low-gains imply smooth response and low cost of control, but they do not provide the needed maneuverability when it is required. "Optimal" gains may sometimes provide a suitable compromise, but may also not be satisfactory for any of these cases. Changing the gains "adaptively," or in accord with the specific situation in order to maintain the performance in the various operational environments is an attractive idea, but then we use nonstationary control and stability is not necessarily guaranteed in general, even if stability of the correspondent linear system was. In the past it was shown for linear systems, that **SAC** guarantees stability with nonlinear adaptive controllers over the entire region that could have been used by fixed gains one at a time. This article extends this result for a class of nonlinear systems linear in control. While it has no intention to present some general solution for the general problem, this approach may thus be offering a simple and robust adaptive solution for many difficult realistic problems, and a completion of preliminary linear designs based only on vague knowledge about the plant to be controlled.

If a controlled system is **ASP**, then simple adaptive control procedures find the necessary gains that can stabilize and control the system without having to know or use this fictitious static feedback or any prior knowledge about the plant. This property can be very useful if one attempts to present some solution when the order of the plant is very large and unknown, because in this case very low-order adaptive controllers may even control large flexible structures which are passive under idealistic assumptions, or can be made passive in realistic environments(Bar-Kana, Kaufman and Balas, 1983; Bar-Kana and Kaufman, 1984; Balas, Kaufman and Wen, 1984; Ih, Wang and Leondes, 1987) .

While the reader with the "general problem" in his mind may still wonder on the basic assumption about the prior availability of some stabilizing configuration, this result is useful in particular control designs, when some prior knowledge on the plant to be controlled, usually available (not more than to know that it is a motor, a manipulator, or a flexible structure), is sufficient to stabilize the plant, even though not necessarily sufficient to obtain the desired performance. In many cases, not necessarily proper controllers may be desired, like the PD controllers with the general (matrical) transfer function $H(s)=K(1+s/s_0)$ (Kidd, 1986; Slotine and Li, 1988; Bar-Kana, Fischl, and Kalata, 1989). However, while Kidd (1986) does not use this desired controller because of the need of derivatives, and while Slotine and Li (1988) try to cope with the noisy tachometer, we only exploit the assumed existence of the *fictitious* improper controller and actually use its proper inverse in parallel with the plant. This way, although we only use position measurement sensors, we still get the effect of the desired differentiators without really differentiating (Bar-Kana, 1987a, 1988c). Since it was proven that global stability and robustness of the nonlinear adaptive controller operating on the augmented plant is guaranteed, the adaptive controller is now called to fit the right gains to the right situation in order to minimize the tracking errors (Bar-Kana, 1987b).

An objective evaluation of this approach as compared with others, although still based on its first results, can be found in the paper of Mattern and Shoureshi (1987). Meanwhile, new papers and reports may better explain some aspects that were not clear enough in the past. Along with new developments in the theory and applications, attempts are made to write some explanation that would intuitively give the motivation behind our approach. Section 4 of Bar-Kana (1987a) may be the best introduction to this approach, and is mainly concerned with SISO systems. One may also note the nonlinear manipulator application of this paper. A new tutorial (Bar-Kana, 1989d) which may be the most complete presentation of the theory and applications so far, gives the necessary explanation and motivation for the various steps and parts of **SAC** in multivariable systems, using the minimal amount of mathematics possible.

Good challenge for this simple adaptive control algorithm in difficult practical applications, can be offered by adaptive control of robots, missiles, or planes. The missiles (Guez, 1980), manipulators (Guez and Dritsas, 1987), and planes (Morse and Ossman, 1989), like the flexible structures, are in fact stable or easily stabilizable configurations. Yet, they are difficult control objects because stability does not necessarily imply performance. In various operational environments the control parameters (angles, loads, inputs, disturbances) change quite a bit and fixed controllers may have a difficult task trying to maintain good performance. Here, **SAC** may provide a solution: the necessary gains are computed automatically such that the performance (small tracking errors) is maintained because the adaptive controller may use the entire domain of admissible gains, the right gain at the right time. We mention in particular the flight control reconfiguration after multiple control surface failure (Morse and Ossman, 1989). The adaptive controllers moves rapidly from some values of the adaptive gains to new value, more fitted to the new situation after failure, thus maintaining a stable flight instead of what could have been total catastrophe. Therefore, this approach is quite different from the other mainstream adaptive control methodologies, who try to use adaptation in order to reach some fixed "optimal" controller. **SAC** may be considered a first, modest, or even primitive example of intelligent control, because it only uses a single control configuration, but attempts to fit the right controller gains to the right environmental and operational condition, in order to maintain the required performance, represented by small tracking errors.

However, although **SAC** was successfully applied in nonstationary and nonlinear systems, the profs of stability only cover linear systems. The state-space representation of the systems and the Lyapunov stability analysis of the adaptive algorithms allowed in the past for a unified presentation of SISO and multivariable systems (Bar-Kana, 1987b, 1989c), and the first attempts at nonlinear systems show that the techniques can be rigorously extended to the classes of nonlinear systems that include, for example, robots and missiles.

This article extends these concepts to nonlinear multivariable systems. Almost strict passivity is defined in Section II, and Section III shows that it is a direct result of the stabilizability properties of the system. The robust model reference adaptive control algorithm for almost passive systems is presented in Section IV, the robot application in Section V, and some conclusions in Section VI.

II. ALMOST PASSIVITY

Let us consider a dynamic nonlinear plant (Fig. 1) given in the following state-space representation

$$\dot{x}_p(t) = A_p(x_p)x_p(t) + B_p(x_p)u_p(t) \tag{1}$$

$$y_p(t) = C_p(x_p)x_p(t) \tag{2}$$

$$y_a(t) = y_p(t) + D_p(x_p)u_p(t) = C_p(x_p)x_p(t) + D_p(x_p)u_p(t) \tag{3}$$

where $x_p(t) \in \mathbb{R}^n$, $y_p(t) \in \mathbb{R}^m$, $y_a(t) \in \mathbb{R}^m$, $u_p(t) \in \mathbb{R}^m$, and where the matrices $A_p(x_p)$, $B_p(x_p)$, $C_p(x_p)$, $D_p(x_p)$ are uniformly bounded. We use the not necessarily strictly causal form $\{A_p(x_p), B_p(x_p), C_p(x_p), D_p(x_p)\}$ only for the sake of completion of the treatment, and the following results remain perfectly valid if the systems are strictly causal ($D_p(x_p) = 0$). Furthermore, we denote the output in (2) by $y_p(t)$ and the output in (3) by $y_a(t)$ to emphasize that usually the actual plant is strictly causal with output $y_p(t) = C_p(x_p)x_p(t)$, and the term $D_p(x_p)u_p(t)$ might be added to obtain some useful properties of the augmented system.

It is easy to see that by using the controller

$$u_p(t) = -K_e y_a(t) + u_{pc}(t) \tag{4}$$

we get the following closed-loop system (Fig. 2)

$$\dot{x}_p(t) = A_{pc}(x_p)x_p(t) + B_{pc}(x_p)u_{pc}(t) \tag{5}$$

$$y_a(t) = C_{pc}(x_p)x_p(t) + D_{pc}(x_p)u_{pc}(t) \tag{6}$$

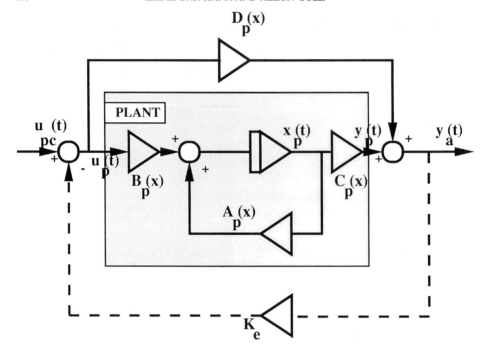

Fig.1: The first closed-loop configuration.

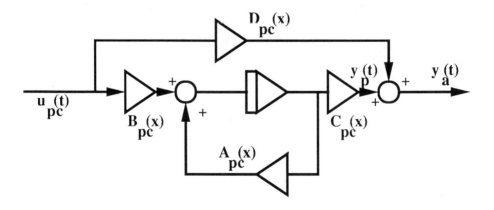

Fig. 2: The second equivalent closed-loop system.

where

$$A_{pc}(x_p) = A_p(x_p) - B_p(x_p)K_{ec}(x_p)C_p(x_p) \tag{7}$$

$$K_{ec}(x_p) = K_e[\ I + D_p(x_p)K_e\ ]^{-1} = [\ I + K_eD_p(x_p)\ ]^{-1}K_e \tag{8}$$

$$B_{pc}(x_p) = B_p(x_p)[\ I + K_eD_p(x_p)\ ]^{-1} \tag{9}$$

$$C_{pc}(x_p) = [\ I + D_p(x_p)K_e\ ]^{-1}C_p(x_p) \tag{10}$$

$$D_{pc}(x_p) = [\ I + D_p(x_p)K_e\ ]^{-1}D_p(x_p) = D_p(x_p)[\ I + K_eD_p(x_p)\ ]^{-1} \tag{11}$$

and where $D_p(x_p)$ and $D_{pc}(x_p)$ are nonsingular. Relations (7)-(11) show that the closed loop system is also equivalent to the system represented in Fig. 3, which will be used later.

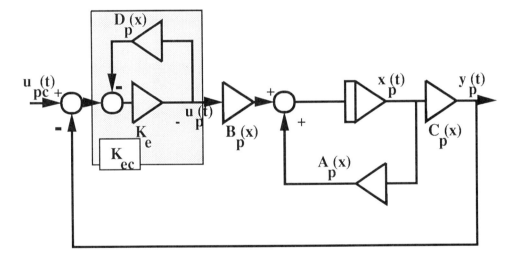

Fig. 3: The third equivalent closed-loop system.

We want now to introduce the passivity property in a form that is convenient for the development of this article. For additional information on passivity, the reader is referred to the references (Willems, 1972; Anderson and Moylan, 1974; Hill and Moylan, 1980; Desoer and Vidyasagar, 1975) for a better understanding of passivity. We define strict passivity in the folowing way:

DEFINITION: The system (5)-(6) is called strictly passive (**SP**) if there exist some uniformly bounded and positive definite matrices, $P(x_p)$ and $Q(x_p)$, and some matrices $L(x_p) \in R^{m \times n}$ and $W(x_p) \in R^{m \times m}$, where $\dot{P}(x_p) = \dfrac{dP(x_p)}{dt} = \dfrac{\partial P(x_p)}{\partial x} \dfrac{dx_p}{dt}$,

such that the following relations are satisfied:

$$\dot{P}(x_p) + P(x_p)A_{pc}(x_p) + A_{pc}^T(x_p)P(x_p) = -Q(x_p) - L^T(x_p)L(x_p) < 0 \tag{12}$$

$$P(x_p)B_{pc}(x_p) = C_{pc}^T(x_p) - L^T(x_p)W(x_p) \tag{13}$$

$$D_{pc}(x_p) + D_{pc}^T(x_p) = W^T(x_p)W(x_p) \tag{14}$$

It is easy to see (Willems, 1972) that relations (12) - (14) are equivalent to the following differential Riccati equation

$$\dot{P}(x_p) + P(x_p)A_{pc}(x_p) + A_{pc}^T(x_p)P(x_p) + [P(x_p)B_{pc}(x_p) - C_{pc}^T(x_p)] \cdot$$
$$\cdot [D_{pc}(x_p) + D_{pc}^T(x_p)]^{-1} [B_{pc}^T(x_p)P(x_p) - C_{pc}(x_p)] + Q(x_p) = 0 \tag{15}$$

Therefore, if the closed-loop system (5)-(6) satisfies relations (12)-(14), we call it "strictly passive" (**SP**). Because only a constant feedback separates between the open-loop plant (1)-(3) and strict passivity, we call (1)-(3) "almost strictly passive" (**ASP**):

DEFINITION 1: The system (1) - (3) is called "almost strictly passive (**ASP**)" if there exists a positive definite static feedback matrix K_e (unknown and not neded for implementation) such that the resulting closed-loop system (5)-(6) is strictly passive (**SP**) (Fig. 1).

The strict negativity of (12) is very useful when one wants to prove stability with nonlinear or adaptive controllers. It will be shown subsequently that it is a direct result of the basic stabilizability properties of the plants. The next section will also show that **ASP** systems are high gain stabilizable, property which is essential for the robustness of the nonlinear adaptive controllers.

III. THE "ALMOST PASSIVITY" LEMMAS

Though very desirable, the almost passivity relations are not usually satisfied by real-world plants. Therefore, a way has been proposed to obtain almost passive forms by using parallel feedforward based only on the fundamental stabilizability properties of the systems. Almost passive forms were recently obtained (Bar-Kana, 1986, 1987a, 1988b, 1989b) in stationary and nonstationary linear systems, based on "almost stabilizability" (Schumacher, 1984) (output stabilizability via static or dynamic output feedback). This article extends this relation to almost passivity" of nonlinear systems.

While in stationary systems, passivity and stability could be intuitively related using the pole-zero description of the systems (MacFarlane, 1982), the state-space representation must be used in the nonstationary or nonlinear case. We also use exponential stability as defined by Hahn (1967) with aplication to our specific structures. We will assume that if the plant is stabilizable the resulting stable configuration is "**exponentially stable**" as defined below:

DEFINITION 2 (Hahn, 1967): Let the general nonlinear system be represented by the n -order vectorial equation $\dot{x}(t) = f(x,t)$ and let x=0 be an equilibrium point The equilibrium point is called "**exponentially stable**" if all solutions x(t) satisfy the relation $|x(t)| \leq \alpha\, |x(0)|\, e^{-\beta t}$ for some scalars $\alpha > 0$ and $\beta > 0$.

THEOREM 1 (Hahn, 1967): Let the right hand of the $\dot{x}(t) = f(x,t)$ have bounded continuous first order partial derivatives. Let the equilibrium be **exponentially**

stable. Then there exists a Lyapunov function V(x,t) which satisfies estimates of

the form

1.1 $\alpha_1 |x(t)|^2 \leq V(x, t) \leq \alpha_2 |x(t)|^2$

1.2 $\dot{V}(x, t) \leq \alpha_3 |x(t)|^2$

1.3 $|\frac{\partial V(x,t)}{\partial x_i}| \leq \alpha_4 |x(t)|^2$, i=1,2,....,n

Because this article deals with nonlinear systems of the form (1)-(2), we cannot

expect to be able to find or even show existence, like in linear time-invariant

systems, of positive definite quadratic Lyapunov function of the form $V(x_p) =$

$x_p^T(t)Px_p(t)$ where P is constant and positive definite. However, after some

experience with specific nonlinear systems like robots, and because we restrict our

discussion to nonlinear systems linear in control of the form (1)-(2), we *assume*

that exponential stability of the *autonomous* system (1), with $u_p(t) \equiv 0$, implies

existence of nonlinear Lyapunov functions V(x) which are not explicit functions of

time. Furthermore, since we can always write $V(x_p) = x_p^T(t)P(x_p)x_p(t)$, and because

then $\dot{V}(x_p) = x_p^T(t)[\dot{P}(x_p) + P(x_p)A(x_p)+A^T(x_p)P(x_p)]x_p(t) = - x_p^T(t)Q(x_p)x_p(t)$, we

will restrict the subsequent discussion to systems that satisfy the following

assumption:

ASSUMPTION 1: Exponential stability of the *autonomous* system (1) ,with

$u_p(t) \equiv 0$, implies existence of Lyapunov functions of the form $V(x_p) =$

$x_p^T(t)P(x_p)x_p(t)$ and derivative of the form $\dot{V}(x_p) = x_p^T(t)Q(x_p)x_p(t)$, where $P(x_p)$ and

$Q(x_p)$ are positive definite for all $x_p \in R^n$. By using the same Lyapunov function

with the nonautonomous equation (1), it is easy to see that exponential stability

implies BIBO stability.

In order to introduce the new concepts and to make as clear as possible the relation

between passivity and the bounds of stability of the systems, we first prove a result

which is only valid if the plant is stabilizable by some static output feedback.

LEMMA 1: Let us consider the strictly causal plant

$$\dot{x}_p(t) = A_p(x_p)x_p(t) + B_p(x_p)u_p(t) \tag{16}$$

$$y_p(t) = C_p(x_p)x_p(t) \tag{17}$$

and let K_y be any positive definite static (constant or time-varying) output feedback matrix such that the closed loop system (Fig. 4) with the system matrix

$$A_y(x_p) = A_p(x_p) - B_p(x_p)K_yC_p(x_p) \tag{18}$$

is exponentially stable according with assumption 1. Then the augmented open-loop system (1)-(3) is almost strictly passive (**ASP**) if $D_p = K_y^{-1}$.

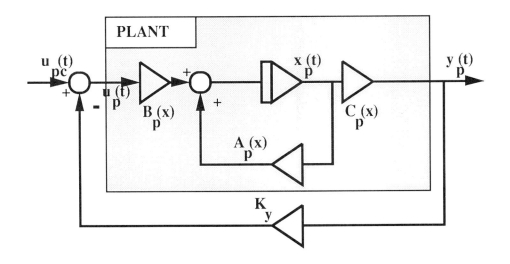

Fig. 4: The stabilized closed-loop system.

PROOF: Exponential stability stability of $A_y(x_p)$, implies that there exist two uniformly bounded and positive definite matrices, $P(x_p)$ and $Q_1(x_p)$, and a positive scalar α, such that if we select the quadratic Lyapunov function $V(x_p) =$

$x_p^T(t)P(x)x_p(t)$, its derivative along the trajectories of (16) is a negative definite quadratic function which satisfies the following differential Lyapunov equation (Hahn, 1967; Willems, 1970)

$$\dot{V}(x_p) = x_p^T(t)[\ \dot{P}(x_p) + P(x_p)A_y(x_p) + A_y^T(x_p)P(x_p)\ ]x_p(t) =$$
$$= - x_p^T(t)Q_1(x_p)x_p(t) < -\alpha\|x_p(t)\|^2 \qquad (19)$$

Let us address now the closed-loop system represented in Figure 1 and Figure 2. According with the assumptions of the lemma, the plant is augmented with constant parallel feedforward $D_p(x) = D_p = K_y^{-1}$. For the proof of lemma 1 we must show that there exists some K_e such that the fictitious closed-loop system is **SP**. For the present time, however, we need the formulas (5)-(11) for any K_e. Since the right term in (19) is positive definite, we can subtract a small positive semidefinite term, and the remainder will still be positive definite. For convenience of the proof, we consider some positive definite matrix K_{yc} and some positive scalar β, both sufficiently small, such that

$$x_p^T(t)Q(x_p)x_p(t) = x_p^T(t)\Big\{ Q_1(x_p) -$$

$$- \frac{1}{2}[P(x_p)B_{pc}(x_p) + C_{pc}^T(x_p)]\ K_{yc}\ [B_{pc}^T(x_p)P(x_p) + C_{pc}(x_p)]\ \Big\}x_p(t) > \beta\|x_p(t)\|^2 \quad (20)$$

because $P(x_p)$ is bounded , and $B_{pc}(x_p)$, and $C_{pc}(x_p)$ defined in (9)-(10) are bounded for any bounded K_e. We make use of these matrices defined in (6)-(10) as an useful step towards obtaining the passivity relations (11)-(13).

Substituting $Q_1(x)$ in (19) and moving all terms to the left side gives

$$x_p^T(t)\Big\{ \dot{P}(x_p) + P(x_p)A_y(x_p) + A_y^T(x_p)P(x_p) +$$

$$+\frac{1}{2}[P(x_p)B_{pc}(x_p) + C_{pc}^T(x_p)]\ K_{yc}\ [B_{pc}^T(x_p)P(x_p) + C_{pc}(x_p)] + Q(x_p)\ \Big\}x_p(t)\ 0 \quad (21)$$

which must be satisfied satisfied for any x_p, and therefore

$$\dot{P}(x_p) + P(x_p)A_y(x_p) + A_y^T(x_p)P(x_p) +$$

$$+\frac{1}{2}[P(x_p)B_{pc}(x_p) + C_{pc}^T(x_p)]\, K_{yc}\, [B_{pc}^T(x_p)P(x_p) + C_{pc}(x_p)] + Q(x_p) = 0 \qquad (22)$$

We try to select K_{yc} in terms of K_y and K_{ec} as follows

$$K_y = K_{yc} + K_{ec} \qquad (23)$$

where we use again the definition of K_{ec} in (7) to get

$$K_{ec} = [\, I + K_e D_p\,]^{-1}K_e = [\, I + K_e K_y^{-1}\,]^{-1}K_e \qquad (24)$$

We define similarly

$$K_{yc}\ [\, I + K_e D_p\,]^{-1}K_y = [\, I + K_e K_y^{-1}\,]^{-1}K_y \qquad (25)$$

and it is easy to see that indeed

$$K_{yc} + K_{ec} = [\, I + K_e K_y^{-1}\,]^{-1}[K_y + K_e] = K_y\,[K_y + K_e\,]^{-1}[K_y + K_e] = K_y \qquad (26)$$

Note that the only condition for this selection results from K_{yc} in (25). K_{yc} is a function of K_y and K_e. K_y is given and fixed, and the unknown K_e must be only sufficiently large such that K_{yc} be small enough. By substituting $A_v(x_p)$ from (18) and K_y from (23) into (22) we get

$$\dot{P}(x_p) + P(x_p)[A_p(x_p) - B_p(x_p)K_{ec}C_p(x_p)] + [A_p(x_p) - B_p(x_p)K_{ec}C_p(x_p)]^T P(x_p) -$$

$$- P(x_p)\, B_p(x_p)K_{yc}C_p(x_p) - C_p^T(x_p)\, K_{yc}\, B_p^T(x_p)P(x_p) +$$

$$+\frac{1}{2}[P(x_p)B_p(x_p) + C_p^T(x_p)]\, K_{yc}\, [B_p^T(x_p)P(x_p) + C_p(x_p)] + Q(x_p) = 0 \qquad (27)$$

or

$$\dot{P}(x_p) + P(x_p)A_{pc}(x_p) + A_{pc}^T(x_p)\,P(x_p) +$$

$$+\frac{1}{2}[P(x_p)B_p(x_p) - C_p^T(x_p)]\,K_{yc}\,[B_p^T(x_p)P(x_p) - C_p(x_p)] + Q(x_p) = 0 \qquad (28)$$

We rewrite K_{yc} as

$$K_{yc} = [\,I + K_eD_p\,]^{-1}K_y = [\,I + K_eD_p\,]^{-1}\left[\frac{D_p}{2} + \frac{D_p^T}{2}\right]^{-1}$$

$$= [\,I + K_eD_p\,]^{-1}\left[\frac{D_p}{2} + \frac{D_p^T}{2}\right]^{-1}[\,I + K_eD_p\,][\,I + K_eD_p\,]^{-1} \qquad (29)$$

and substitute it into (27) to get finally

$$\dot{P}(x_p) + P(x_p)A_{pc}(x_p) + A_{pc}^T(x_p)P(x_p) +$$

$$[P(x_p)B_{pc}(x_p) - C_{pc}^T(x_p)]\,[D_{pc} + D_{pc}^T]^{-1}[B_{pc}^T(x_p)P(x_p) - C_{pc}(x_p)] + Q(x_p) = 0 \quad (30)$$

which is the strict passivity relation (15) and thus proves the lemma.

In conclusion, if the plant (16)-(17) is exponentially stabilizable by some feedback K_y, we can use the inverse of K_y in parallel with the plant in order to get the augmented **ASP** system (1)-(2). Then there exists some other output feedback gain K_e *(unknown and not needed for any implementation)* which makes the closed loop system (4)-(5) strictly passive.

We will now show that **ASP** systems are high-gain stabilizable. We also want to emphasize the fact that implementation of **ASP** systems is a convenient use of the basic stabilizability properties of the plants that we want to control, first of all for plants that can be controlled by constant feedback, and later for more general plants that require dynamic controllers in order to reach stability. Since we only assume that some idea about stabilizability of the plant may be available, but do *not* assume that real prior knowledge about a good stabilizing feedback is available, let K_{max} be some vague estimate for the upper bound of the admissible class of stabilizing *fixed* gains K_y, and let $K_{ec_{min}} = K_{max} - K_{yc}$ be the minimal value of the gain K_{ec} that satisfies (28). Then, select $D_p = K_{max}^{-1}$, to get from (28)

$$\dot{P}(x_p) + P(x_p)[A_p(x_p) - B_p(x_p)K_{ec_{min}} C_p(x_p)] + [A_p(x_p) - B_p(x_p)K_{ec_{min}} C_p(x_p)]^T P(x_p)$$

$$= - Q(x_p) - \frac{1}{2}[P(x_p)B_p(x_p) - C_p^T(x_p)][K_{max} - K_{ec_{min}}] [B_p^T(x_p)P(x_p) - C_p(x_p)] \qquad (31)$$

For any other value of K_{ec}, constant or time-varying, that we denote for generality by $K_{ec}(t)$, we check the differential Lyapunov equation for the system closed by $K_{ec}(t)$ to get

$$\dot{P}(x_p) + P(x_p)[A_p(x_p) - B_p(x_p)K_{ec}(t)C_p(x_p)] + [A_p(x_p) - B_p(x_p)K_{ec}(t)C_p(x_p)]^T P(x_p)$$

$$= \dot{P}(x_p) + P(x_p)[A_p(x_p) - B_p(x_p)K_{ec_{min}} C_p(x_p)] + [A_p(x_p) - B_p(x_p)K_{ec_{min}} C_p(x_p)]^T P(x_p)$$

$$- P(x_p)B_p(x_p)[K_{ec}(t) - K_{ec_{min}}]C_p(x_p) - C_p^T(x_p)[K_{ec}(t) - K_{ec_{min}}]^T B_p^T(x_p)P(x_p) =$$

$$= - Q(x_p) - \frac{1}{2}[P(x_p)B_p(x_p) - C_p^T(x_p)][K_{max} - K_{ec_{min}}] [B_p^T(x_p)P(x_p) - C_p(x_p)] -$$

$$- P(x_p)B_p(x_p)[K_{ec}(t) - K_{ec_{min}}]C_p(x_p) - C_p^T(x_p)[K_{ec}(t) - K_{ec_{min}}]^T B_p^T(x_p)P(x_p) =$$

$$= - Q(x_p) - \frac{1}{2}[P(x_p)B_p(x_p) - C_p^T(x_p)][K_{max} - K_{ec}(t)] [B_p^T(x_p)P(x_p) - C_p(x_p)]$$

$$- \frac{1}{2}[P(x_p)B_p(x_p) - C_p^T(x_p)][K_{ec}(t) - K_{ec_{min}}] [B_p^T(x_p)P(x_p) - C_p(x_p)]$$

$$- P(x_p)B_p(x_p)[K_{ec}(t) - K_{ec_{min}}]C_p(x_p) - C_p^T(x_p)[K_{ec}(t) - K_{ec_{min}}]^T B_p^T(x_p)P(x_p) =$$

$$= - Q(x_p) - \frac{1}{2}[P(x_p)B_p(x_p) - C_p^T(x_p)][K_{max} - K_{ec}(t)] [B_p^T(x_p)P(x_p) - C_p(x_p)]$$

$$- \frac{1}{2}[P(x_p)B_p(x_p) + C_p^T(x_p)][K_{ec}(t) - K_{ec_{min}}] [B_p^T(x_p)P(x_p) + C_p(x_p)] \qquad (32)$$

Since the right term in (32) is negative definite for any value of $K_{ec}(t)$ which satisfies $K_{ec_{min}} \leq K_{ec}(t) \leq K_{max}$, exponential stability of the closed loop system is guaranteed for arbitrary variation of $K_{ec}(t)$ within these boundaries. Now, from (25) we get $K_e = [I - D_p K_{ec}]^{-1} K_{ec}$, and one can see that the stability equation (32) holds for any K_e, constant or time-varying, which is greater than some minimal value given by the relation $K_e \geq K_{e_{min}} = [I - D_p K_{ec_{min}}]^{-1} K_{ec_{min}}$,which shows that

ASP systems (Fig. 1) are high-gain stabilizable.

In this simple case it is also easy to see from (25) that if the gain K_e which stabilizes

the *augmented* **ASP** plant, varies between 0 and ∞, then the correspondent gain

K_{ec} which actually operates on the *original* plant (Fig. 3), varies between 0 and

K_{max}. This relation shows that the high-gain stabilizability of the augmented ASP

plants is a useful utilization of the basic stabilizability bounds of the controlled

plants. It should be mentioned that *any* stabilizing gain may play the role of K_{max}.

However, if we have a good estimate of the *maximal* admissible gain K_{max}, we can

use smaller D_p. This is important because we now control $y_a(t) = y_p(t) + D_p u_p(t)$,

although we are still interested in $y_p(t)$ (Fig. 1). Therefore, we would like to have

$y_a(t) \approx y_p(t)$ for all practical purposes.

The assumption on output stabilizability via *static* feedback was only assumed as a

first step, to illustrate the relation between stabilizability and passivity. Since it may

apparently restrict the applicability of "almost passivity," we now extend its

meaning to general nonlinear systems that need some dynamic output feedback in

order to reach stability.

LEMMA 2 (The almost passivity lemma): Let G_p be any strictly causal

nonlinear system of the form (16)-(17). Let H be some stabilizing *linear*

configuration for G_p and assume that we use the inverse of H in parallel with G_p,

such that the augmented system $G_a = G_p + H^{-1}$ has the form G_a: $\{A_p(x_p), B_p(x),$

$C_p(x_p), D_p(x_p)\}$. Then G_a is "almost strictly passive" (**ASP**). In other words, the

augmented system G_a satisfies almost strictly passivity relations of the form (12)-

(14) (Appendix A).

Notice that the proof of Appendix A holds also if augmented plants G_a is strictly causal $(D_a(x)=0)$. In *strictly causal* **ASP** systems the passivity relations (12)-(14) become:

$$\dot{P}(x_p) + P(x_p)[A_p(x_p) - B_p(x_p)K_eC_p(x_p)] + [A_p(x_p) - B_p(x_p)K_eC_p(x_p)]^T P(x_p) =$$

$$= - Q(x_p) < 0 \tag{33}$$

$$P(x_p) B_p(x_p) = C_p^T(x_p) \tag{34}$$

Relations (33)-(34) can be obtained directly from (15). When $D_p(x_p)$ is singular, the passivity relation (15) can be satisfied only if (34) holds, and then (33) follows immediately. Let us assume that $B_p(x_p)$ is maximal rank. In this case, we get from relation (34)

$$B_p^T(x_p)P(x_p)B_p(x_p) = B_p^T(x_p)C_p^T(x_p) > 0 \tag{35}$$

which requires that $C_p(x_p)$ be also maximal rank. This is the analog of the Linear Time-Invariant conditions which require that strictly causal **ASP** systems have *n-m* minimum-phase finite zeros and *n* arbitrary poles (Shaked, 1977; Bar-Kana, 1987a, 1989b). The almost passivity lemmas can be used for the proof of stability of robust nonlinear adaptive controllers in nonlinear systems (Bar-Kana and Guez, 1990).

The most attractive and direct applications of the almost passivity lemma can be implemented if some raw prior information about the controlled plant is given and some estimate of the maximal stabilizing output feedback, K_{max}, is known. If it can be evaluated, then by using the parallel feedforward $D_p = K_{max}^{-1}$, we get an augmented system that can be easily and reliably controlled by almost any reasonable control method. Since instead of controlling the plant output $y_p(t)=C_p(x_p)x_p(t)$ we now control the augmented output $y_a(t) = y_p(t) + D_pu_p(t)$, the

lemma is especially useful in systems that maintain stability with high feedback gains. In this case, the supplementary gain may be very small and may not essentially affect the controlled plant output. For example, in very common cases, if the gain of the plant is 10 and the largest admissible gain is evaluated to be about 100, then we only add $1/100 = 0.01$ in parallel with the plant. It is amazing how such a small parallel term changes the stability properties of the plant. (Bar-Kana and Kaufman, 1987).

To end these remarks, we note that K_{max} (or H, in general), is indeed required to guarantee the stability of the fictitious closed-loop system, and thus the almost passivity of the augmented open-loop system (15)-(16), but only in a very week sense. The fictitious stable system is by no means required to be good. Thus, stability is only a sufficient condition that enables the subsequent non-linear control to impose upon the (large) plant the desired behavior of, for example, a well designed low-order reference model.

The proof of stability of the subsequent multivariable adaptive control systems extends the applicability of simple adaptive controllers to nonlinear systems, and some references at the end of this article may give a good illustration about what can be done in stationary, nonstationary, or nonlinear systems when appropriate feedforward can be selected a priori (Bar-Kana and Kaufman, 1983, 1984, 1985a, b, 1988; Bar-Kana, Kaufman, and Balas, 1983; Bar-Kana, 1987a, b) or if it must also be computed adaptively (Bar-Kana, 1986b, 1987b).

IV. THE MODEL FOLLOWING PROBLEM FOR NONLINEAR ASP SYSTEMS

In the next section we show how to select the necessary parallel configurations in practical design. This section assumes that the adaptive control is applied to the

augmented **ASP** system. In realistic environments we must take into consideration
the input and output disturbances, and therefore our augmented ASP controlled
plant has the following representation

$$\dot{x}_p(t) = A_p(x_p)x_p(t) + B_p(x_p)u_p(t) + d_i(x_p, t) \tag{36}$$

$$y_a(t) = C_p(x_p)x_p(t) + D_p(x_p)u_p(t) + d_o(x_p, t) \tag{37}$$

where the matrices $A_p(x_p)$, $B_p(x_p)$, $C_p(x_p)$, and $D_p(x_p)$ are uniformly bounded, and
where $d_i(x_p, t)$ and $d_o(x_p, t)$ are some input and output disturbances, not
necessarily bounded, that can also represent some inaccuracies of the
representation of the real plant. Since they are of the same dimension as the output
vector $y_a(t)$, we will assume for the following robustness analysis, that there exist
some finite positive coefficients α and β such that $\|d_i(x_p, t)\|$, $\|d_o(x_p, t)\| \le \alpha +$
$\beta\|y_a(t)\|$. The plant is assumed to be ASP, however it may also be very large and
basically unknown. The output of the plant is required to follow the output of the
(possibly) very low-order model

$$\dot{x}_m(t) = A_m(x_m)x_m(t) + B_m(x_m)u_m(t) \tag{38}$$
$$y_m(t) = C_m(x_m)x_m(t) + D_m(x_m)u_m(t) \tag{39}$$

The model represents the desired behavior of the plant, but is free otherwise and
does not have to be the result of some prior modeling of the plant. It is also
allowed to be of any arbitrarily high or low order. It will be, usually, a linear time-
invariant model, and we represent it in the general, nonlinear form, only for the
sake of generality of the solution.

We define now the output tracking error

$$e_y(t) = y_m(t) - y_p(t) \tag{40}$$

and use the following simple multivariable adaptive control algorithm (Fig. 5) (Bar-Kana and Kaufman, 1983 and 1985b, Bar-Kana, 1987a)

$$u_p(t) = K_{e_y}(t)e_y(t) + K_{x_m}(t)x_m(t) + K_{u_m}(t)u_m(t) = K(t)r(t) \tag{41}$$

$$K(t) = [\ K_{e_y}(t) \qquad K_{x_m}(t) \qquad K_{u_m}(t)\ ] \tag{42}$$

$$r^T(t) = [\ e_y^T(t) \qquad x_m^T(t) \qquad u_m^T(t)\ ] \tag{43}$$

where the adaptive gains are a combination of "proportional" and "integral" gains

$$K_p(t) = [\ e_y(t)e_y^T(t)\bar{\Gamma}_{e_y} \qquad e_y(t)x_m^T(t)\bar{\Gamma}_{x_m} \qquad e_y(t)u_m^T(t)\bar{\Gamma}_{u_m}\ ] = e_y(t)r^T(t)\bar{\Gamma} \tag{44}$$

$$\dot{K}_I(t) = e_y(t)r^T(t)\Gamma - \sigma K_I(t) \tag{45}$$

$$K(t) = K_p(t) + K_I(t) \tag{46}$$

where Γ and $\bar{\Gamma}$ are selected positive definite scaling matrices.

The basic algorithm (41) - (46), without the σ - term in (45), was introduced by Sobel, Kaufman, and Mabius (1979, 1982) and extended by Bar-Kana and Kaufman (1983, 1984, 1985a). The gain $K_{e_y}(t)$ takes care of the stability of the controlled plant, while $K_{x_m}(t)$ and $K_{u_m}(t)$ help improving the performance of the adaptive model following system and even achieve perfect following in idealistic clean environments (Bar-Kana and Kaufman, 1985a).

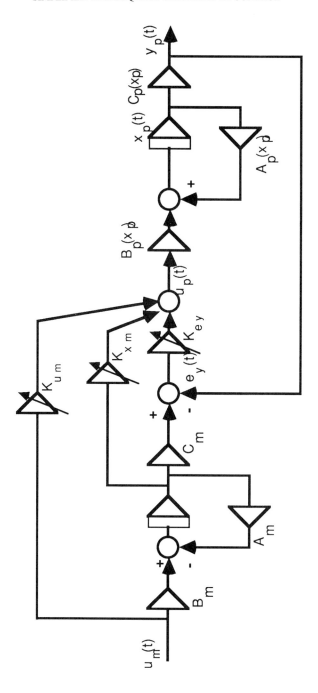

Fig. 5: The complete adaptive control system.

The σ-term is a "forgetting" factor (Ioannou and Kokotovic, 1983) which avoids divergence of the integral gain in the presence of disturbances. Since it is only a low-pass filtering of $e_y(t)r^T(t)\Gamma$, this gain will not diverge unless the output tracking error diverges. Furthermore, the adaptive gain now move up and down according to the specific situation (errors), as adaptive gains should do. The proportional gain $K_p(t)$ is added for its immediate action. Due to this action, it heavily punishes high tracking errors and quickly brings the system towards tracking with small errors (Landau, 1979). The efficiency of the integral term $K_{e_y}(t)$ was also recently demonstrated (Nussbaum, 1983; Willems and Byrnes, 1984, 1984; Morse, 1984; Bar-Kana, 1988d) with respect to the problem of the unknown sign of the high frequency gain.

The subsequent theorem of robust adaptive stability extends the applicability of this simple adaptive algorithm to "almost passive" multivariable nonlinear systems, and gives the necessary theoretical background for the successful applications of the algorithm in nonlinear or nonstationary systems (Bar-Kana, 1987a, b; 1988b; Bar-Kana and Kaufman, 1988).

THEOREM 2: The model reference adaptive controller (41)-(46) guarantees robust adaptive stabilization of the ASP plant (36) - (37) in the presence of any bounded input commands and input or output disturbances.

PROOF: See Appendix B for proper and Appendix C for strictly proper ASP systems.

"Robust adaptive stabilization" means that all values involved in the adaptation process namely, states, gains and errors, are bounded in the presence of any bounded input commands and input or output disturbances. Asymptotic perfect following or asymptotic perfect regulation are obtained in idealistic conditions.

Although theorem 2 guarantees boundness of all dynamic values involved in the adaptation process, some undesired phenomena where observed in adaptive algorithms with forgetting factors (Åström, 1983; Anderson, 1985; Bar-Kana, 1989b, c; Fortesque and colleagues, 1981; Hsu and Costa, 1987) when no external excitation is present. In our configuration , these effects are apparent in particular if the controlled plant is unstable (Hsu and Costa, 1987; Bar-Kana, 1989c). In this case, the adaptive gain may initially increase due to the initial errors, and reach some stabilizing value. As a result, the states and outputs move towards their zero value, and the decrease of the output leads to the decrease of the gains towards their zero value. However, since the equilibrium point $(y_a(t)=0, K(t) = 0)$ is unstable, and since on the other hand all values are bounded, we can expect that the system will reach some other limiting trajectories or equilibrium points (Willems, 1971).

Since the gain has a stabilizing effect initially, the states and outputs come very close to their zero value. Therefore, the gains must go well into the unstable region before their destabilizing effect is felt. Then we have a sudden "burst" of error that brings the gains into the stable region again, and so on (Mareels and Bitmead, 1986; Bar-Kana, 1989b, c). If we use fast adaptation, which in our case means using high adaptation coefficients Γ and $\bar{\Gamma}$, then the phenomena can be reduced to less then noticeable dimensions (Bar-Kana, 1989c). However, when the phenomenon appears, it finally settles at or about the minimum stabilizing gain K_{min}, and this value can now be used to totally eliminate the bursts.

The bursting phenomena can be eliminated by replacing $K_{e_y}(t)$ in (42) with

$$K_{e_y} + K_0 \tag{47}$$

where K_0 is any stabilizing gain

$$K_0 \geq K_{min} \tag{48}$$

because if either $\|y_a(t)\|$, $\|K(t)\|$ or $\|K(t)y_a(t)\|$ becomes small, the adaptive system

enters the domain of attraction of the equilibrium point $(y_a(t) = 0, K(t) = K_0)$

(Appendix D). Since it can be shown that the eventual existence of other

equilibrium points is excluded, the equilibrium point $(x_p(t) = 0, K(t) = K_0)$ is both

asymptotically stable (attractive) and unique (Appendix D).

In conclusion, although in general the prior knowledge on the bounds of stability

$K_{min} \div K_{max}$ that are admissible with fixed controlers does not guarantee stability

with nonlinear controllers (Aizerman and Gantmacher, 1964), stability is

guaranteed in the specific case of this particular nonlinear adaptive algorithm.

V. ADAPTIVE CONTROL OF MANIPULATORS

A state space representation of the manipulator is (Slotine and Li, 1988):

$$\dot{x}_p(t) = A_p(x_p)x_p(t) + B_p(x_p)u_p(t) + d(x_p \tag{49}$$

$$y_p(t) = x_p(t) = \begin{bmatrix} x_1(t) \\ x_2(t) \end{bmatrix} \tag{50}$$

where $x_p \in R^{2n}$, $y_p \in R^{2n}$, $u_p \in R^n$, and where $x_1(t)$ is the n position-state
vector and $x_2(t)$ is the n velocity-state vector. The various matrices of
corresponding dimensions in (49)-(50) are:

$$A_p(x_p) = \begin{bmatrix} 0 & I \\ 0 & -H^{-1}(x_1)C(x_1,x_2) \end{bmatrix} \tag{51}$$

$$B_p(x_p) = \begin{bmatrix} 0 \\ H^{-1}(x_1) \end{bmatrix} \tag{52}$$

$$d(x_p) = \begin{bmatrix} 0 \\ -H^{-1}(x_1)g(x_1) \end{bmatrix} \tag{53}$$

where $d(x_p)$ is considered to be a state-dependent disturbance generated by the gravity $g(x_1)$. We first neglect the gravity term in order to establish some basic relations and show that the ideal system can be stabilized by feedback controller of the form

$$u_p(t) = -K(x_1 + \alpha x_2) \tag{54}$$

for any positive definite matrix K and positive coefficient α. In other words, the closed-loop system

$$\dot{x}_p(t) = A_{CL}(x_p)x_p(t) \tag{55}$$

where

$$A_{CL}(x_p) = \begin{bmatrix} 0 & I \\ -H^{-1}(x_1)K & -H^{-1}(x_1)[\ C(x_1,x_2) + \alpha K\ ] \end{bmatrix} \tag{56}$$

is uniformly asymptotically stable. To this end we select the positive definite Lyapunov function

$$V(x_p) = x_p^T(t)P(x_p)x_p(t) \tag{57}$$

where

$$P(x_p) = \begin{bmatrix} K & 0 \\ 0 & H(x_1) \end{bmatrix} \tag{58}$$

and the derivative of the Lyapunov function along the trajectories of (51) is

$$\dot{V}(x_p) = x_p^T(t)[\ \dot{P}(x_p) + P(x_p)A_{CL}(x_p) + A_{CL}^T(x_p)P(x_p)\ ]x_p(t) =$$

$$= x_p^T(t)\begin{bmatrix} 0 & K - K^T \\ \\ K^T - K & \dot{H}(x_1) - C(x_1,x_2) - C^T(x_1,x_2) - \alpha(K + K^T) \end{bmatrix} x_p(t) \qquad (59)$$

and where we use (Slotine and Li, 1988) the manipulator property that $H(x_1)$ is positive definite and that $\dot{H}(x_1) = C(x_1,x_2) + C^T(x_1,x_2)$, to finally get

$$\dot{V}(x_p) = -2\alpha x_2^T(t)Kx_2(t) \qquad (60)$$

Although $\dot{V}(x_p)$ is only negative semidefinite rather than negative definite, uniform asymptotic stability of (55) is guaranteed by the following theorems, which we formulate in a form that fits our specific problem.

THEOREM 3 (Lyapunov): Let $V(x_p)$ be a positive definite Lyapunov function of $x_p(t)$. Then, the system (55) is globally asymptotically stable if and only if the derivative of the Lyapunov function along the trajectories of (55) is a negative definite function of $x_p(t)$ (Vidyasagar, 1978).

THEOREM 4 (LaSalle): Under the assumptions of theorem 3, the system (51) is stable if the derivative of the Lyapunov function along the trajectories of (65) is negative semidefinite. The trajectories of the system finally reach the region defined by $\dot{V}(x_p) \equiv 0$ (LaSalle, 1981). The system (55) is asymptotically stable if the derivative $\dot{V}(x_p)$ is *not* identically zero along any nontrivial solution of (55). In our case, since $\dot{V}(x_p)$ is quadratic, the limiting trajectories condition $\dot{V}(x_p) \equiv 0$ is equivalent with $x_2(t) = \dot{x}_1(t) \equiv 0$, which implies $x_1(t) \equiv 0$, and therefore $x_p(t) \equiv 0$.

Although stabilizability of the robotic manipulator through some PD controller was thus established using a negative semidefinite derivative of the Lyapunov function, the passivity relations needed for the convergence of the nonlinear adaptive controller require that the underlying $\dot{V}(x_p)$ be negative definite, or have a negative definite upper bound with respect to $x_p(t)$. Now, since for any $x_2(t) \neq 0$ we have $V(x_p) > 0$ (strictly positive) and $\dot{V}(x_p) < 0$ (strictly negative), which implies that $V(x_p)$ is strictly decreasing, then $\dot{V}(x_p)/V(x_p) \leq -\beta$ for some scalar $\beta > 0$, sufficiently small. Then $\dot{V}(x_p) \leq -\beta V(x_p)$ and

$$\dot{V}(x_p) = -2\alpha x_2^T(t)Kx_2(t) \leq -\beta x_p^T(t)P(x_p)x_p(t) = -\beta V(x_p) \tag{61}$$

and therefore, all passivity lemmas, as well as the proofs of robust stability of the adaptive systems are valid in the case of the robot manipulators.

The stabilizability of the robot via PD controllers was only needed as an underlying knowledge. The implementation of the simplified adaptive controller does not actually make use of the derivative of the output and we can only use position sensors. Instead of the fictitious stabilizing PD controller $H(s)$ given by (54) we use the configuration $H^{-1}(s)$ given by

$$y_s(s) = \frac{K_{max}^{-1}}{1+\alpha s} u_p(s) \tag{62}$$

in parallel with the controlled plant (Bar-Kana, 1977a), where K_{max} is some reasonable evaluation of the maximal admissible gain K in (54), which theoretically may be arbitrarily large. Since we treat the gravity term as some unknown and nonmeasured bounded disturbance, and since we do not use any identification algorithm in order to identify and eliminate this term, the adaptive controller can only guarantee stability with respect to boundness of all states, gains, and tracking errors, in general. However, since the adaptive algorithm moves the gains up and down in order to reduce the errors, this residual error may be negligible (Bar-

Kana, 1987a; Laniado and colleagues, 1989), and may obviate the need for more complex algorithms using identifiers in the closed loop.

VI.   CONCLUSIONS

In this article the basic stabilizability properties of systems were used in order to achieve "almost passive" configurations for a class of nonlinear continuous-time systems. It was also shown that the "almost passivity" property of systems can be used to extend the applicability of simple nonlinear adaptive algorithms to nonlinear systems, since it guarantees global stability and robustness of the adaptive control systems in the presence of bounded disturbances. Other usually needed prior knowledge and conditions, like: the order of the plant, the relative degree, inverse stability, stationarity, external excitation, etc.. are immaterial in this context. The article then suggests a simple and robust adaptive controller for nonlinear systems with unknown parameters, with particular application for robot manipulators.

APPENDIX A. PROOF OF LEMMA 2

Let us assume that the nonlinear plant $G_p(\cdot)$ is stabilizable by the linear configuration H. In other words, the system (Fig. 6) $y(t) = G_{CL}(u(t))$, given by the implicit relation

$$y(t) = H \cdot [u(t) - G_p(y(t))] \qquad (A.1)$$

is stable. The stability assumption will be needed only later. From (A.1) we first find that the inverse of the closed-loop system $G_{CL}(\cdot)$ which gives $u(t) = G_{CL}^{-1}(y(t))$ can be expressed as

$$u(t) = H^{-1} \cdot y(t) + G_p(y(t)) \qquad (A.2)$$

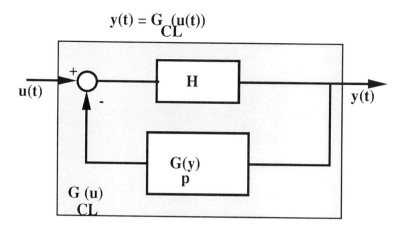

Fig. 6. The closed-loop system

Therefore the inverse system whose output y(t) and input u(t) would be related by the equation $y(t) = G_{CL}^{-1}(u(t))$, is given by

$$y(t) = H^{-1} \cdot u(t) + G_p(u(t)) \tag{A.3}$$

and is represented in Fig. 7. Note that u(t) and y(t) are only general notations, u(t) representing inputs and y(t) representing outputs of the various systems and inverse systems, rather that relations between specific functions.

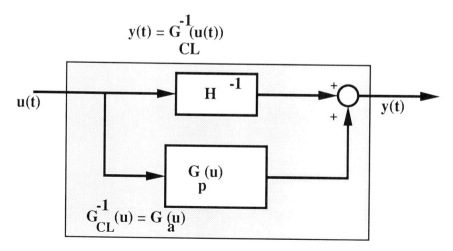

Fig. 7. The inverse of the closed-loop system

Formally, we can say that if H is some linear stabilizing configuration such that

$G_{CL}(\cdot)=[I+HG_p(\cdot)]^{-1}H=[G_p(\cdot)+H^{-1}]^{-1}=G_a^{-1}(\cdot)$ is a bounded nonlinear operator,

then there exists some constant matrix K_e, sufficiently large (positive definite) such

that the operator $G_1(\cdot)=G_{CL}(\cdot)+K_e$ is strictly positive (Fig. 8). We will also

present below the direct proof that it satisfies the desired passivity conditions.

$$y(t) = G_1(u(t))$$

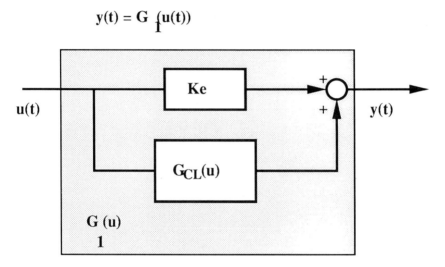

Fig. 8. A strictly passive system

By using the same argument as above, we get (Fig. 9) the inverse

$$G_s(\cdot) \overset{\Delta}{=} G_1^{-1}(\cdot) = G_{CL}(\cdot) + K_e^{-1}.$$

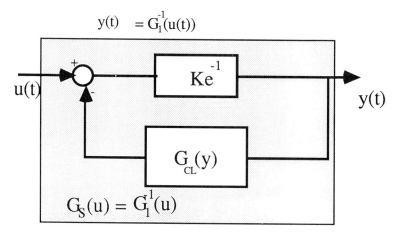

Fig. 9. The inverse of the strictly passive system

In this case, $G_s(\cdot) = G_1^{-1}(\cdot) = [G_{cl}(\cdot) + K_e]^{-1} = [G_a^{-1}(\cdot) + K_e]^{-1} = [I + G_a(\cdot)K_e]^{-1}G_a(\cdot)$ represented in Fig.10 is also SP, and therefore $G_a(\cdot) = G_p(\cdot) + H^{-1}$ is **ASP**, where $G_a(\cdot)$ may be either causal or, more important, strictly causal.

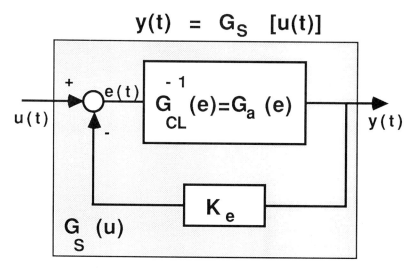

Fig. 10. Another inverse of the strictly passive system.

Now, in order to avoid any doubt about the satisfaction of the passivity relations (12)-(14), we will obtain them directly from the stability properties of the plant.

Assume that the fictitious stabilized plant $G_{CL}(\cdot)$ in Fig. 6 has the general representation

$$\dot{x}(t) = A_{CL}(x) \, x(t) + B_{CL}(x)u(t) \qquad (A.4)$$

$$y(t) = C_{CL}(x)x(t) + D_{CL}(x)u(t) \qquad (A.5)$$

We assume that $G_{CL}(\cdot)$ is exponentially stable, and therefore, that there exists some positive definite symmetric matrices $P(x)$ and $Q_1(x)$ such that if we select the Lyapunov function $V(t)=x^T(t)P(x)x(t)$, its derivative along the trajectories of (A.4)-(A.5) satisfies the following equation:

$$\dot{V}(t)=x^T(t)[\dot{P}(x)+P(x)A_{CL}(x)+A_{CL}^T(x)P(x)]x(t) = - \, x^T(t)Q_1(x)x(t) \le -\epsilon\|x(t)\|^2 \quad (A.6)$$

for some positive scalar ϵ, where $\dot{P}(x){=}\dfrac{dP(x)}{dt} = \dfrac{\partial P(x)}{\partial x}\dfrac{dx}{dt}$.

It is then easy to see that there exists a (unknown) positive definite gain matrix $\tilde{K}_e$, sufficiently large, such that

$$x^T(t)\{Q_1(x) - [P(x)B_{CL}(x) - C_{CL}^T(x)][D_{CL}(x)+D_{CL}^T(x)+\tilde{K}_e +\tilde{K}_e^T]^{-1}[B_{CL}^T(x)P(x)$$

$$-C_{CL}(x)]\}x(t)$$

$$= x^T(t)Q(x)x(t) \ge \delta\|x(t)\|^2 \qquad (A.7)$$

for some positive scalar $\delta \le \epsilon$. Relation (A.6) becomes

$$x^T(t)[\dot{P}(x) + P(x)A_{CL}(x)+ A_{CL}^T(x)P(x)]x(t) =$$

$$= - \, x^T(t)\{Q(x)+[P(x)B_{CL}(x) - C_{CL}^T(x)][D_{CL}(x)+D_{CL}^T(x)+\tilde{K}_e +\tilde{K}_e^T]^{-1}[B_{CL}^T(x)P(x)$$

$$-C_{CL}(x)]\}x(t) \qquad (A.8)$$

which is the strict passivity relation for the system G_1 in Fig. 8 with the representation

$$\dot{x}(t) = A_{CL}(x) \, x(t) + B_{CL}(x)u(t) \qquad (A.9)$$

$$y(t) = C_{CL}(x)x(t) + [D_{CL}(x) + \tilde{K}_e]u(t) \qquad (A.10)$$

The inverse system G_1^{-1} in Fig. 9 and Fig. 10 has the representation

$$\dot{x}(t) = A_2(x)x(t) + B_2(x)u(t) \qquad (A.11)$$

$$y(t) = C_2(x)x(t) + D_2(x)u(t) \qquad (A.12)$$

were

$$A_2(x) = A_{CL}(x) - B_{CL}(x)[D_{CL}(x) + \tilde{K}_e]^{-1}C_{CL}(x) \qquad (A.13)$$

$$B_2(x) = B_{CL}(x)[D_{CL}(x) + \tilde{K}_e]^{-1} \qquad (A.14)$$

$$C_2(x) = - [D_{CL}(x) + \tilde{K}_e]^{-1}C_{CL}(x) \qquad (A.15)$$

$$D_2(x) = [D_{CL}(x) + \tilde{K}_e]^{-1} \qquad (A.16)$$

Then

$$
\begin{aligned}
& x^T(t)[\dot{P}(x) + P(x)A_2(x)+ A_2^T(x)P(x)]x(t) = \\
& = x^T(t)[\dot{P}(x) + P(x)A_{CL}(x)+ A_{CL}^T(x)P(x)]x(t) - \\
& \quad - x^T(t)\{P(x)B_{CL}(x)[D_{CL}(x)+\tilde{K}_e]^{-1}C_{CL}(x) \\
& \quad + C_{CL}^T(x)[D_{CL}^T(x)+\tilde{K}_e^T]^{-1}B_{CL}^T(x)P(x)\}x(t) \\
& = - x^T(t)\{Q(x)+[P(x)B_{CL}(x) - C_{CL}^T(x)]\cdot \\
& \quad \cdot[D_{CL}(x)+D_{CL}^T(x)+\tilde{K}_e +\tilde{K}_e^T]^{-1}[B_{CL}^T(x)P(x)-C_{CL}(x)] + \\
& \quad + P(x)B_{CL}(x)[D_{CL}(x)+\tilde{K}_e]^{-1}C_{CL}(x) + C_{CL}^T(x)[D_{CL}^T(x)+\tilde{K}_e^T]^{-1}B_{CL}^T(x)P(x)\}x(t) = \\
& = -x^T(t)\{Q(x)+[P(x)B_{CL}(x)+C_{CL}^T(x)]\cdot \\
& \quad \cdot[D_{CL}(x)+D_{CL}^T(x)+\tilde{K}_e+\tilde{K}_e^T]^{-1}[B_{CL}^T(x)P(x)+C_{CL}(x)]\}x(t)= \\
& = - x^T(t)\{Q(x)+[P(x)B_{CL}(x) +C_{CL}^T(x)][D_{CL}(x) + \tilde{K}_e]^{-1}[D_{CL}(x) + \tilde{K}_e]\cdot \\
& \quad \cdot[D_{CL}(x)+D_{CL}^T(x)+\tilde{K}_e +\tilde{K}_e^T]^{-1}[B_{CL}^T(x)P(x)+C_{CL}(x)] \}x(t) = \\
& = - x^T(t)\{Q(x)+[P(x)B_2(x) - C_2^T(x)]\cdot \\
& \quad \cdot[D_{CL}(x)+D_{CL}^T(x)+\tilde{K}_e +\tilde{K}_e^T][B_2^T(x)P(x) - C_2(x)]\}x(t) = \\
& = - x^T(t)\{Q(x)+[P(x)B_2(x) - C_2^T(x)][D_2(x)+D_2^T(x)]^{-1}[B_2^T(x)P(x) - C_2(x)]\}x(t)
\end{aligned}
$$

$$(A.17)$$

and finally

$$\dot{P}(x) + P(x)A_2(x) + A_2^T(x)P(x) +$$

$$+[P(x)B_2(x) - C_2^T(x)][D_2(x) + D_2^T(x)]^{-1}[B_2^T(x)P(x) - C_2(x)] + Q(x) = 0 \qquad (A.18)$$

which shows that the inverse system (Fig. 10)also satisfies a strict passivity relation, and therefore the augmented system $G_a(\cdot)$ is almost strictly passive (ASP).

We used here the quadratic *functional* relations, rather than the usual matrical quadratic equations, because this way we could use, in general, $x^T(t)Mx(t) = x^T(t)M^Tx(t)$, were M is a general notation, even if M is not symmetric.

APPENDIX B. THE PROOF OF STABILITY OF THE ADAPTIVE CONTROL ALGORITHM (41)-(46)

We shall question first the existence of some "ideal" bounded trajectories $x_p^*(t)$ and "ideal" control $x_p^*(t)$ that satisfy some unknown differential equation

$$\dot{x}_p^*(t) = A_p^*(x_p^*)x_p^*(t) + B_p^*(x_p^*)u_p^*(t) \qquad (B.1)$$

$$y_p^*(t) = C_p(x_p)x_p^*(t) + D_p(x_p)u_p^*(t) \qquad (B.2)$$

where the matrices $A_p^*(x_p^*)$, $B_p^*(x_p^*)$, $C_p(x_p)$, and $D_p(x_p)$ are uniformly bounded. Note that the ideal system (B. 1) may be entirely different from the plant to be controlled, and that only their output equations (B. 2) are assumed to be equal.

We now try to represent $x_p^*(t)$ and $u_p^*(t)$ as linear combinations (Bar-Kana and Kaufman, 1985a) of $x_m(t)$ and $u_m(t)$

$$x_p^*(t) = Xx_m(t) + Uu_m(t) \qquad (B.3)$$

$$u_p^*(t) = \tilde{K}_{x_m} x_m(t) + \tilde{K}_{u_m} u_m(t) \qquad (B.4)$$

It is now required that

$$y_p^*(t) = y_m(t) \tag{B.5}$$

Substituting (B.2) - (B.4) and (36) into (B.5) gives (Bar-Kana and Guez, 1990)

$$[C_p(x_p)X + D_p(x_p)\tilde{K}_{x_m} - C_m(x_m)]x_m(t) +$$
$$+ [C_p(x_p)U + D_p(x_p)\tilde{K}_{u_m} - D_m(x_m)]u_m(t) = 0 \tag{B.6}$$

Equation (B.7) has much more variables than equations, therefore it has (at least) a solution for the unknown matrices X, U, $\tilde{K}_{x_m}$, and $\tilde{K}_{u_m}$, in general.

Notice that none of the values that were calculated above are needed for implementation of the adaptive algorithm (37) - (42). The conditions of existence that were established for the idealistic perfect following solution will be subsequently used for the proof of stability.

Since we want the plant states to reach the "ideal" states that allow perfect tracking we define the state error as

$$e_x(t) = x_p^*(t) - x_p(t) \tag{B.7}$$

and the output error then becomes

$$e_y(t) = y_m(t) - y_p(t) = y_p^*(t) - y_p(t) \tag{B.8}$$

$$e_y(t) = C_p(x_p) x_p^*(t) + D_p(x_p) u_p^*(t) - C_p(x_p) x_p(t) + D_p(x_p) u_p(t) - d_0(t) \tag{B.9}$$

or (Bar-Kana and Kaufman, 1990)

$$e_y(t) = C_{pc}(x_p) e_x(t) - D_{pc}(x_p)[K(t) - \tilde{K}]r(t) - [I + D_p(x_p)\tilde{K}_{e_y}]^{-1}d_0(t) \tag{B.10}$$

where

$$\tilde{K} = [\tilde{K}_{e_y} \qquad \tilde{K}_{x_m} \qquad \tilde{K}_{u_m}] \tag{B.15}$$

and where $C_{pc}(x_p)$ and $D_{pc}(x_p)$ were defined in (9)-(10).

Differentiating $e_x(t)$ in (B.7) gives

$$\dot{e}_x(t) = \dot{x}\,^*_p(t) - \dot{x}_p(t) - \ddot{x}\,^*_p(t) - A_p(x_p)x^*_p(t) + A_p(x_p)x^*_p(t)\ \dot{x}_p(t) \tag{B.12}$$

$$\dot{e}_x(t) = A^*_p(x^*_p)\,x^*_p(t) + B^*_p(x^*_p)\,u^*_p(t) - A_p(x_p)x^*_p(t) +$$
$$+ A_p(x_p)x^*_p(t) - A_p(x_p)x_p(t) - B_p(x_p)K(t)r(t) - d_i(t) \tag{B.13}$$

and after some algebra we get (Bar-Kana and Guez, 1990)

$$\dot{e}_x(t) = A_{pc}(x_p)e_x(t) - B_{pc}(x_p)[\ K(t) - \tilde{K}\ ]r(t) + F(t) \tag{B.14}$$

where $A_{pc}(x_p)$, $B_{pc}(x_p)$, $K_{ec}(x_p)$ were defined in (4) - (10) and where $F(t)$ is the residual uniformly bounded term

$$F(t) = \{[\ A^*_p(x^*_p) - A_p(x_p)]X + [\ B^*_p(x^*_p) - B_p(x_p)]\tilde{K}_{x_m}\ \}x_m(t) +$$
$$+ \{[\ A^*_p(x^*_p) - A_p(x_p)]U + [\ B^*_p(x^*_p) - B_p(x_p)]\tilde{K}_{u_m}\ \}u_m(t) -$$
$$- B_p(x_p)\tilde{K}_{ec}d_0(t) - d_i(t) \tag{B.15}$$

The following quadratic Lyapunov function will be used for the proof of stability of the adaptive system (B.14) and (45):

$$V(t) = e_x^T(t)P(x_p)e_x(t) + tr\ \{\ [K_I(t) - \tilde{K}\ ]\Gamma^{-1}[K_I(t) - \tilde{K}\ ]^T\ \} \tag{B.16}$$

where $P(x_p)$ is the positive definite matrix defined in (12)-(14) and Γ is the positive definite scaling factor defined in (45). The derivative of $V(t)$ "along the trajectories" of (B.14) and (45) gives:

$$\dot{V}(t) = e_x^T(t)\dot{P}(x_p)e_x(t) + \dot{e}_x^T(t)P(x_p)e_x(t) + e_x^T(t)P(x_p)\dot{e}_x(t) +$$
$$+ 2tr\ \{\ [K_I(t) - \tilde{K}\ ]\Gamma^{-1}\dot{K}_I^T(t)\ \} \tag{B.17}$$

Substituting $\dot{e}_x(t)$ from (B.14) and $\dot{K}_I(t)$ from (45) gives

$$\dot{V}(t) = e_x^T(t)\dot{P}(x_p)e_x(t) + \dot{e}_x^T(t)A_{pc}^T(x_p)P(x_p)e_x(t) -$$

$$-r^T(t)[K(t) - \tilde{K}]^T B_{pc}^T(x_p)P(t)e_x(t) + F^T(t)P(t)e_x(t) +$$

$$+ e_x^T(t)P(x_p)A_{pc}(x_p)e_x(t) - e_x^T(t)P(x_p)B_{pc}(x_p)[K(t) - \tilde{K}]r(t) +$$

$$+ e_x^T(t)P(x_p)F(t) + 2tr\{[K_I(t) - \tilde{K}]\Gamma^{-1}[e_y(t)r^T(t)\Gamma - \sigma K_I(t)]\} \tag{B.18}$$

Substituting

$$K_I(t) = K(t) - K_p(t) = K(t) - e_y(t)r^T(t)\overline{\Gamma} \tag{B.19}$$

and using (12) - (14) and (B.14) finally gives after some algebra

$$\dot{V}(t) = - e_x^T(t)Q(x_p)e_x(t) -$$

$$- \{L(x_p)e_x(t) - W(x_p)[K(t) - \tilde{K}]r(t)\}^T\{L(x_p)e_x(t) - W(x_p)[K(t) - \tilde{K}]r(t)\}$$

$$- 2e_y^T(t)e_y(t)r^T(t)\overline{\Gamma}r(t)$$

$$- 2\sigma tr\{[K_I(t) - \tilde{K}]\Gamma^{-1}[K_I(t) - \tilde{K}]^T\}$$

$$- 2\sigma tr\{[K_I(t) - \tilde{K}]\Gamma^{-1}\tilde{K}^T\}$$

$$- 2r^T(t)[K(t) - \tilde{K}]^T[I + D_p(x_p)\tilde{K}_{e_y}]^{-1}d_0(t)$$

$$+ 2e_x^T(t)P(x_p)F(t) \tag{B.20}$$

It is easy to see that there exist some positive coefficients $\alpha_1 \div \alpha_{10}$ such that

$$\dot{V}(t) \le - \alpha_1\|e_x(t)\|^2 - \alpha_2\|[K(t) - \tilde{K}]r(t)\|^2 - \alpha_3\|e_y(t)\|^4 - \alpha_4\|e_y(t)\|^2\|x_m(t)\|^2 -$$

$$- \alpha_5\|e_y(t)\|^2\|u_m(t)\|^2 - \alpha_6\sigma\|K_I(t) - \tilde{K}\|^2 + \alpha_7\sigma\|K_I(t) - \tilde{K}\| +$$

$$+ \alpha_8\|[K(t) - \tilde{K}]r(t)\| + \alpha_9\|e_x(t)\| + \alpha_{10}\|e_x(t)\|\cdot\|e_y(t)\| \tag{B.21}$$

If either $\|e_x(t)\|$, $\|[K(t) - \tilde{K}]r(t)\|$, or $\|K_I(t) - \tilde{K}\|$ increases beyond some bound, the negative definite quadratic terms in (B.21) become dominant, and thus $\dot{V}(t)$ becomes negative. The quadratic form of the Lyapunov function $V(t)$ then guarantees that all the dynamic values, namely $e_x(t)$, $K_I(t)$, and $e_y(t)$ are bounded. In idealistic situations with $d_i(t)=0$ and $d_0(t)=0$, F(t) may vanish and thus allow perfect following (Bar-Kana and Kaufman, 1985a) . If, however, the σ-term were missing in (41), then the adaptive controller could not have avoided those situations when $\| K_I(t)-\tilde{K}\|$ might increase without bound in spite of $\| [K(t) - \tilde{K}]r(t) \|$ being small. As shown in Section II, the properties of ASP systems guarantee that theoretically, the adaptive system remains stable, even if the gains increase without bound. However, the gains may then reach nonnecessarily or nonrealistically large value. The negative quadratic σ-term in (B.21) shows that such a situation is not possible and all values are therefore bounded.

As already mentioned, only the $K_{e_y}(t)$ terms are needed for the stability of the adaptive control system. The effect of the other terms, like the proportional adaptive gains or the terms built on $x_m(t)$ and $u_m(t)$ is expressed by the corespondent supplementary negative terms in (B.20) and (B.21), which increase the rate of convergence and also reduce the bounds of the final bounded error (where $\dot{V}(t)$ is not necessarily negative) thus improving the performance of the adaptive controller.

APPENDIX C. ADAPTIVE CONTROL OF STRICTLY CAUSAL ALMOST PASSIVE SYSTEMS

The adaptive controller (41) - (46) can also be applied to *strictly causal* ASP systems. Such systems satisfy the following relations:

$$\dot{P}(x_p) + P(x_p)[A_p(x_p) - B_p(x_p)K_eC_p(x_p)] + [A_p(x_p) - B_p(x_p)K_eC_p(x_p)]^T P(x_p) =$$
$$= - Q(x_p) < 0 \tag{C.1}$$
$$P(x_p) B_p(x_p) = C_p^T(x_p) \tag{C.2}$$

Relations (C.1)-(C.2) can be obtained directly from (15) which, in case $D_p(x_p)$ is singular, cannot hold unless (C.2) holds, and then (C.1) follows immediately. Let us also assume that $B_p(x_p)$ is maximal rank. In this case, relation (C.2) requires that $C_p(x_p)$ be also maximal rank and that

$$B_p^T(x_p)C_p^T(x_p) = B_p^T(x_p)P(x_p)B_p(x_p) > 0 \qquad (C.3)$$

which is the equivalent of the Linear Time-Invariant conditions which require that strictly causal ASP systems have $n\text{-}m$ minimum-phase finite zeros and n arbitrary poles (Shaked, 1977; Bar-Kana, 1987a).

The proof of stability in such a case follows immediately from the proof of Appendix B, if we substitute $D_p(x_p)=0$, $L(x_p)=0$, $W(x_p)=0$. Then

$$\dot{V}(t) = - e_x^T(t)Q(x_p)e_x(t)$$
$$- 2e_y^T(t)e_y(t)r^T(t)\overline{\Gamma}r(t)$$
$$- 2\sigma tr \{ [K_I(t) - \tilde{K}]\Gamma^{-1}[K_I(t) - \tilde{K}]^T \}$$
$$- 2\sigma tr \{ [K_I(t) - \tilde{K}]\Gamma^{-1} \tilde{K}^T\}$$
$$- 2r^T(t)[K(t) - \tilde{K}]^T d_0(t)$$
$$+ 2e_x^T(t)P(x_p)F(t) \qquad (C.3)$$

and the stability results following (B.20) - (B.21) are also perfectly valid here.

APPENDIX D. THE ROLE OF THE STABILIZING GAIN K_0

Let us assume that we used the fixed gain K_0 and control the plant

$$\dot{x}_p(t) = A_0(x_p)x_p(t) + B_p(x_p)u_p(t) \qquad (D.1)$$
$$y_a(t) = C_p(x_p)x_p(t) + D_p(x_p)u_p(t) \qquad (D.2)$$

where

$$A_0(x_p) = A_p(x_p) - B_p(x_p)K_0C_p(x_p) \qquad (D.3)$$

Let us assume that no external input command or disturbance is present, and that therefore $u_m(t)=0$; $x_m(t)=0$; $y_m(t)=0$; $K_x(t)=0$; $K_y(t)=0$, and for convenience, $K_p(t)=0$; $K(t)=K_{e_y}(t)=K_{e_y}I(t)$. Then

$$\dot{K}(t) = y_a(t)y_a^T(t)\Gamma_I - \sigma K(t) \qquad (D.4)$$

$$u_p(t) = - K(t)y_a(t) = - K(t)C_p(x_p)x_p(t) - K(t)D_p(x_p)u_p(t) \qquad (D.5)$$

or

$$u_p(t) = - [I + K(t)D_p(x_p)]^{-1}K(t)C_p(x_p)x_p(t) = - K_c(t)C_p(x_p)x_p(t) \qquad (D.6)$$

where

$$K_c(t) = [I + K(t)D_p(x_p)]^{-1}K(t) = K(t)[I + D_p(x_p)K(t)]^{-1} \qquad (D.7)$$

and define

$$B_c(x_p) = B_p(x_p) [I + K(t)D_p(x_p)]^{-1} \qquad (D.8)$$

$$C_c(x_p) = [I + D_p(x_p)K(t)]^{-1} C_p(x_p) \qquad (D.9)$$

From (D.2)

$$y_a(t) = C_p(x_p)x_p(t) + D_p(x_p)u_p(t)$$

$$= C_p(x_p)x_p(t) - D_p(x_p)[I + K(t)D_p(x_p)]^{-1}K(t)C_p(x_p)x_p(t) \qquad (D.10)$$

$$y_a(t) = \{I - D_p(x_p)[I + K(t)D_p(x_p)]^{-1}K(t)\}C_p(x_p)x_p(t)$$

$$= [I + D_p(x_p)K(t)]^{-1} C_p(x_p)x_p(t) \qquad (D.11)$$

$$y_a(t) = C_c(x_p)x_p(t) \qquad (D.12)$$

Since (D.1) is, by assumption, uniformly asymptotically stable, there exists some uniformly positive definite matrices, $P_0(x_p)$, $Q_0(x_p)$, $Q_1(x_p)$, and Δ (all unknown and not needed for any implementation) such that the following stability Lyapunov relation is satisfied:

$$\dot{P}_0(x_p) + P_0(x_p)A_0(x_p) + A_0^T(x_p)P_0(x_p) = -Q_1(x_p)$$

$$= -Q_0(x_p) - [P_0(x_p)B_p(x_p) - 2C_c^T(x_p)]\Delta[B_p^T(x_p)P_0(x_p) - 2C_c(x_p)] < 0 \quad (D.13)$$

where $\Delta \in \mathbb{R}^{n*m}$ is "sufficiently small" (but not necessarily small) such that $Q_0(x_p) > 0$.

We use the following positive definite quadratic Lyapunov function

$$V(t) = x_p^T(t)P_0(x_p)x_p(t) + \text{tr}\,[K^T(t)\Gamma_I^{-1}K(t)] \quad (D.14)$$

and the derivative of $V(t)$ along the trajectories of (D.1) and (D.4) is

$$\dot{V}(t) = x_p^T(t)[\,\dot{P}_0(x_p) + P_0(x_p)A_0(x_p) + A_0^T(x_p)P_0(x_p)\,]x_p(t)$$

$$+ x_p^T(t)P_0(x_p)B_p(x_p)u_p(t) + u_p^T(t)B_p^T(x_p)P_0(x_p)x_p(t)$$

$$+ \text{tr}\,[K^T(t)\Gamma_I^{-1}\dot{K}(t)] + \text{tr}\,[\dot{K}^T(t)\Gamma_I^{-1}K(t)] \quad (D.15)$$

Substituting $u_p(t)$ from (D.6) and using (D.4) gives

$$\dot{V}(t) = -x_p^T(t)Q_0(x_p)x_p(t)$$

$$-x_p^T(t)[P_0(x_p)B_p(x_p) - 2C_c^T(x_p)]\Delta[B_p^T(x_p)P_0(x_p) - 2C_c(x_p)]x_p(t)$$

$$- x_p^T(t)P_0(x_p)B_p(x_p)K(t)C_c(x_p)x_p(t) - x_p^T(t)C_c^T(x_p)K^T(t)B_p^T(x_p)P_0(x_p)x_p(t)$$

$$+ y_a^T(t)K(t)y_a(t) + y_a^T(t)K^T(t)y_a(t)$$

$$- 2\sigma\,\text{tr}\,[K^T(t)\Gamma_I^{-1}K(t)] \quad (D.16)$$

$$\dot{V}(t) = -x_p^T(t)Q_0(x_p)x_p(t)$$

$$-x_p^T(t)[P_0(x_p)B_p(x_p) - 2C_c^T(x_p)]\Delta[B_p^T(x_p)P_0(x_p) - 2C_c(x_p)]x_p(t)$$

$$- x_p^T(t)P_0(x_p)B_p(x_p)K(t)C_c(x_p)x_p(t) - x_p^T(t)C_c^T(x_p)K^T(x_p)B_p^T(x_p)P_0(x_p)x_p(t)$$

$$+ x_p^T(t)C_c^T(x_p)K(t)C_c(x_p)x_p(t) + x_p^T(t)C_c^T(x_p)K^T(t)C_c(x_p)x_p(t)$$
$$- 2\sigma \text{ tr } [K^T(t)\Gamma_I^{-1}K(t)] \tag{D.17}$$

Substituting $\Delta = [\Delta - K(t)] + K(t)$ gives

$$\dot{V}(t) = -x_p^T(t)Q_0(x_p)x_p(t)$$
$$-x_p^T(t)[P_0(x_p)B_p(x_p) - 2C_c^T(x_p)][\Delta - K(t)][B_p^T(x_p)P_0(x_p) - 2C_c(x_p)]x_p(t)$$
$$-x_p^T(t)[P_0(x_p)B_p(x_p) - 2C_c^T(x_p)]K(t)[B_p^T(x_p)P_0(x_p) - 2C_c(x_p)]x_p(t)$$
$$- x_p^T(t)P_0(x_p)B_p(x_p)K(t)C_c(x_p)x_p(t) - x_p^T(t)C_c^T(x_p)K^T(t)B_p^T(x_p)P_0(x_p)x_p(t)$$
$$+ x_p^T(t)C_c^T(x_p)K(t)C_c(x_p)x_p(t) + x_p^T(t)C_c^T(x_p)K^T(t)C_c(x_p)x_p(t)$$
$$- 2\sigma \text{ tr } [K^T(t)\Gamma_I^{-1}K(t)] \tag{D.18}$$

$$\dot{V}(t) = -x_p^T(t)Q_0(x_p)x_p(t)$$
$$-x_p^T(t)[P_0(x_p)B_p(x_p) - 2C_c^T(x_p)][\Delta - K(t)][B_p^T(x_p)P_0(x_p) - 2C_c(x_p)]x_p(t)$$
$$-x_p^T(t)P_0(x_p)B_p(x_p) K(t)B_p^T(x_p)P_0(x_p)x_p(t)$$
$$-2x_p^T(t)P_0(x_p)B_p(x_p)K(t)C_c(x_p)x_p(t) - 2x_p^T(t)C_c^T(x_p)K(t)B_p^T(x_p)P_0(x_p)x_p(t)$$
$$-4x_p^T(t)C_c^T(x_p)K(t)C_c(x_p)x_p(t)$$
$$- x_p^T(t)P_0(x_p)B_p(x_p)K(t)C_c(x_p)x_p(t) - x_p^T(t)C_c^T(x_p)K^T(t)B_p^T(x_p)P_0(x_p)x_p(t)$$
$$+ x_p^T(t)C_c^T(x_p)K(t)C_c(x_p)x_p(t) + x_p^T(t)C_c^T(x_p)K^T(t)C_c(x_p)x_p(t)$$
$$- 2\sigma \text{ tr } [K^T(t)\Gamma_I^{-1}K(t)] \tag{D.19}$$

$$\dot{V}(t) = -x_p^T(t)Q_0(x_p)x_p(t)$$
$$-x_p^T(t)[P_0(x_p)B_p(x_p) - 2C_c^T(x_p)][\Delta - K(t)][B_p^T(x_p)P_0(x_p) - 2C_c(x_p)]x_p(t)$$
$$-x_p^T(t)P_0(x_p)B_p(x_p) K(t)B_p^T(x_p)P_0(x_p)x_p(t)$$
$$-x_p^T(t)P_0(x_p)B_p(x_p)K(t)C_c(x_p)x_p(t) - x_p^T(t)C_c^T(x_p)K(t)B_p^T(x_p)P_0(x_p)x_p(t)$$
$$-2x_p^T(t)C_c^T(x_p)K(t)C_c(x_p)x_p(t)$$
$$- 2\sigma \text{ tr } [K^T(t)\Gamma_I^{-1}K(t)] \tag{D.20}$$

$$\dot{V}(t) = -x_p^T(t)Q_0(x_p)x_p(t)$$
$$-x_p^T(t)[P_0(x_p)B_p(x_p) - 2C_c^T(x_p)][\Delta - K(t)][B_p^T(x_p)P_0(x_p) - 2C_c(x_p)]x_p(t)$$
$$-x_p^T(t)[P_0(x_p)B_p(x_p) - C_c^T(x_p)]K(t)[B_p^T(x_p)P_0(x_p) - C_c(x_p)]x_p(t)$$
$$-x_p^T(t)C_c^T(x_p)K(t)C_c(x_p)x_p(t)$$
$$- 2\sigma \, \mathrm{tr} \, [K^T(t)\Gamma_I^{-1}K(t)] \qquad\qquad\qquad\qquad (D.21)$$

We mention again that stability with respect to boundness is guaranteed (Appendix A and B) and that we are interested to test what happens if because of lack of excitations, the adaptive gains tend to decrease. We will show that the equilibrium point $\{x_p = 0, K(t) = 0\}$ is unique, and use (D.21) only to test the behavior of the system in the neighborhood of $\{x_p = 0, K(t) = 0\}$. From (D.21) it is clear that, for any value of $x(t)$, $\dot{V}(t)$ becomes negative whenever $K(t) < \Delta$, and the system is thus led to the equilibrium point given by $\dot{V}(t) \equiv 0$, which is equivalent to $x_p(t) \equiv 0$, and thus $y_p(t) \equiv 0$ and $K(t) \equiv 0$.

We will show now that for any $K_0 > K_{min}$, the point $\{x_p = 0, K(t) = 0\}$ is also the only equilibrium point of the system. To this end, we restrict the discussion to stationary systems and to those nonlinear systems for which uniform asymptotic stability guarantees that the system matrix is nonsingular (for example, because all its eigenvalues belong to the open left halfplane).

We then assume that before applying the adaptive controller, we used some preliminary constant matrix K_0 in (D.3), such that the maximal admissible (supplementary) feedback gain is then

$$K_{max} - K_0 \qquad\qquad\qquad\qquad\qquad\qquad (D.22)$$

and therefore

$$D_p = [K_{max} - K_0]^{-1} \qquad\qquad\qquad\qquad\qquad (D.23)$$

From (D.1) we get

$$\dot{x}_p(t) = A_K(x_p)x_p(t) \tag{D.24}$$

where

$$A_K(x_p) = A_0(x_p) - B_p(x_p)K_c(t)C_p(x_p) \tag{D.25}$$

From (D.7) it can be seen again that

$$0 < K(t) < \infty \quad \text{implies} \quad 0 < K_c(t) < D_p^{-1}(t) = K_{max} - K_0 \tag{D.26}$$

The equilibrium points of (D.4) and (D. 24) are obtained from

$$A_K(x_p)x_p(t) = 0 \tag{D.27}$$

$$y_a(t)y_a^T(t)\Gamma_I - \sigma K(t) = 0 \tag{D.28}$$

If $K_0 > K_{min}(t)$, then $A_K(x_p)$ is stable for any value of $K_c(t)$ and it is therefore, nonsingular. In this case, from (D.27) we get the only solution $x_p(t) = 0$, and then, from (D.28) we get $K_I(t) = 0$.

In conclusion, if $K_0 > K_{min}$, then the equilibrium point $\{x_p(t) = 0, K_I(t) = 0\}$ is both stable and unique.

ACKNOWLEDGMENT

This article is based in part on earlier works (Bar-Kana, ; Bar-Kana and Guez, 1990). The authors are grateful to Howard Kaufman for valuable comments and suggestions. The main support of Drexel University's Stein Fellowship Fundation and the partial support of AFOSR grant No. 890010 are gratefully acknowledged.

REFERENCES

Aizerman, M. A., and Gantmacher, F.R. (1964) *Absolute Stability of Regulator Systems*, Holden Day, San Fransisco.

Anderson, B.D.O. (1968), *Quadratic Minimization, Positive-Real Matrices and Spectral Factorization,* Technical Report EE-6812, University of Newcastle, Australia.

Anderson, B.D.O., and Moylan, P. (1974), "Synthesis of Time-Varying Passive Networks," *IEEE Transactions on Circuits and Systems,* CAS-21, 4, 678-687.

Åström, K.J. (1983), "Theory and Applications of Adaptive Control - A Survey," *Automatica*, 19, 471-481.

Åström, K.J., Borisson, U., Ljung, L. and Wittenmark, B. (1977) "Theory and Applications of Self Tuning Regulators," *Automatica*, 13, 457-476.

Balas, M., Kaufman, H. and Wen, J.(1984) "Stable Direct Adaptive Control of Linear Infinite Dimensional Systems Using Command Generator Approach," *Workshop on Identification and Control of Flexible Space structures*, San Diego, CA, 1984.

Bar-Kana, I. (1986a) "Positive Realness in Discrete-Time Adaptive Control Systems," *International Journal of Systems Sciences*, 17, 7, 1001-1006; also *Proceedings of American Control Conference*, Seattle, WA, June 1986, 1440-1443.

Bar-Kana, I. (1986b) "Extension of a Continuous-Time Multivariable Adaptive Control Algorithm," *Proceedings of American Control Conference*, Seattle, WA, June 1986, 1081-1086.

Bar-Kana, I. (1987a) "Parallel Feedforward and Simplified Adaptive Control," *International Journal of Adaptive Control and Signal Processing*, 1, 2, 95-109.

Bar-Kana, I. (1987b) "Adaptive Control - A Simplified Approach," in C. Leondes (Ed.) *Control and Dynamic Systems - Advances in Theory and Applications*, 25, 187-235.

Bar-Kana, I. (1988a) *Rohrs Examples and Robustness of Simplified Adaptive Control*, Technical Report, Rafael.

Bar-Kana, I. (1988b) *Reduction of Bursting without External Excitation*, Technical Report, Rafael ; also *Proceedings of 1990 Conference on Informational Sciences and Systems*, Princeton University, Princeton, New Jersey.

Bar-Kana, I. (1988c) "Comments on a paper by Kidd," *International Journal of Control*, 48, 3, 1011-1023.

Bar-Kana, I. (1988d) "On the Lur'e Problem and Stability of Nonlinear Controllers," *Journal of The Franklin Institute*, 325, 6, 687-693.

Bar-Kana, I. (1989a) *On Passivity of a Class of Nonlinear Systems,* Technical Report, Drexel University.

Bar-Kana, I. (1989b) "Positive Realness in Multivariable Stationary Linear Systems," *Proceedings of 1989 Conference on Informational Sciences and Systems*, The Johns Hopkins University, Baltimore, Maryland, pp. 383-388; also (full report) Technical Report, Drexel University.

Bar-Kana, I. (1989c) "Absolute Stability and Robust Discrete Adaptive Control of Multivariable Systems," in C. Leondes (Ed) *Control and Dynamic Systems - Advances in Theory and Applications*, Vol. 31, pp. 157-183.

Bar-Kana, I. (1989d) "Robust Simplified Adaptive Stabilization of Not Necessarily Minimum-Phase Systems," *Transactions of ASME, Journal of Dynamic Systems, Measurements, and Control* , pp. 364-370.

Bar-Kana, I. (1990) "Almost Passivity and Simple Adaptive Control of Nonstationary Continuous Linear Systems," *International Journal of Systems Science,* (forthcoming).

Bar-Kana, I., Fischl, R., and Kalata, P., (1989) "Direct Position-plus-Velocity Feedback Control of Large Flexible Space Structures," *Proceedings of 1989 Conference on Informational Sciences and Systems*, The Johns Hopkins University, Baltimore, Maryland, pp. 574-577.

Bar-Kana, I., and Guez, A.(1990) "Simple Adaptive Control for a Class of Nonlinear Systems with Application to Robotics," *International Journal of Control* (forthcoming).

Bar-Kana, I. and Kaufman, H. (1982a) "Model Reference Adaptive Control for Time-Variable Input Commands," *Proceedings of 1982 Conference on Informational Sciences and Systems*, Princeton, New Jersey, 208-211.

Bar-Kana, I. and Kaufman, H. (1982b) "Multivariable Direct Adaptive Control for a General Class of Time-Variable Commands," *Proceedings of 21st IEEE Conference on Decision and Control*, Orlando, Florida, 750-751.

Bar-Kana, I. and Kaufman, H. (1983) "Direct Adaptive Control with Bounded Tracking Errors," *Proceedings of 22nd Conference on Decision and Control*, San Antonio, TX, USA, 181-182.

Bar-Kana, I., and Kaufman, H. (1984) "Some Applications of Direct Adaptive Control to Large Structural Systems," *AIAA Journal of Guidance, Control and Dynamics,* 7, 6, 717-724; also, *Aerokosmicheskaia Technika*, 3, 6, 88-96 (in Russian).

Bar-Kana, I. and Kaufman, H. (1985a) "Global Stability and Performance of a Simplified Adaptive Algorithm," *International Journal of Control,* 42, 6, 1491-1505.

Bar-Kana, I., and Kaufman, H. (1985b) "Robust Simplified Adaptive Control for a Class of Multivariable Continuous-Time Systems," *Proceedings of 24th Conference on Decision and Control,* 141-146.

Bar-Kana, I., and Kaufman, H. (1987) "Robust Simplified Adaptive Control of Large Flexible Space Structures," in H.E. Rauch (Ed) *Control of Distributed Parameter Systems,* IFAC Proceedings Series, 3, 121-126; also (full report) *Technical Report,* Drexel University, 1989.

Bar-Kana, I., and Kaufman, H., (1988) "Simple Adaptive Control of Uncertain Systems," *International Journal of Adaptive Control and Signal Processing,* 2, 133-143.

Bar-Kana, I., Kaufman, and Balas, M., (1983) "Model Reference Adaptive Control of Large Structural Systems," *AIAA Journal of Guidance, Control and Dynamics,* 6, 112-118; also, *Aerokosmicheskaia Technika,* 1, 163-171 (in Russian).

Chen, C. T. (1984) *Linear Systems Theory and Design,* Holt, Rinehart and Winston, New York.

Corless, M., and Leitman, G. (1984) "Adaptive Control for Uncertain Dynamical Systems," in *Dynamical Systems and Microphysics - Control Theory and Mechanics,* Academic Press, 91-158.

Desoer, C.A., and Vidyasagar, M., (1975), *Feedback Systems: Input - Output Properties,* Academic Press, Orlando, FL.

Donaldson, D. D., and Leondes, C. T. (1963) "A Model Referenced Parameter Tracking Technique for Adaptive Control Systems, Part I and II," *J. ASME.*

Fortesque, R.R., Kershenbaum, L.S., and Ydstie, B.D.E. (1981) "Implementation of Self-Tuning Regulators with Variable Forgetting Factors," *Automatica,* 17, 831-835.

Guez, A. (1980) "Application of Singular Perturbation Methods to Air-to-Air Missile Control," *Technical Report, System Dynamics*.

Guez, A., and Dritsas, L. (1987) Ultimate Boundness of Robot Tracking," *Proceedings of 2nd International IEEE Intelligent Control Conference,* Philadelphia, PA.

Gutman, S. (1979) "Uncertain Dynamical Systems - Lyapunov Min-Max Approach," *IEEE Transactions on Automatic Control,* AC-24, 437-443.

Hahn, W. (1967) *Stability of motion,* Springer Verlag, New York.

Hill, D. J., and Moylan, P. (1980), "Dissipative Dynamical Systems," Basic Input-Output Properties," *Journal of The Franklin Institute,* 309, 5, 327-357.

Hsu, L., and Costa, R.R. (1987) "Bursting Phenomena in Continuous - Time Adaptive Systems with σ - Modification," *IEEE Transactions on Automatic Control,* AC-32, 2, 84-85.

Ih, C.H.C., Wang, S. J. and Leondes, C.T. (1987) "Adaptive Control for the Space Station," *Control Systems Magazine,* 7,1, 29-34.

Ioannou, P.A. (1986) "Adaptive Stabilization of Not Necessarily Minimum Phase Plants," *Systems and Control Letters,* 7, 281-287.

Ioannou, P.A. and Kokotovic, P.V. (1983) *Adaptive Systems with Reduced Models,* Springer Verlag, New York.

Kalman, R. E. (1960) "Contributions to the Theory of Optimal Control," *Bol. Soc. Mat. Mexicana,* 102-119.

Kalman, R. E. (1964) "When is a Linear System Optimal?" *Transactions of ASME, Journal of Basic Engineering,* 51- 60.

Kaufman, H., Balas, M., Bar-Kana, I, and Rao, L. (1981) "Model Reference Adaptive Control of Large Scale Systems," *Proceedings of 20th IEEE Conference of Decision and Control,* San Diego, CA, 984-989.

Kidd, P. T. (1985) "Comparison of the Performance of a Model Reference Adaptive System and a Classical Linear Control System under Nonideal Conditions," *International Journal of Control,* 42, 671-694.

Landau, I. (1979) *Adaptive Control - The Model Reference Approach,* Marcel Decker, New York.

Laniado, I., Kreindler, E., and Bar-Kana, I. (1989) "Simple Adaptive Control of a Robot Arm," *Proceedings of 1989 IEEE Conference on Control and Applications - ICCON,* Jerusalem, Israel, April 3-6, 1989.

LaSalle, J. P. (1981) "Stability of Nonautonomous Systems," *Nonlinear Analysis Theory, Methods and Applications,* 1, 1, 83-91.

MacFarlane, A. (1982) "Complex Variable Methods in Feedback Systems Analysis and Design," in *Design of Modern Control Systems,* Peter Peregrinus, London.

Mareels, I.M.Y., and R.R. Bitmead (1986) "Nonlinear Dynamics in Adaptive Control - Chaotic and Periodic Stabilization," *Automatica,* 22, 3, 641-655.

Mattern, D. L., and Shoureshi, R. (1987) "Model Reference Adaptive Control - A Perspective," *Proceedings 1987 American Control Conference*, 201-206.

Monopoli, R. V.(1974) "Model Reference Adaptive Control with an Augmented Error Signal," IEEE Transactions on Automatic Control, AC-25, 433-439.

Morse, A.S. (1984) "New Directions in Parameter Adaptive Control Systems*," Proc. of 23rd Conference of Decision and Control,* Las Vegas, NV, 1566-1568.

Morse, W., and Ossman, K. (1989) "Flight Control Reconfiguration Using Model Reference Adaptive Control," *Proceedings of 1989 American Control Conference,* Pittsburgh, Pennsylvania, pp. 159-164.

Narendra, K. S., and Valavani, L. (1979) "Direct and Indirect Adaptive Control," Automatica, 15, 653-661.

Nussbaum, R.O. (1983) "Some Remarks on a Conjecture in Parameter Adaptive Control," *Systems and Control Letters,* 3, 243-246.

Osburn, P. V., Whitaker, H. P., and Kezer, A., (1961) "New Developments in the Design of Model Reference Adaptive Control Systems," *Inst. Aeronautical Sciences,* Paper 61-39.

Ortega, R., and Spong, M. (1988) "Adaptive Motion Control of Rigid Robots: A Tutorial," *Proceedings of 27th Conference on Decision and Control,* Austin, TX, 1575-1584.

Parks, P. C. (1966) "Lyapunov Redesign of Model Reference Adaptive Control Systems," IEEE Transactions on Automatic Control, AC-11, 362-367.

Popov, V.M. (1964), "Hyperstability and Optimality of Automatic Systems with Several Control Functions," *Rev. Roumaine Sci. Tech. Electrotechn. et Energy,* 9, 4, 629-690.

Rohrs, C., Valavani, L., Athans, M. and Stein, G. (1982) "Robustness of Adaptive Control Algorithms in the Presence of Unmodelled Dynamics," *Proc. of 21st IEEE Conference on. Decision and Control,* Orlando, FL, 3-11.

Rohrs, C., Valavani, L., Athans, M. and Stein, G., (1985) "Robustness of Continuous-Time Adaptive Control Algorithms in the Presence of Unmodelled Dynamics," *IEEE Transactions on Automatic Control,* AC-30, 9, 881-889.

Salam, F.M.A., and Bai, S. (1986) "Disturbance-Generated Bifurcations in a Simple Adaptive System: Simulation Evidence," *Systems and Control Letters,* 7, 4, 269-280.

Schumacher, J.M. (1984) "Almost Stabilizability Subspaces and High Gain Feedback," *IEEE Transactions on Automatic Control,* AC-29, 620-628.

Seraji, H. (1989) "Decentralized Adaptive Control of Manipulators: Theory, Simulation, and Experimentation," *IEEE Transactions on Robotics and Automation,* 5, 2, 183-201.

Shaked, U. (1977) "The Zero Properties of Linear Passive Systems," *IEEE Transactions on Automatic Control,* AC-22, 6, 973-976.

Slotine, J.-J. E., and Li, W. (1988) "Adaptive Manipulator Control: A Case Study," *IEEE Transactions on Automatic Control,* 33, 11, 995-1003.

Sobel, K., Kaufman, H., and Mabius, L. (1979) "Model Reference Output Control Systems Without Parameter Identification," *Proceedings of 18th Conference on Decision and Control,* Fort Lauderdale, Florida.

Sobel, K., Kaufman, H. and Mabius, L. (1982) "Implicit Adaptive Control for a Class of MIMO Systems," *IEEE Transactions on Aerospace and Electronic Systems,* AES-18, 576-590.

Steinberg, A., and Corless, M. (1985) "Output Feedback Stabilization of Uncertain Dynamical Systems," *IEEE Transactions on Automatic Control,* AC-30, 1025-1027.

Vidyasagar, M. (1978) *Nonlinear Systems Analysis,* Prentice Hall, Engelwood Cliffs, NJ.

Whitaker, H. P. (1959) "An Adaptive System for Control of the Dynamics Performance of Aircraft and Spacecraft," *Inst. Aeronautical Sciences,* Paper 59-100.

Willems, J.C. (1971) *The Analysis of Feedback Systems,* MIT Press, Cambridge, MA.

Willems, J. C. (1972) "Dissipative Dynamical Systems," in *Archive for Rational Mechanics and Analysis,* 45, 5, 321-393.

Willems, J. C., and Byrnes, C.I. (1984) "Global Adaptive Stabilization in the Absence of Information on the High Frequency Gain," *Proceedings of Sixth International Conference on Analysis and Optimization of Systems,* Springer Verlag, 49-57.

Willems, J. L. (1970) *Stability Theory of Dynamical Systems,* Nelson, London.

THEORY AND APPLICATIONS OF CONFIGURATION CONTROL FOR REDUNDANT MANIPULATORS

HOMAYOUN SERAJI

Jet Propulsion Laboratory
California Institute of Technology
Pasadena, CA 91109

ABSTRACT

Previous investigations of redundant manipulators have often focussed on *local* optimization for redundancy resolution by using the Jacobian pseudoinverse to solve the instantaneous relationship between the joint and end-effector velocities. This paper presents a new and simple approach to control the manipulator configuration over the *entire* motion based on augmentation of the manipulator forward kinematics. A set of kinematic functions is defined in Cartesian or joint space to reflect the desirable configuration that will be achieved in addition to the specified end-effector motion. The user-defined kinematic functions and the end-effector Cartesian coordinates are combined to form a set of task-related configuration variables as generalized coordinates for the manipulator. A task-based adaptive scheme is then utilized to directly control the configuration variables so as to achieve tracking of some desired reference trajectories throughout the robot motion. This accomplishes the basic task of desired end-effector motion, while utilizing the redundancy to achieve any additional task through the desired time variation of the kinematic functions. The present formulation can also be used for optimization of any kinematic objective function, or for satisfaction of a set of kinematic inequality constraints, as in the obstacle avoidance problem. In contrast to pseudoinverse-based methods, the configuration control scheme ensures cyclic motion of the manipulator, which is an essential requirement for repetitive operations. The control law is simple and computationally very fast, and does not require either the complex manipulator dynamic model or the complicated inverse kinematic transformation. The configuration control scheme can alternatively be implemented in joint space. Simulation results of a direct-drive three-link arm are given to illustrate the proposed control scheme. The scheme is also implemented for real-time control of three links of a PUMA 560 industrial robot and experimental results are presented and discussed. The simulation and experimental results validate the configuration control scheme, and demonstrate its capabilities for performing various realistic tasks.

1. INTRODUCTION

The remarkable dexterity and versatility that the human arm exhibits in performing various tasks can be attributed largely to the kinematic redundancy of the arm, which provides the capability of reconfiguring the arm without affecting the hand position. A robotic manipulator is called (kinematically) "redundant" if it possesses more degrees-of-freedom than is necessary for performing a specified task. Redundancy of a robotic manipulator is therefore determined relative to the particular task to be performed. For example, in the two-dimensional space, a planar robot with three joints is redundant for achieving any end-effector position, whereas the robot is non-redundant for tasks involving both position and orientation of the end-effector. In the three-dimensional space, a manipulator with seven or more joints is redundant since six degrees-of-freedom are sufficient to position and orient the end-effector in any desired configuration. In a non-redundant manipulator, a given position and orientation of the end-effector corresponds to a limited set of joint angles and associated robot configurations with distinct poses (such as elbow up or down). Therefore, for a prescribed end-effector trajectory and a given pose, the motion of the robot is uniquely determined. When this motion is undesirable due to collision with obstacles, approaching kinematic singularities or reaching joint limits, there is no freedom to reconfigure the robot so as to reach around the obstacles, or avoid the singularities and joint limits.

Redundancy in the manipulator structure yields increased dexterity and versatility for performing a task due to the infinite number of joint motions which result in the same end-effector trajectory. However, this richness in choice of joint motions complicates the manipulator control problem considerably. In order to take full advantage of the capabilities of redundant manipulators, effective control schemes should be developed to utilize the redundancy in some useful manner. During recent years, redundant manipulators have been the subject of considerable research, and several methods have been suggested to resolve the redundancy [1-26]. Whitney [1] suggests the use of Jacobian pseudoinverse for the control of redundant manipulators. Liégeois [2] proposes a modification to the pseudoinverse approach to resolve manipulator redundancy. Nakamura and Yoshikawa [3-6] develop a scheme based on task priority using pseudoinverses. Baillieul [7-11] proposes the extended Jacobian method to minimize or maximize an objective function. Walker and Marcus [12] suggest a method based on the pseudoinverse approach to impose a constraint relationship on the manipulator. A comprehensive review of the pseudoinverse approach to redundant manipulators is given by Klein and Huang [13]. Oh, Orin and Bach [14,15] describe a numerical procedure for solving the inverse kinematic problem which uses constraints on the manipulator. Khatib [16] gives a method for the resolution of redundancy using the robot dynamics in the operational space. Klein

[17-19] addresses obstacle avoidance and dynamic simulation of redundant robots. Baker and Wampler [20] study the kinematic properties of redundant manipulators. The problems of robot design and torque optimization are addressed by Hollerbach [21,22]. Egeland [23] describes a method for Cartesian control of a hydraulic redundant manipulator. Sciavicco and Siciliano [24] give a dynamic solution to the inverse kinematic problem for redundant robots. Hsu, Hauser and Sastry [25] discuss the resolution of redundancy using the manipulator dynamics. Dubey, Euler and Babcock [26] describe a gradient projection optimization scheme for 7 dof robots.

Over the past two decades, investigations of redundant manipulators [1-26] have often been explicitly or implicitly based on the Jacobian pseudoinverse approach for the utilization of redundancy through local optimization of some objective function. Furthermore, most proposed methods resolve the redundancy in joint space and are concerned solely with solving the inverse kinematic problem for redundant manipulators.

In this paper, a new and conceptually simple approach to *configuration control* of redundant manipulators is presented, which takes a complete departure from the conventional pseudoinverse methods. In this approach, the redundancy is utilized for control of the manipulator configuration directly in *task space*, where the task is performed, thus avoiding the complicated inverse kinematic transformation. A set of kinematic functions in Cartesian or joint space is chosen to reflect the desired additional task that will be performed due to the redundancy. The kinematic functions can be viewed as a parameterization of the manipulator "self-motion," in which the internal movement of the links does not move the end-effector. In other words, given the end-effector position and orientation, the kinematic functions are used to "shape" the manipulator configuration. The end-effector Cartesian coordinates and the kinematic functions are combined to form a set of "configuration variables" which describe the physical configuration of the entire manipulator in the task space. The control scheme then ensures that the configuration variables track some desired trajectories as closely as possible, so that the manipulator motion meets the task requirements. The control law is adaptive and does not require knowledge of the complex dynamic model or parameter values of the manipulator or payload. The scheme can be implemented in either a centralized or a decentralized control structure, and is computationally very fast, making it specially suitable for real-time control of redundant manipulators.

The paper is organized as follows. In Section 2, the configuration control approach is formulated and a task-based control scheme is proposed. The configuration control scheme is illustrated in Section 3 by simulation of a direct-drive three-link arm and experimentation on three links of an industrial robot. Section 4 discusses the results of the paper and draws some conclusions.

2. DEVELOPMENT OF CONFIGURATION CONTROL SCHEME

The mechanical manipulator under consideration consists of a linkage of rigid bodies with n revolute or prismatic joints. Let T be the $n \times 1$ vector of torques or forces applied at the joints and θ be the $n \times 1$ vector of the resulting relative joint rotations or translations. The dynamic equation of motion of the manipulator which relates T to θ can be represented in the general form [27]

$$M(\theta)\ddot{\theta} + N(\theta, \dot{\theta}) = T \qquad (1)$$

where the matrices M and N are highly complex nonlinear functions of θ, $\dot{\theta}$, and the payload. Let the $m \times 1$ vector Y (with $m < n$) represent the position and orientation of the end-effector (last link) with respect to a fixed Cartesian coordinate system in the m-dimensional task space where the task is to be performed. The $m \times 1$ end-effector coordinate vector Y is related to the $n \times 1$ joint angle vector θ by the forward kinematic model

$$Y = Y(\theta) \qquad (2)$$

where $Y(\theta)$ is an $m \times 1$ vector whose elements are nonlinear functions of the joint angles and link parameters, and which embodies the geometry of the manipulator. For a redundant manipulator, a Cartesian coordinate vector (such as Y) that specifies the end-effector position and orientation does not constitute a set of generalized coordinates to completely describe the manipulator dynamics. Nonetheless, equations (1) and (2) form a valid dynamic model that describes the *end-effector* motion itself in the task space. The desired motion of the end-effector is represented by the reference position and orientation trajectories denoted by the $m \times 1$ vector $Y_d(t)$, where the elements of $Y_d(t)$ are continuous twice-differentiable functions of time. The vector $Y_d(t)$ embodies the information on the "*basic task*" to be accomplished by the end-effector in the task space.

We now discuss the definition of configuration variables and the task-based control of redundant manipulators in the subsequent sections.

2.1 Definition of Configuration Variables

Let $r = n - m$ be the "degree-of-redundancy" of the manipulator, i.e. the number of "extra" joints. Let us define a set of r kinematic functions $\{\phi_1(\theta), \phi_2(\theta), \ldots, \phi_r(\theta)\}$ in Cartesian or joint space to reflect the "*additional task*" that will be performed due to the manipulator redundancy. Each ϕ_i can be a function of the joint angles $\{\theta_1, \ldots, \theta_n\}$ and the link geometric parameters. The choice of the kinematic functions can be made in several ways to represent, for instance, the Cartesian coordinates of any point on the manipulator, or any combination of the joint angles. The kinematic functions parameterize the "self-motion" of the manipulator, in which the internal movement of the links does not move the end-effector.

For illustration, consider the planar three-link arm shown in Figure 1(i). The basic task is to control the end-effector position coordinates $[x, y]$ in the base frame. Suppose that we fix the end-effector position and allow internal motion of the links so that the arm takes all possible configurations. It is found that the locus of point A is an arc of a circle with center O and radius ℓ_1 which satisfies the distance constraint $AC \leq (\ell_2 + \ell_3)$. Likewise, the locus of point B is an arc of a circle with center C and radius ℓ_3 which satisfies $OB \leq (\ell_1 + \ell_2)$. The loci of A and B are shown by solid arcs in Figure 1(i), and represent the self-motion of the arm. Now, in order to characterize the self-motion, we can select a kinematic function $\phi(\theta)$ to represent, for instance, the terminal angle $\phi = \theta_3$ in joint space, or alternatively we can designate the wrist height y_B as the kinematic function $\phi = \ell_1 \sin\theta_1 + \ell_2 \sin\theta_2$ in Cartesian space. The choice of ϕ clearly depends on the particular task that we wish to perform by the utilization of redundancy, in addition to the end-effector motion. Let us now consider the spatial 7 dof arm [21] shown in Figure 1(ii), in which the end-effector position and orientation are of concern. With the end-effector frame fixed in space, the self-motion of the arm consists of the elbow transcribing a circle with center C and radius CA, as shown in Figure 1(ii). This circle is the intersection of a sphere with center O and radius OA, and another sphere with center B and radius BA. In order to parameterize the self-motion, we can now define the kinematic function $\phi(\theta) = \alpha$, where α is the angle between the plane OAB and the vertical plane passing through OB, as shown in Figure 1(ii). The kinematic function ϕ then succinctly describes the redundancy and gives a simple characterization of the self-motion.

Once a set of r task-related kinematic functions $\Phi = \{\phi_1, \phi_2, \ldots, \phi_r\}$ is defined, we have partial information on the manipulator configuration. The set of m end-effector position and orientation coordinates $Y = \{y_1, y_2, \ldots, y_m\}$ provides the remaining information on the configuration. Let us now combine the two sets Φ and Y to obtain a complete set of n configuration variables as

$$X = \{Y, \Phi\} = \{y_1, y_2, \ldots, y_m \vdots \phi_1, \phi_2, \ldots, \phi_r\}$$
$$= \{x_1, x_2, \ldots, x_n\} \tag{3}$$

The $n \times 1$ vector X is referred to as the *"configuration vector"* of the redundant manipulator and the elements of X, namely $\{x_1, \ldots, x_n\}$, are called the *"configuration variables."* The configuration variables define an n-dimensional *"task space"* whose coordinate axes are $\{x_1, x_2, \ldots, x_n\}$. The task space is composed of two subspaces; namely, the m-dimensional end-effector subspace with axes $\{x_1, \ldots, x_m\}$ and the r-dimensional subspace due to kinematic functions with axes $\{x_{m+1}, \ldots, x_n\}$. The configuration variables $\{x_1, \ldots, x_n\}$ constitute a set of generalized coordinates for the redundant manipulator in a given region of the workspace. Using the configuration vector

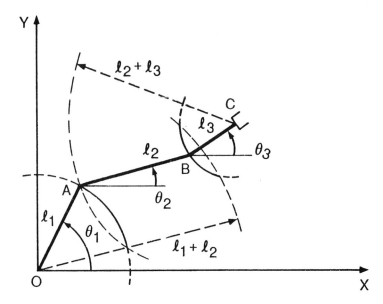

Figure 1(i). Self-motion of Planar 3 dof Arm

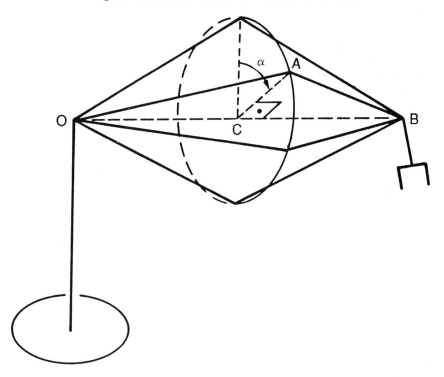

Figure 1(ii).   Self-motion of Spatial 7 dof Arm

X, the manipulator is fully specified kinematically and is no longer redundant in this representation. It is noted that in some applications, certain end-effector coordinates are not relevant to the task, for instance, in a spot welding task the orientation of the end-effector is not important. In such cases, the present approach allows the designer to replace the insignificant end-effector coordinates with additional kinematic functions which are more relevant to the particular application. In fact, if $m'(< m)$ end-effector coordinates are specified, then $n - m' = r'(> r)$ kinematic functions can be defined.

The *augmented forward kinematic model* which relates the configuration vector X to the joint angle vector θ is now given by

$$X = \begin{pmatrix} Y(\theta) \\ \dots \\ \Phi(\theta) \end{pmatrix} = X(\theta) \tag{4}$$

From equation (4), the differential kinematic model which relates the rates of change of X and θ is obtained as

$$\dot{X}(t) = J(\theta)\dot{\theta}(t) \tag{5}$$

where

$$J(\theta) = \begin{pmatrix} J_e(\theta) \\ \dots \\ J_c(\theta) \end{pmatrix} = \begin{pmatrix} \frac{\partial Y}{\partial \theta} \\ \dots \\ \frac{\partial \Phi}{\partial \theta} \end{pmatrix} \tag{6}$$

is the $n \times n$ *augmented Jacobian matrix*. The $m \times n$ submatrix $J_e(\theta) = \frac{\partial Y}{\partial \theta}$ is associated with the end-effector, while the $r \times n$ submatrix $J_c(\theta) = \frac{\partial \Phi}{\partial \theta}$ is related to the kinematic functions. The combination of the two non-square submatrices J_e and J_c forms the square Jacobian matrix J.

The augmented Jacobian matrix J can be used to test the functional independence of the kinematic functions $\{\phi_1, \dots, \phi_r\}$ and the end-effector coordinates $\{y_1, \dots, y_m\}$. For the set of configuration variables $X = \{x_1, \dots, x_n\}$ to be functionally independent throughout the workspace, it suffices to check that det $[J(\theta)]$ is not identically zero for *all* θ, [28]. In other words, when the augmented Jacobian matrix J is rank-deficient for *all* values of θ, the kinematic functions chosen are functionally dependent on the end-effector coordinates and a different choice of Φ is necessary. For example, consider the three-link planar arm in Figure 1(i) with the end-effector coordinates $[x, y]$ and suppose that we define the end-effector distance from the origin, i.e., OC, as the kinematic function $\phi(\theta)$, that is,

$$\phi(\theta) = (OC)^2 = x^2 + y^2$$

The third row of the resulting Jacobian matrix is

$$
\begin{aligned}
J_c &= \left[\frac{\partial \phi}{\partial \theta_1} \quad \frac{\partial \phi}{\partial \theta_2} \quad \frac{\partial \phi}{\partial \theta_3} \right] \\
&= \left[2x\frac{\partial x}{\partial \theta_1} + 2y\frac{\partial y}{\partial \theta_1} \quad , \quad 2x\frac{\partial x}{\partial \theta_2} + 2y\frac{\partial y}{\partial \theta_2} \quad , \quad 2x\frac{\partial x}{\partial \theta_3} + 2y\frac{\partial y}{\partial \theta_3} \right] \\
&= [2x \quad , \quad 2y] \begin{bmatrix} \frac{\partial x}{\partial \theta_1} & \frac{\partial x}{\partial \theta_2} & \frac{\partial x}{\partial \theta_3} \\ \frac{\partial y}{\partial \theta_1} & \frac{\partial y}{\partial \theta_2} & \frac{\partial y}{\partial \theta_3} \end{bmatrix} = [2x, 2y]J_e
\end{aligned}
$$

It is seen that the third row is a linear combination of the first and second rows, and hence det $[J] = \det \left(\begin{smallmatrix} J_e \\ J_c \end{smallmatrix} \right) \equiv 0$ for all θ. This implies that the particular choice of $\phi(\theta)$ is not independent of the end-effector coordinates, as expected, and is therefore unacceptable.

When det $[J(\theta)]$ is not identically zero, the configuration variables $\{x_1, \ldots, x_n\}$ are not functionally dependent for all θ. Nonetheless, there can be certain joint configurations $\theta = \theta_o$ at which det $[J(\theta_o)] = 0$, i.e., the augmented Jacobian matrix J is rank-deficient. This implies that the rows J^i of J satisfy the linear relationship $\sum_{i=1}^{n} c_i J^i = 0$, where c_i are some constants which are not all zero. Since the changes of the configuration variables and joint angles are related by $\Delta x = J(\theta)\Delta\theta$, we conclude that at $\theta = \theta_o$, $\sum_{i=1}^{n} c_i \Delta x_i = 0$. Therefore at a Jacobian singularity, the changes in the configuration variables $\{\Delta x_1, \ldots, \Delta x_n\}$ must satisfy the constraint relationship $\sum_{i=1}^{n} c_i \Delta x_i = 0$, and hence the configuration vector X cannot be changed arbitrarily. This also implies that $\sum_{i=1}^{n} c_i \dot{x}_i = 0$, i.e., no combination of joint velocities will produce motion along certain directions in the task space. From expression (6), it is clear that the Jacobian matrix J will be singular at any joint configuration for which the submatrix J_e is rank-deficient; i.e., at any end-effector singular configuration. These are referred to as *kinematic singularities* of the manipulator. Due to the additional task requirements, *algorithmic singularities* may further be introduced into the Jacobian matrix J due to the submatrix J_c, [8]. These singularities occur when either J_c is rank-deficient or some rows of J_c and J_e are linearly dependent. However, by a judicious choice of the kinematic functions, some algorithmic singularities may be avoided. Further discussion of the augmented Jacobian singularities can be found in [29].

For the sake of illustration, let us reconsider the planar three-link arm shown in Figure 1(i). Let x and y represent the Cartesian coordinates of the hand (C) position to be controlled, and suppose that the hand orientation is not of concern so that the arm is redundant with the degree-of- redundancy equal to one. Suppose that we wish to utilize the redundancy in order to control the elbow (A) position or equivalently the shoulder angle θ_1, *in addition to* the hand position. For this purpose, the kinematic function ϕ can be defined in a number of ways such as

$$\phi = \ell_1 \sin \theta_1 = \text{elbow vertical position}$$
$$\phi = \ell_1 \cos \theta_1 = \text{elbow horizontal position}$$
$$\phi = \theta_1 \qquad = \text{shoulder angle}$$

The above choices of ϕ will serve the same purpose. Nonetheless, each choice of ϕ will yield a different augmented Jacobian matrix J and hence different singular configurations. Let us take the last choice of ϕ and, from Figure 1(i), obtain the configuration variables for the arm as

$$x = \ell_1 \cos \theta_1 + \ell_2 \cos \theta_2 + \ell_3 \cos \theta_3$$
$$y = \ell_1 \sin \theta_1 + \ell_2 \sin \theta_2 + \ell_3 \sin \theta_3$$
$$\phi = \theta_1$$

From these equations, the augmented Jacobian matrix is found to be

$$J = \begin{pmatrix} J_e \\ \cdots \\ J_c \end{pmatrix} = \begin{pmatrix} -\ell_1 \sin \theta_1 & -\ell_2 \sin \theta_2 & -\ell_3 \sin \theta_3 \\ \ell_1 \cos \theta_1 & \ell_2 \cos \theta_2 & \ell_3 \cos \theta_3 \\ \cdots\cdots\cdots\cdots\cdots\cdots\cdots\cdots \\ 1 & 0 & 0 \end{pmatrix}$$

and hence

$$\det J = \ell_2 \ell_3 \sin(\theta_3 - \theta_2)$$

The singular configurations are obtained from $\det J = 0$ as $\theta_3 = \theta_2$ and $\theta_3 = \theta_2 + 180°$. It is seen that for our particular choice of ϕ, the arm "appears" to be a two-link arm (ℓ_2, ℓ_3) with a moveable base (A). As for any two-link arm, the two singular configurations are when the arm is either fully extended $(\theta_3 = \theta_2)$ or fully folded $(\theta_3 = \theta_2 + 180°)$. It is noted that if, instead, the elbow vertical or horizontal position is selected as the kinematic function ϕ, additional *algorithmic* singularities will be introduced at $\theta_1 = \pm 90°$ or $\theta_1 = 0°$, $180°$, respectively. Therefore, the choice of the kinematic function ϕ is a critical factor in determining the algorithmic singularities of the arm. From this example, we see that it is preferable to choose a combination of joint angles as a kinematic function, if feasible, since the resulting Jacobian matrix J_c will then consist of a constant row which does not depend on the robot configuration. Otherwise, additional algorithmic singularities may be introduced due to J_c.

An alternative way to utilize the redundancy is to ensure avoidance of certain kinematic singularities, in addition to desired end-effector motion. For the three-link arm, the square of "manipulability measure" [6] associated with the end-effector can be defined as the kinematic function ϕ, namely

$$\phi(\theta) = \det [J_e(\theta) J_e'(\theta)]$$
$$= \ell_1^2 \ell_2^2 \sin^2(\theta_2 - \theta_1) + \ell_1^2 \ell_3^2 \sin^2(\theta_3 - \theta_1) + \ell_2^2 \ell_3^2 \sin^2(\theta_3 - \theta_2)$$

where $J_e(\theta)$ is the end-effector Jacobian matrix, and the prime denotes transposition. The function $\phi(\theta)$ will be zero at the kinematic singularities related to the end-effector. In order to avoid these singularities, the redundancy is used to ensure that the manipulability measure $\phi(\theta)$ is a non-zero constant or tracks some desired positive time function $\phi_d(t)$.

2.2 Task-Based Configuration Control

Suppose that a user-defined *"additional task"* can be expressed by the following kinematic equality constraint relationships

$$
\begin{aligned}
\phi_1(\theta) &= \phi_{d1}(t) \\
\phi_2(\theta) &= \phi_{d2}(t) \\
&\ \ \vdots \qquad \vdots \\
\phi_r(\theta) &= \phi_{dr}(t)
\end{aligned}
\tag{7}
$$

where $\phi_{di}(t)$ denotes the desired time variation of the kinematic function ϕ_i and is a user-specified continuous twice-differentiable function of time so that $\dot{\phi}_{di}(t)$ and $\ddot{\phi}_{di}(t)$ are defined. The kinematic relationships (7) can be represented collectively in the vector form

$$
\Phi(\theta) = \Phi_d(t)
\tag{8}
$$

where Φ and Φ_d are $r \times 1$ vectors. Equation (8) represents a set of "kinematic constraints" on the manipulator and defines the task that will be performed *in addition to* the basic task of desired end-effector motion. The kinematic equality constraints (8) are chosen to have physical interpretations and are used to formulate the desirable characteristics of the manipulator configuration in terms of motion of other members of the manipulator. The utilization of redundancy to provide *direct* control over the entire robot configuration enables the user to specify the evolution of the robot configuration while the end-effector is travelling a prescribed path. This control strategy is particularly useful for maneuvering the robot in a constricted workspace or a cluttered environment. Since the robot "shape" is controlled directly, it can be configured to fit within a restricted space or keep clear of workspace objects. For instance, in the 7 dof arm of Figure 1(ii), the user can ensure that the arm assumes a desired angle relative to the vertical, when the hand reaches the goal location. Alternatively, by controlling the elbow height as well as the hand coordinates, we can ensure that the elbow reaches over vertical obstacles (such as walls) in the workspace while the hand tracks the desired trajectory. The proposed formulation appears to be a highly promising approach to the additional task performance in comparison with the previous approaches which attempt to minimize or maximize objective functions, since we are now able to make a more specific statement about the evolution of the manipulator configuration. The present approach also covers the intuitive solution to redundant arm control in which certain joint angles are held constant for a portion of the task in order to resolve the redundancy. The functional forms of the kinematic functions ϕ_i and their desired behavior ϕ_{di} may vary widely for different additional tasks, making the approach unrestricted to any particular type of application.

Based on the foregoing formulation, we can now consider the manipulator with the $n \times 1$ configuration vector $X = \binom{Y}{\Phi}$ and the augmented forward kinematic model $X = X(\theta)$. Once the desired motion of the end-effector $Y_d(t)$ is specified for the particular basic task and the required evolution of the kinematic functions $\Phi_d(t)$ is specified to meet the desired additional task, the $n \times 1$ desired configuration vector $X_d(t) = \binom{Y_d(t)}{\Phi_d(t)}$ is fully determined. The configuration control problem for the redundant manipulator is to devise a dynamic control scheme as shown in Figure 2 which ensures that the manipulator configuration vector $X(t)$ tracks the desired trajectory vector $X_d(t)$ as closely as possible. In the control system shown in Figure 2, the actual end-effector position $Y(t)$ and the current value of the kinematic functions $\Phi(t)$ are first computed based on the joint positions $\theta(t)$ using the augmented forward kinematic model (4). This information is then fed back to the controller which also receives the commanded end-effector motion $Y_d(t)$ and the desired time variation $\Phi_d(t)$ to compute the driving torques $T(t)$. These torques are applied at the manipulator joints so as to meet the basic and additional task requirements simultaneously.

Once the forward kinematic model of the manipulator is augmented to include the kinematic functions, different control strategies can be improvised to meet the above tracking requirement, taking into account the dynamics of the manipulator given by equation (1). There are two major techniques for the design of tracking controllers in task space, namely model-based control and adaptive control. For the model-based control [16], the manipulator dynamics is first expressed in task space as

$$M_x(\theta)\ddot{X} + N_x(\theta, \dot{\theta}) = F \tag{9}$$

where F is the $n \times 1$ "virtual" control force vector in the task space, and M_x and N_x are obtained from equations (1)-(6). The centralized control law which achieves tracking through global linearization and decoupling is given by

$$F = M_x(\theta)\left[\ddot{X}_d(t) + K_v\left(\dot{X}_d(t) - \dot{X}(t)\right) + K_p\left(X_d(t) - X(t)\right)\right] + N_x(\theta, \dot{\theta}) \tag{10}$$

where K_p and K_v are constant position and velocity feedback gain matrices. This control formulation requires a precise knowledge of the full dynamic model and parameter values of the manipulator and the payload. Furthermore, since M_x and N_x depend on the definition of the kinematic functions Φ,

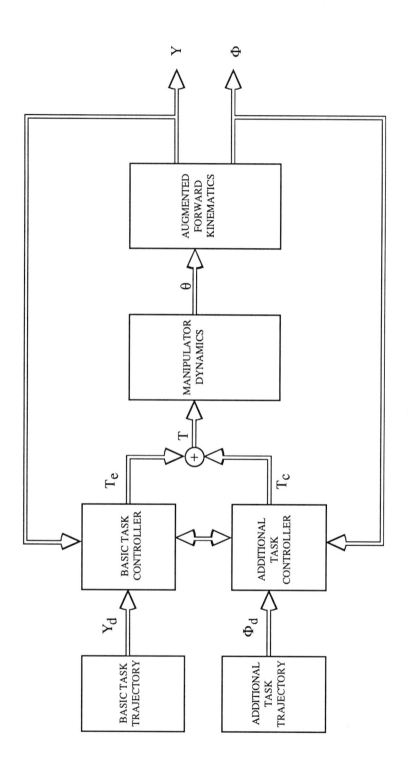

Figure 2. Architecture of Configuration Control Scheme

any change in the additional task specifications requires re-evaluation of the robot dynamic model (9). The alternative approach is the adaptive control technique in which the on-line adaptation of the controller gains eliminates the need for the complex manipulator dynamic model. In this section, we adopt an adaptive control scheme which has been developed recently and validated experimentally on a PUMA industrial robot [30-32]. The adaptive controller produces the control signal based on the observed performance of the manipulator and has therefore the capability to operate with minimal information on the manipulator/payload and to cope with unpredictable gross variations in the payload. The proposed adaptive control scheme is developed within the framework of Model Reference Adaptive Control (MRAC) theory, and the centralized adaptive tracking control law in the task space is given by [30]

$$
\begin{aligned}
F(t) = d(t) &+ [K_p(t)E(t) + K_v(t)\dot{E}(t)] \\
&+ [C(t)X_d(t) + B(t)\dot{X}_d(t) + A(t)\ddot{X}_d(t)]
\end{aligned}
\tag{11}
$$

as shown in Figure 3. This control force is composed of three components,[1] namely:

(i) The *auxiliary signal* $d(t)$ is synthesized by the adaptation scheme and improves transient performance while resulting in better tracking and providing more flexibility in the design.

(ii) The term $[K_p(t)E(t) + K_v(t)\dot{E}(t)]$ is due to the PD *feedback controller* acting on the position tracking-error $E(t) = X_d(t) - X(t)$ and the velocity tracking-error $\dot{E}(t) = \dot{X}_d(t) - \dot{X}(t)$.

(iii) The term $[C(t)X_d(t) + B(t)\dot{X}_d(t) + A(t)\ddot{X}_d(t)]$ is the contribution of the PD^2 *feedforward controller* operating on the desired position $X_d(t)$, the desired velocity $\dot{X}_d(t)$, and the desired acceleration $\ddot{X}_d(t)$.

The required auxiliary signal and feedback/feedforward controller gains are updated based on the $n \times 1$ "weighted" error vector $q(t)$ by the following simple adaptation laws [30]:

$$
q(t) = W_p E(t) + W_v \dot{E}(t)
\tag{12}
$$

$$
d(t) = d(0) + \delta_1 \int_0^t q(t)dt + \delta_2 q(t)
\tag{13}
$$

$$
K_p(t) = K_p(0) + \alpha_1 \int_0^t q(t)E'(t)dt + \alpha_2 q(t)E'(t)
\tag{14}
$$

$$
K_v(t) = K_v(0) + \beta_1 \int_0^t q(t)\dot{E}'(t)dt + \beta_2 q(t)\dot{E}'(t)
\tag{15}
$$

---

[1] Note that since the desired velocity $\dot{X}_d(t)$ and acceleration $\ddot{X}_d(t)$ are directly available from the trajectory generator, it is not necessary to perform differentiation in implementing the control law (11).

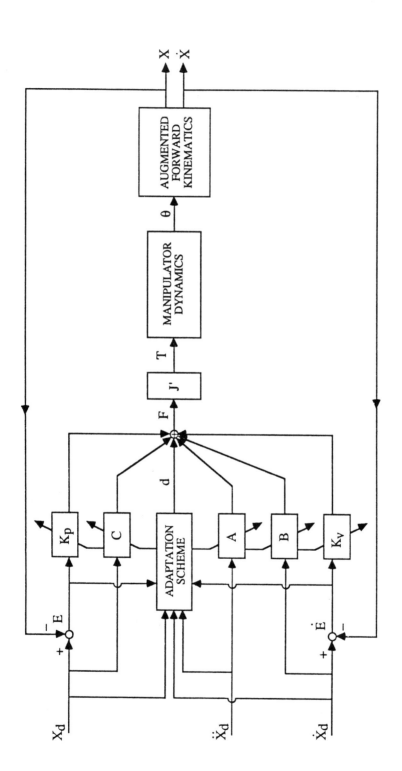

Figure 3. Adaptive Manipulator Control System

$$C(t) = C(0) + \nu_1 \int_0^t q(t)X_d'(t)dt + \nu_2 q(t)X_d'(t) \qquad (16)$$

$$B(t) = B(0) + \gamma_1 \int_0^t q(t)\dot{X}_d'(t)dt + \gamma_2 q(t)\dot{X}_d'(t) \qquad (17)$$

$$A(t) = A(0) + \lambda_1 \int_0^t q(t)\ddot{X}_d'(t)dt + \lambda_2 q(t)\ddot{X}_d'(t) \qquad (18)$$

In equations (13)-(18), $\{\delta_1, \alpha_1, \beta_1, \nu_1, \gamma_1, \lambda_1\}$ are any positive scalar integral adaptation gains, and $\{\delta_2, \alpha_2, \beta_2, \nu_2, \gamma_2, \lambda_2\}$ are zero or any positive scalar proportional adaptation gains.[2] In equation (12), $W_p = \text{diag}_i\{w_{pi}\}$ and $W_v = \text{diag}_i\{w_{vi}\}$ are $n \times n$ weighting matrices chosen by the designer to reflect the relative significance of the position and velocity errors E and $\dot{E}$ in forming the vector q. The values of the adaptation gains and weighting matrices determine the rate at which the tracking-errors converge to zero.

Since the control actuation is at the manipulator joints, the control force F is implemented as the joint torque T where

$$T(t) = J'(\theta)F(t) \qquad (19)$$

The augmented Jacobian matrix $J(\theta)$ is used in equation (19) to map the task-space forces $F(t)$ to the joint-space torques $T(t)$. Equation (19) represents the fundamental relationship between the task and joint spaces and is the basis for implementation of any task-based control scheme [16]. Equation (19) can be rewritten as

$$T(t) = \left[J_e'(\theta) \vdots J_c'(\theta) \right] \begin{bmatrix} F_e(t) \\ \dots \\ F_c(t) \end{bmatrix} = J_e'(\theta)F_e(t) + J_c'(\theta)F_c(t) \qquad (20)$$

where F_e and F_c are the $m \times 1$ and $r \times 1$ control force vectors corresponding to the basic task and the additional task, respectively. It is seen that the total control torque is the sum of two components: $T_e = J_e'F_e$ for the end-effector motion (basic task), and $T_c = J_c'F_c$ for the kinematic constraints (additional task). Equation (20) shows distinctly the contributions of the basic and the additional tasks to the overall control torque. Under the joint control law (20), the desired end-effector trajectory $Y_d(t)$ is tracked, and the "extra" degrees-of-freedom are appropriately used to control the evolution of the manipulator configuration through tracking of the desired functions

---

[2] Note that when the adaptation gains are all set to zero, the control scheme reduces to the constant PD feedback controller $\{K_p(0), K_v(0)\}$, the constant PD^2 feedforward controller $\{C(0), B(0), A(0)\}$, and the constant bias term $\{d(0)\}$.

$\Phi_d(t)$. In other words, the self-motion of the manipulator is controlled by first characterizing this motion in terms of user-defined kinematic functions and then controlling these functions through trajectory tracking.

In the foregoing centralized adaptive controller (11), each task-space control force F_i is generated on the basis of all configuration variables $\{x_1, \ldots, x_n\}$. The controller complexity and computations can be reduced significantly if the adaptive control scheme is implemented in a decentralized control structure [32]. In this case, each task-space control force F_i is responsible only for the corresponding configuration variable x_i, and each x_i is controlled *independently* of the others by a local adaptive controller. The couplings between the configuration variables then appear as "disturbances" and the adaptation laws are modified slightly to compensate for the unmodeled disturbances [33]. The local control scheme for the configuration variable x_i is [32]

$$\begin{aligned} F_i(t) = {} & d_i(t) + k_{pi}(t)e_i(t) + k_{vi}(t)\dot{e}_i(t) + c_i(t)x_{di}(t) \\ & + b_i(t)\dot{x}_{di}(t) + a_i(t)\ddot{x}_{di}(t) \end{aligned} \tag{21}$$

where $e_i = x_{di} - x_i$ is the tracking-error of the ith configuration variable, and the modified adaptation laws are:

$$q_i(t) = w_{pi}e_i(t) + w_{vi}\dot{e}_i(t) \tag{22}$$

$$d_i(t) = d_i(0) - \sigma_i \int_0^t d_i(t)dt + \delta_1 \int_0^t q_i(t)dt + \delta_2 q_i(t) \tag{23}$$

$$k_{pi}(t) = k_{pi}(0) - \sigma_i \int_0^t k_{pi}(t)dt + \alpha_{1i} \int_0^t q_i(t)e_i(t)dt + \alpha_{2i}q_i(t)e_i(t) \tag{24}$$

$$k_{vi}(t) = k_{vi}(0) - \sigma_i \int_0^t k_{vi}(t)dt + \beta_{1i} \int_0^t q_i(t)\dot{e}_i(t)dt + \beta_{2i}q_i(t)\dot{e}_i(t) \tag{25}$$

$$c_i(t) = c_i(0) - \sigma_i \int_0^t c_i(t)dt + \nu_{1i} \int_0^t q_i(t)x_{di}(t)dt + \nu_{2i}q_i(t)x_{di}(t) \tag{26}$$

$$b_i(t) = b_i(0) - \sigma_i \int_0^t b_i(t)dt + \gamma_{1i} \int_0^t q_i(t)\dot{x}_{di}(t)dt + \gamma_{2i}q_i(t)\dot{x}_{di}(t) \tag{27}$$

$$a_i(t) = a_i(0) - \sigma_i \int_0^t a_i(t)dt + \lambda_{1i} \int_0^t q_i(t)\ddot{x}_{di}(t)dt + \lambda_{2i}q_i(t)\ddot{x}_{di}(t) \tag{28}$$

where σ_i is a positive scalar design parameter. Using the local control law (21), each end-effector coordinate y_i and each kinematic function ϕ_i is controlled independently of the remaining configuration variables. The decentralized adaptive control scheme (21)-(28) can be computed much faster than the centralized control law (11)-(18), since the number of mathematical operations is reduced considerably due to decentralization. Note that although the task-space control law (21) is decentralized, this property is lost in the transformation from task space to joint space $T = J'F$ for implementation, since all joint torques must be applied simultaneously to control each configuration variable.

The centralized and decentralized adaptive control schemes presented in this section are extremely simple since the auxiliary signal and controller gains are evaluated from equations (12)-(18) or (22)-(28) by simple numerical integration by using, for instance, the trapezoidal rule. Thus the computational time required to calculate the adaptive control law (11) or (21) is extremely short. As a result, the scheme can be implemented for on-line control of redundant manipulators with high sampling rates, resulting in improved dynamic performance. This is in contrast to most existing approaches which require time-consuming optimization processes unsuitable for fast on-line control implementation. It is important to note that the adaptation laws (12)-(18) or (22)-(28) are based solely on the observed performance of the manipulator rather than on any knowledge of the complex dynamic model or parameter values of the manipulator and the payload.

We now discuss briefly the implications of Jacobian singularities on configuration control. Suppose that the transposed augmented Jacobian matrix J' has n distinct eigenvalues $\{\lambda_1, \ldots, \lambda_n\}$, and n right and left eigenvectors $\{u_1, \ldots, u_n\}$ and $\{v_1, \ldots, v_n\}$ respectively, where $J'u_i = \lambda_i u_i$ and $v_i J' = \lambda_i v_i$ for $i = 1, \ldots, n$, and $v_i u_j = 0$ for all $i \neq j$. Then J' can be expressed by the modal decomposition form

$$J' = \sum_{i=1}^{n} \lambda_i u_i v_i$$

Hence, the force-torque transformation becomes

$$T = J'F = \sum_{i=1}^{n}(\lambda_i v_i F)u_i \tag{29}$$

It is seen that along each eigenvector u_i, the contribution of the control force F to the joint torque T is equal to $(\lambda_i v_i F)u_i$. Now, suppose that $\lambda_j = 0$ for some j; i.e. the Jacobian matrix J is singular. Then, any control force F in the task-space direction u_j will make no contribution to the joint torque T. Thus, the ability to control the manipulator in a certain task-space direction is impaired. However, the proposed method allows complete control of the manipulator motion in the remaining directions for which λ_i is non-zero; since the corresponding control force F_i does not map into zero joint torque T. This is in contrast to pseudoinverse-based methods requiring inversion of the matrix $(J_e J_e')$ which becomes rank-deficient when the end-effector Jacobian matrix J_e is singular.

The following points are noted about the proposed adaptive configuration control scheme:

1. By controlling the manipulator directly in task space, the complicated and time-consuming inverse kinematic computations are not required. This makes the scheme computationally efficient as a real-time control algorithm.

2. Since the control problem is formulated in task space, the method can be extended readily to hybrid force and configuration control, and also to the control of redundant multi-arm robots.

3. Using this dynamic control scheme, accurate tracking of desired trajectories for the basic and the additional tasks can be achieved simultaneously. Furthermore, in contrast to local methods based on Jacobian pseudoinverse, the proposed scheme provides *direct* control of the manipulator motion over the entire trajectory.

4. Redundant manipulators are often composed of many joints to enhance their dexterity and versatility, and will therefore have highly complicated dynamic models. When model-based control schemes such as the Computed Torque Technique are used, the on-line computation of the full dynamic model may make it impractical to implement fast control loops. On the other hand, adaptive control schemes provide a practical alternative, since on-line adaptation eliminates the need for the complex dynamic model and thus allows fast control loops to be implemented.

5. In contrast to pseudoinverse-based methods, the proposed approach does not require the assignment of priorities to the basic and additional task specifications, since the additional task requirements are met *independently* through the kinematic functions. In other words, the performance of the basic task (end-effector motion) is not sacrificed due to the presence of the additional task (kinematic constraints), provided the task trajectories avoid the augmented Jacobian singularities. In our formulation, the weighting matrices $\{W_p, W_v\}$ are used to assign some degree of relative significance to position and velocity errors of all task variables. These matrices need not be constant throughout the robot motion, and can be varied according to the task requirements.

6. The proposed formulation provides the capability of satisfying multiple objectives through the definition of basic and additional task requirements. These requirements have simple tangible physical interpretations in terms of the manipulator configuration, rather than abstract mathematical goals. The task requirements are achieved by means of a simple control law which can be implemented as a real-time algorithm for on-line control with high sampling rates. Using the proposed formulation, the task to be performed by the redundant manipulator can be decomposed into a number of subtasks with different kinematic constraints. In the execution of each subtask, the appropriate kinematic constraint is satisfied, in addition to the specified end-effector motion.

7. A distinctive feature of the proposed control scheme is its applicability to the shared operator/autonomous mode of operation for performing a given task. This is due to the fact that the basic task and the additional task appear as two distinct and separate entities in the proposed control scheme, as shown in Figure 2. The present formulation allows the operator to specify the basic task of desired end-effector motion using an input device such as a hand controller. The autonomous system can then invoke the AI spatial planner to specify the additional task of desired kinematic constraints that will be performed simultaneously through the utilization of redundancy. In this way, the operator and the autonomous system can share the execution of a complex task.

2.3 Special Case: Kinematic Optimization

In this section, the configuration control scheme is used to optimize any desired kinematic objective function.

Let $g(\theta)$ denote the scalar kinematic objective function to be optimized by the utilization of redundancy. In order to optimize $g(\theta)$ subject to the end-effector constraint $\dot{Y} = J_e\dot{\theta}$, we apply the standard gradient projection optimization theory [34] to obtain the optimality criterion for the constrained optimization problem as

$$\left(I - J_e^+ J_e\right)\frac{\partial g}{\partial \theta} = 0 \tag{30}$$

where $J_e^+ = J_e'(J_e J_e')^{-1}$ is the pseudoinverse of J_e. The $n \times n$ matrix $(I - J_e^+ J_e)$ is of rank r and therefore equation (30) reduces to

$$N_e \frac{\partial g}{\partial \theta} = 0 \tag{31}$$

where N_e is an $r \times n$ matrix formed from r linearly independent rows of $(I - J_e^+ J_e)$. The rows of N_e span the r-dimensional null-space of the end-effector Jacobian J_e, since $J_e(I - J_e^+ J_e) = 0$ and $(I - J_e^+ J_e)$ is symmetric. Equation (31) implies that the projection of the gradient of the objective function $g(\theta)$ onto the null-space of the end-effector Jacobian matrix J_e must be zero. This is the result used by Baillieul [7-11] in the extended Jacobian method. Using the configuration control approach, we define the r kinematic functions as $\Phi(\theta) = N_e \frac{\partial g}{\partial \theta}$ and the desired trajectory as $\Phi_d(t) \equiv 0$ to represent equation (31). Therefore, the configuration vector X and the augmented Jacobian matrix J are now

$$X = \begin{pmatrix} Y \\ \cdots \\ N_e\frac{\partial g}{\partial \theta} \end{pmatrix} \quad ; \quad J = \begin{pmatrix} J_e \\ \cdots\cdots\cdots \\ \frac{\partial}{\partial \theta}\left(N_e\frac{\partial g}{\partial \theta}\right) \end{pmatrix}$$

The adaptive control scheme of Section 2.2 can now be applied directly to ensure that $X(t)$ tracks the desired trajectory $X_d(t) = \left(\begin{smallmatrix} Y_d(t) \\ \Phi_d(t) \end{smallmatrix}\right)$. Therefore, we have shown that the kinematic optimization problem can be reformulated as a special case of the configuration control problem, and the proposed solution does not require the complicated inverse kinematic transformation.

It must be noted that for a general objective function $g(\theta)$, equation (30) is only a necessary condition for optimality, and not a sufficient condition. As a result, the satisfaction of the kinematic constraint (31) does not always represent a truly optimal solution.

2.4 Inequality Constraints

In Section 2.2, the additional task to be performed is formulated as the trajectory tracking problem $\Phi(\theta) = \Phi_d(t)$. In some applications, however, the additional task requirements are expressed naturally as a set of kinematic *inequality* constraints $\Phi(\theta) \geq C$, where C is a constant vector. For instance, when the redundancy is utilized to avoid collision with a workspace object, the distance between the object and the closest robot link should exceed a certain threshold, which leads to an inequality constraint [35]. Similarly, inequality constraints can represent avoidance of joint limits and kinematic singularities.

The formulation of the configuration control scheme allows kinematic inequality constraints to be incorporated directly into the control law. For instance, suppose that the additional task requires the inequality constraint $\phi_i(\theta) \geq c_i$, where c_i is some constant, to be satisfied by the kinematic function ϕ_i. Then, for $\phi_i(\theta) < c_i$, the tracking-errors are defined as $e_i = c_i - \phi_i(\theta)$ and $\dot{e}_i = 0 - \dot{\phi}_i(\theta)$; while for $\phi_i(\theta) \geq c_i$ we have $e_i = \dot{e}_i = 0$. Therefore, we can choose $\phi_{di}(t) = \dot{\phi}_{di}(t) = \ddot{\phi}_{di}(t) = 0$ and use the adaptive feedback control law $F_i = d_i + k_{pi}e_i + k_{vi}\dot{e}_i$ to achieve the additional task requirement.

2.5 Joint-Based Configuration Control

The configuration control scheme described in Section 2.2 is "task-based," in the sense that the tracking-errors are formed and the control actions are generated in the task space. In some applications, it is preferable to use a joint-based control system, for instance, to compensate joint frictions more effectively. In this section, we describe briefly the implementation of the configuration control scheme in a joint-based robot control system.

For joint-based control, we first need to determine the $n \times 1$ desired joint angular position vector $\theta_d(t)$ that corresponds to the $n \times 1$ desired configuration vector $X_d(t)$. This can be done by solving the augmented inverse kinematic equations pertaining to the manipulator [29]. The inverse kinematic solution produces a finite set of joint angles with distinct poses, and often the solution corresponding to the initial pose is selected. Alternatively, the differential kinematic equations can be solved recursively using the augmented Jacobian. The Jacobian approach can be computationally more efficient than the inverse kinematic solution, and can also be used when there is no closed-form analytical solution to the inverse kinematic problem, e.g., for robots with non-spherical wrists.

Once the desired joint trajectory $\theta_d(t)$ is computed, it is used as a setpoint for the joint servo loops. A joint-based control scheme such as [31,32] can then be employed to achieve trajectory tracking in joint space. Following [31], the adaptive joint control law is given by

$$T(t) = d(t) + [K_p(t)E(t) + K_v(t)\dot{E}(t)]$$
$$+ [C(t)\theta_d(t) + B(t)\dot{\theta}_d(t) + A(t)\ddot{\theta}_d(t)]$$

where $E(t) = \theta_d(t) - \theta(t)$ is the position tracking-error in joint space, and the gains $\{d(t), K_p(t), K_v(t), C(t), B(t), A(t)\}$ are updated on-line in real-time using adaptation laws similar to equations (12)-(18). This control scheme ensures that the actual joint positions $\theta(t)$ track $\theta_d(t)$, and consequently the robot configuration vector $X(t)$ follows the desired trajectory $X_d(t)$.

For practical implementation of the control schemes discussed so far, we require the capability of sending the control torques computed by the controllers directly to the joint motors. In some applications, the joint servo loops provided by the robot manufacturer cannot be modified easily to accept torque commands. In such cases, the desired joint trajectories $\theta_d(t)$ computed from the inverse kinematics or Jacobian are simply sent as setpoints to the joint servo loops, which ensure that trajectory tracking is achieved.

2.6 Cyclicity of Motion

Robot manipulators are often employed in repetitive operations. For greater efficiency and reliability, it is highly desirable that at the end of each operation cycle the robot returns to the same configuration. This property is known as "cyclicity" of motion. For a redundant robot, it is possible for the end-effector to return to the same task space position and yet the robot to be in a completely different configuration. In fact, this is generally the result obtained when methods based on Jacobian pseudoinverse are used to control the robot motion [13].

The configuration control approach has the attractive feature of cyclicity of motion, since through definition of the n configuration variables $\{x_1, \ldots, x_n\}$, the kinematic representation of the manipulator is no longer redundant. Therefore, it can be viewed as a non-redundant manipulator which, in general, possesses the property of cyclicity, provided the singularity boundaries are not crossed [20,36]. In other words, when the configuration variables traverse a closed path in task space, the joint angles will, in general, traverse a unique closed curve in joint space, and hence the initial and final manipulator configurations will be identical. Therefore, the configuration control scheme meets the essential requirement of cyclicity for repetitive operations.

3. APPLICATIONS OF CONFIGURATION CONTROL SCHEME

In this section, we present some applications of the configuration control scheme through simulations of a direct-drive manipulator and experiments with an industrial robot.

3.1 Simulations of a direct-drive manipulator

The configuration control scheme described in Section 2 will now be applied to a direct-drive three-link manipulator in a series of computer simulations. The results presented here are samples selected from a fairly comprehensive computer simulation study which was carried out to test the performance of the proposed control scheme. These examples are chosen for presentation because they illustrate the flexibility and versatility of the configuration control approach to redundant manipulators.

Consider the planar three-link manipulator in a horizontal plane shown in Figure 4. The manipulator parameters are link lengths $l_1 = l_2 = l_3 = 1.0$ meter, and link masses $m_1 = m_2 = m_3 = 10.0$kg; the link inertias are modeled by thin uniform rods. The manipulator dynamic equation which relates joint torques T and joint angles θ is given by [37]

$$T = M(\theta)\ddot{\theta} + N(\theta, \dot{\theta}) + V\dot{\theta} \qquad (32)$$

where the mass matrix $M = [m_{ij}]$, Coriolis and centrifugal torque vector $N = [n_i]$, and viscous friction coefficient matrix $V = \text{diag}\{v_i\}$ have the following representations:

$$
\begin{aligned}
m_{11} &= 40 + 30\cos\theta_2 + 10\cos\theta_3 + 10\cos(\theta_2 + \theta_3) \\
m_{12} &= m_{21} = 16.67 + 15\cos\theta_2 + 10\cos\theta_3 + 5\cos(\theta_2 + \theta_3) \\
m_{13} &= m_{31} = 3.33 + 5\cos\theta_3 + 5\cos(\theta_2 + \theta_3) \\
m_{22} &= 16.67 + 10\cos\theta_3 \\
m_{23} &= m_{32} = 3.33 + 5\cos\theta_3 \\
m_{33} &= 3.33 \\
n_1 &= -30\dot{\theta}_1\dot{\theta}_2\sin\theta_2 - 10\dot{\theta}_1\dot{\theta}_3\sin\theta_3 - 10\dot{\theta}_1(\dot{\theta}_2 + \dot{\theta}_3)\sin(\theta_2 + \theta_3) \\
&\quad - 15\dot{\theta}_2^2\sin\theta_2 - 10\dot{\theta}_2\dot{\theta}_3\sin\theta_3 - 5\dot{\theta}_2(\dot{\theta}_2 + \dot{\theta}_3)\sin(\theta_2 + \theta_3) - 5\dot{\theta}_3^2\sin\theta_3 \\
&\quad - 5\dot{\theta}_3(\dot{\theta}_2 + \dot{\theta}_3)\sin(\theta_2 + \theta_3) \\
n_2 &= -10\dot{\theta}_1\dot{\theta}_3\sin\theta_3 - 10\dot{\theta}_2\dot{\theta}_3\sin\theta_3 - 5\dot{\theta}_3^2\sin\theta_3 \\
&\quad + (15\sin\theta_2 + 5\sin(\theta_2 + \theta_3))\dot{\theta}_1^2 \\
n_3 &= (5\sin\theta_3 + 5\sin(\theta_2 + \theta_3))\dot{\theta}_1^2 + 10\dot{\theta}_1\dot{\theta}_2\sin\theta_3 + 5\dot{\theta}_2^2\sin\theta_3 \\
v_1 &= v_2 = v_3 = 30.0
\end{aligned}
\qquad (33)
$$

Note that the gravity vector is orthogonal to the plane of motion of the manipulator, so that no gravity torques appear in (32). It must be emphasized

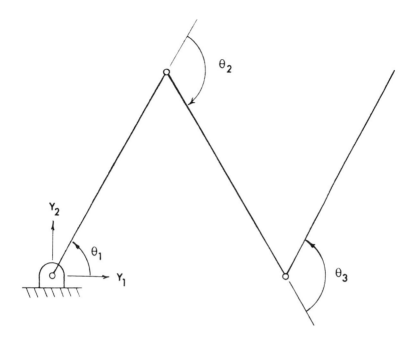

Figure 4. Three-Link Manipulator in Horizontal Plane

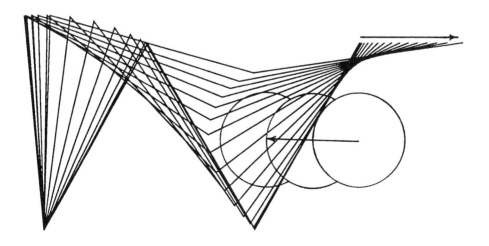

Figure 5(i). Manipulator Trajectory in Obstacle Avoidance Simulation

that the dynamic model (32)-(33) is used only to simulate the manipulator behavior and is *not* used in the control law formulation; i.e., the manipulator dynamics is treated as a "black box" by the controller in the simulations.

The configuration control scheme described in Section 2 will now be applied to the planar redundant manipulator depicted in Figure 4 and described by (32)-(33). In order to demonstrate the versatility of the proposed approach, three different simulation examples will be considered: obstacle avoidance, force control with terminal angle constraint, and manipulability maximization.

Case (i) — Obstacle Avoidance: One of the advantages of a redundant manipulator is the potential to use the "extra" DOFs to maneuver in a congested workspace and avoid collision with obstacles. The capability for automatic obstacle avoidance may be readily incorporated into the configuration control approach. The basic strategy for obstacle avoidance that will be adopted here is intuitively appealing and is somewhat similar to the strategy proposed in [38]. However, the resulting theory is quite different and more general because the strategy is implemented within the framework of configuration control.

It is supposed that all workspace obstacles can be enclosed in convex volumes, and that each volume defines a "space of influence" (SOI) for the control law. For simplicity, it is assumed that these volumes are spheres in three dimensions and disks in two dimensions; the extension to more general volumes is straightforward. The basic strategy for obstacle avoidance may be summarized as follows: if any point on the manipulator enters the SOI of any obstacle, the manipulator redundancy is utilized to inhibit the motion of that point in the direction toward the obstacle. Note that this strategy is simple and intuitively appealing.

The implementation of obstacle avoidance capability within the configuration control method may be achieved by formulating the obstacle avoidance strategy in terms of kinematic inequality constraints. We shall summarize this procedure for the case of the redundant manipulator shown in Figure 4 and a single *moving* workspace obstacle. The extension to the case of a more general manipulator and multiple obstacles is straightforward; however, this extension is beyond the scope of the present paper. Define Y_c as the 2×1 position vector of the "critical point" on the manipulator with respect to the base frame, where the critical point is defined to be that point currently at minimum distance from the obstacle. Let Y_o denote the 2×1 position vector of the obstacle center, also in the base frame. Then, letting $d_c(\theta) = ||Y_c - Y_o|| = [(Y_c - Y_o)'(Y_c - Y_o)]^{1/2}$ be the critical distance and r_o be the radius of the SOI of the obstacle, the criterion for obstacle avoidance is readily expressed as the inequality constraint

$$d_c(\theta) - r_o \equiv g(\theta) \geq 0 \qquad (34)$$

Observe that (34) is simply a special case of the kinematic inequality constraints discussed in Section 2.4. Thus incorporation of the obstacle avoidance criterion (34) into the configuration control scheme should not be difficult. Before turning to that problem, however, one comment is in order. The location of the critical point varies during the motion, so that Y_c (and d_c, of course) must be continually recomputed; therefore computational efficiency is very critical for real-time control applications. It has been found that an effective means of performing these calculations is to compute d_c via the vector norm definition given above, and to structure all calculations in a recursive framework. Additionally, it has proven to be more efficient to compute the required geometric parameters using vector algebra rather than geometric techniques.

Simultaneous tracking of desired end-effector motion and satisfaction of the obstacle avoidance criterion (34) may be achieved with a slightly modified version of the configuration control law (11)-(19). Examination of the control torque equation (19) reveals that only the control torque corresponding to the additional task performance, i.e., $T_c = J_c' F_c$, needs to be modified. The constraint Jacobian J_c can be computed directly from (34) as

$$J_c \;=\; \partial g/\partial\theta \;=\; \partial d_c/\partial\theta \tag{35}$$

where it is assumed that the radius of the SOI is constant. Note that selection of the obstacle avoidance criterion (34) has resulted in a constraint Jacobian (35) which is simply the Jacobian corresponding to the motion of the critical point in terms of the "obstacle frame" (the frame with origin at the obstacle center and axis of interest aligned with the $Y_c - Y_o$ vector). This choice has significant computational advantages, and allows the case of moving obstacles to be handled with no additional effort. Note also that explicit expressions for d_c and J_c are not provided here; these expressions depend on which link of the manipulator contains the critical point and their derivation requires a development that is beyond the intended scope of this paper. A complete development of this theory and the presentation of the resulting expressions may be found in [35].

The constraint control input F_c may be computed from (11) as

$$F_c = d + K_p E + K_v \dot{E} \tag{36}$$

where F_c is a scalar and the feedforward term $C X_d + B \dot{X}_d + A \ddot{X}_d$ is identically zero. The vanishing of the feedforward term for the additional task is a direct consequence of choosing the obstacle avoidance criterion (34), since this choice has the effect of reducing the "constraint tracking" problem to a "constraint regulation" problem. Thus the constraint control force F_c is completely specified by (36) with $E = 0$ when $g \geq 0$ and $E = -(d_c - r_o)$ when $g < 0$.

The performance of the configuration control algorithm with obstacle avoidance capability is tested through a set of computer simulations. The control algorithm is implemented using the control structure shown in Figure 2, where the basic task is end-effector motion and the additional task is obstacle avoidance. The moving obstacle for the simulations is assigned an SOI radius of $r_o = 0.22$m and has its center location defined by the trajectory $Y_o'(t) = [1.0 + 0.5cos0.5t \quad 0.43]$; thus from Figure 5(i) it is seen that the obstacle is initially close to the third link of the manipulator and subsequently moves toward the manipulator base. The initial configuration of the manipulator is $\theta' = [60° \quad -120° \quad 120°]$ and the manipulator is initially at rest.

The desired end-effector trajectory for the simulation is $Y_d'(t) = [2.0 - 0.5cos0.5t \quad \sqrt{3}/2]$. This trajectory is tracked as the basic task in the configuration control algorithm shown in Figure 2 and described in (11)-(19). Each end-effector coordinate Y_1, Y_2 is controlled by simultaneous actuation of both control forces F_{e_1}, F_{e_2} generated by a centralized 2×2 controller. The controller gains d, K_p, K_v, C, B, and A are all set to zero initially. The adaptation gains are assigned the following values: $\delta_2 = \alpha_2 = \beta_2 = \nu_2 = \gamma_2 = \lambda_2 = 0$, $\delta_1 = \nu_1 = \gamma_1 = \lambda_1 = 0.5$, $\alpha_1 = \beta_1 = 2.0$, The weighting matrices are chosen as: $W_p = diag(6000), W_v = diag(700)$.

The desired additional task in the configuration control algorithm shown in Figure 2 is obstacle avoidance. This constraint is quantified in (34) and tracked using (11)-(19), (35)-(36). Therefore, the additional task is controlled independently of the end-effector motion and using the control force F_c generated by a scalar controller. The controller gains are all set to zero initially. The adaptation gains are assigned the following values: $\delta_2 = \alpha_2 = \beta_2 = \nu_2 = \gamma_2 = \lambda_2 = 0$, $\delta_1 = \alpha_1 = \beta_1 = 2.0$, $\nu_1 = \gamma_1 = \lambda_1 = 0$. The weighting matrices are simply scalar constants and are defined as follows: $W_p = 6000, W_v = 700$. The control law is applied to the dynamic model (32)-(33) through computer simulations on a SUN3 computer with a sampling period of one millisecond; all integrals required by the controller are implemented using a simple trapezoidal integration rule with a time step of one millisecond. The results of the simulation are given in Figures 5(i)-5(ii), and indicate that the desired end-effector trajectory is closely tracked (maximum error of less than 1.0 mm) and the moving obstacle is successfully avoided.

Case (ii) — Force Control with Terminal Angle Constraint: Thusfar in this paper, it has been assumed that the manipulator end-effector is unconstrained and is free to move anywhere in the workspace; therefore position control is all that is required. In many practical applications, however, the manipulator end-effector must come into contact with the environment. In these cases, the contact forces must be controlled in the directions constrained by the environment while the position must be controlled in the

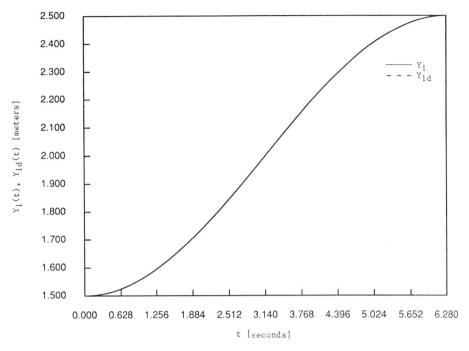

Figure 5(ii). Response of Position $Y_1(t)$ in Obstacle Avoidance Simulation

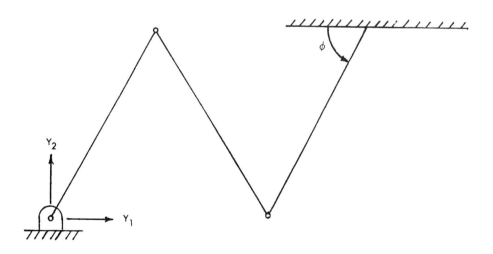

Figure 5(iii). Force Control Task

(orthogonal) directions which are unconstrained. A potential advantage of redundancy is in improving the manipulator performance in those tasks requiring force control capabilities. In the hybrid position/force control framework, the orthogonality of the "position subspace" and the "force subspace" is exploited to allow position and force to be controlled independently [39]. More specifically, within the framework of configuration control, hybrid position/force control prescribes that the end-effector control force F_e be constructed by combining the $\ell \times 1$ position control input F_p and the $j \times 1$ force control input F_f as follows:

$$F_e = \begin{bmatrix} F_p \\ -- \\ F_f \end{bmatrix} \tag{37}$$

where $l + j = m$, and the control inputs F_p and F_f are computed *independently*. The position control input F_p may be computed from (11)-(18) as before, where it is understood that terms such as desired trajectory, trajectory tracking error, and so on, refer to quantities in the l-dimensional position subspace. The force control input F_f may be computed in a manner analogous to that used in calculating F_p. Recently, Seraji [40] has used an improved Model Reference Adaptive Control (MRAC) method to derive an adaptive force controller for non-redundant manipulators which requires no knowledge of the manipulator dynamic model or parameter values for the manipulator or the environment. This control algorithm computes F_f as follows:

$$F_f = P_d + d_f + K_I \int_0^t E_f dt + K_{pf} E_f - K_{vf} \dot{Y}_f \tag{38}$$

where P_d, P are the $j \times 1$ desired and actual contact forces, respectively, $E_f = P_d - P$ is the force tracking error, $\dot{Y}_f$ is the $j \times 1$ end-effector velocity in the force subspace, and $\{d_f, K_I, K_{pf}, K_{vf}\}$ are controller gains which are updated adaptively based on the observed force tracking-error. For the present simulation study, the dimension of the force subspace is $j = 1$; thus the controller gains are all scalar quantities and may be updated using the following simple adaptation laws:

$$q_f = 2 \int_0^t E_f dt + 0.25 E_f - 100 \dot{Y}_f \quad ; \quad d_f = q_f + \int_0^t q_f dt$$

$$K_I = 1000 + 100 \int_0^t q_f \left[\int_0^t E_f dt \right] dt \quad ; \quad K_{pf} = 100 + 10 \int_0^t q_f E_f dt \tag{39}$$

$$K_{vf} = 5000 - 40000 \int_0^t q_f \dot{Y}_f dt$$

It can be seen from the preceding development that force control capability may be readily incorporated into the configuration control scheme. Indeed, this process may be viewed as replacing the j generalized coordinates corresponding to end-effector position in the force subspace with j generalized coordinates which represent end-effector/environment contact force. Note that the inclusion of force control capability alters only the m-dimensional end-effector subspace, so that the manipulator redundancy may still be conveniently utilized to perform some desired additional task.

The hybrid position/force control scheme may be implemented within the configuration control framework shown in Figure 2 by specifying hybrid position/force control of the end-effector as the basic task. The performance of this version of the configuration control scheme is tested via the computer simulation task depicted in Figure 5(iii). The basic task of the simulation study is to track the position trajectory $Y_{1d}(t) = 2.0 - 0.5\cos t$ while simultaneously exerting a 10Nt force normal to the frictionless reaction surface which is located parallel to the Y_1-axis at $Y_2 = \sqrt{3}/2$m and has stiffness coefficient of 10^4Nt/m. The desired additional task is to maintain the following inequality constraint on the terminal angle $\phi(\theta) = \theta_1 + \theta_2 + \theta_3$ shown in Figure 5(iii):

$$\phi(\theta) \geq 1.0 \quad \text{rad}$$

which may be written in standard kinematic inequality constraint form as follows:

$$\phi(\theta) - 1.0 \quad \equiv \quad g(\theta) \quad \geq \quad 0 \tag{40}$$

This additional task is chosen to illustrate the ease with which inequality constraints can be incorporated within the configuration control framework, and also because terminal angle inequality constraints may be important in certain force control applications such as scraping tasks or assembly operations. The initial state of the manipulator is chosen to be the same as in Case (i). The position control input F_p is computed with the one-dimensional version of the adaptive algorithm (11)-(18), where the adaptation gains, the initial values for the controller gains $\{d, K_p, K_v, C, B, A\}$, and the weighting terms W_p and W_v are all set to the values used for the basic task in Case (i). The force control input F_f is computed from (38)-(39) with $P_d = 10$Nt. The constraint control input F_c is computed exactly as in Case (i) with g defined by (40) instead of (34). The control forces $F_e' = [F_p \ \ F_f]$ and F_c are mapped to the joint control torque T using (19) with $J_c = \partial g/\partial \theta = [1\ 1\ 1]$. Note that the control forces F_p, F_f, and F_c are computed using a decentralized control structure; i.e., each of the configuration variables is controlled *independently* by an adaptive control law.

In order to simulate the behavior of the manipulator defined in (32)–(33) for the case in which the end-effector is in contact with the reaction surface, the dynamic model (32) is modified as

$$T = M\ddot{\theta} + N + V\dot{\theta} + J_e' \begin{bmatrix} 0 \\ -- \\ P \end{bmatrix} \tag{41}$$

where $M, N,$ and V are defined in (33) and the contact force P is computed as follows:

$$P = \begin{cases} 0 & \text{if } Y_2 < \sqrt{3}/2 \quad \text{(no contact)} \\ 10^4(Y_2 - \sqrt{3}/2) & \text{if } Y_2 \geq \sqrt{3}/2 \quad \text{(contact)} \end{cases}$$

Note that because the reaction surface is parallel to the Y_1-axis and frictionless, the contact force acts in the Y_2 direction only. The control law outlined in the preceding paragraph is applied to the dynamic model (41) through computer simulation on a SUN3 computer with a sampling period of one millisecond. The results of the simulation are given in Figures 5(iv)-5(vi), and indicate that the desired end-effector position/force trajectories are closely tracked and that the terminal angle inequality constraint is satisfied.

Case (iii)— Manipulability Maximization: An additional advantage of a redundant manipulator is the potential to utilize the redundancy to increase the dexterity of the manipulator. While it is generally agreed that dexterity is a desirable quality in a manipulator, there is less agreement as to the proper way to quantify dexterity, and several dexterity measures have been proposed [41]. One popular measure of dexterity is the "manipulability measure" w defined by Yoshikawa as follows [6]:

$$w(\theta) = (det[J_e J_e'])^{1/2} \tag{42}$$

Thus one means of using manipulator redundancy to increase dexterity is to maximize $w(\theta)$ subject to the constraint imposed by specifying the desired end-effector trajectory.

The manipulability maximization problem summarized above is a special case of the following general constrained optimization problem:

$$\text{maximize} \quad g(\theta)$$

$$\text{subject to the constraint} \quad Y - Y(\theta) = 0 \tag{43}$$

where the scalar g may be constructed to represent a measure of any desired kinematic performance objective and the constraint $Y - Y(\theta) = 0$ is simply the forward kinematics of the manipulator (2). The solution to the optimization problem (43) can be obtained using Lagrange multipliers. Let the augmented scalar objective function $g^*(\theta, \lambda)$ be defined as follows:

$$g^*(\theta, \lambda) = g(\theta) + \lambda'[Y - Y(\theta)] \tag{44}$$

where λ is the $m \times 1$ vector of Lagrange multipliers. The necessary conditions for optimality of (43) may be written using (44) as

$$\partial g^*/\partial \lambda = 0 \quad \Rightarrow \quad Y = Y(\theta) \tag{45}$$

$$\partial g^*/\partial \theta = 0 \quad \Rightarrow \quad \partial g/\partial \theta \;=\; \left(\frac{\partial Y}{\partial \theta}\right)' \lambda \;=\; J_e' \lambda \tag{46}$$

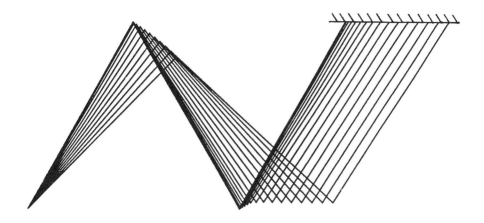

Figure 5(iv). Manipulator Trajectory in Force Control Simulation

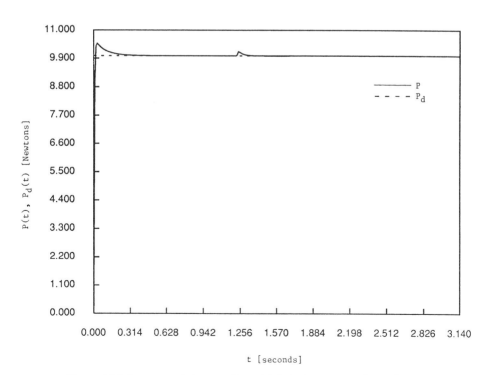

Figure 5(v). Response of Contact Force P(t) in Force Control Simulation

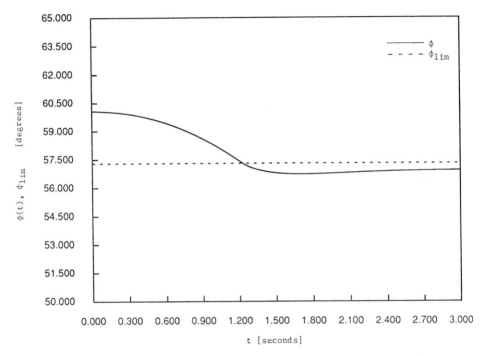

Figure 5(vi). Response of Terminal Angle $\phi(t)$ in Force Control Simulation

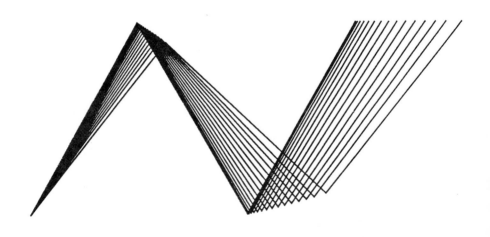

Figure 5(vii).
Manipulator Trajectory in Manipulability Maximization Simulation

From (46), a necessary condition for optimality of (43) is that $\partial g/\partial \theta$ lies in the range space of J_e'. This requirement may be written concisely as

$$N_e \partial g/\partial \theta = 0 \qquad (47)$$

where N_e is any $r \times n$ matrix whose rows form a basis for the null space of J_e. This result is a direct consequence of the fact that the row space and null space of any matrix are orthogonal complements. Note that (47) is the same as (31) given in Section 2.3. When $g(\theta)$ is convex, the condition (47) is both necessary and sufficient for constrained optimization of (43). This is of interest because in robotics applications it is usually possible to construct a convex $g(\theta)$.

Observe that the optimality condition (47) is an r-dimensional kinematic constraint of the form (8) with $\Phi = N_e \partial g/\partial \theta$ and $\Phi_d = 0$. Therefore it is quite convenient within the configuration control framework to specify the additional task necessary to achieve optimization of any desired kinematic objective function $g(\theta)$. Indeed, assuming that $g(\theta)$ is defined (and differentiable), specification of this additional task requires only that an appropriate N_e matrix be constructed and that the calculations specified in (47) be performed. The matrix N_e may be constructed in several ways; one formulation for N_e is [42]

$$N_e = \left[-J_2'(J_1^{-1})' \mid I_r \right] \qquad (48)$$

where J_1 and J_2 are $m \times m$ and $m \times r$ partitions of J_e defined by $J_e = [J_1 \mid J_2]$ with $|J_1| \neq 0$, and I_r is the $r \times r$ identity matrix. Since J has rank m, the m linearly independent columns of J can always be grouped together to form the nonsingular $m \times m$ matrix J_1. In order to see that (48) is a valid construction for N_e, observe by direct calculation that $N_e J_e' = 0$, and that $rank N_e = r$ for all θ due to the I_r partition.

The necessary condition for maximizing manipulability may be obtained from the above results by writing (47) with $g = w^2/2$ as

$$N_e \frac{\partial}{\partial \theta} \left[\frac{w^2}{2} \right] = 0 \qquad (49)$$

where w and N_e are given in (42) and (48), respectively, and the objective function $w^2/2$ is used in place of w for computational convenience (note that maximizing $w^2/2$ and maximizing w are equivalent). For the three DOF manipulator considered in this section, $w^2/2$ may be readily calculated from (42) as

$$\frac{w^2}{2} = \frac{1}{2}[sin^2\theta_2 + sin^2\theta_3 + sin^2(\theta_2 + \theta_3)] \qquad (50)$$

while N_e may be computed from (48) as

$$N_e = [sin\theta_3 \quad - sin\theta_3 - sin(\theta_2 + \theta_3) \quad sin(\theta_2 + \theta_3) + sin\theta_2] \qquad (51)$$

Finally, the gradient vector required by (49) may be calculated from (50) as

$$\frac{\partial[w^2/2]}{\partial\theta} = \begin{bmatrix} 0 \\ sin\theta_2 cos\theta_2 + sin(\theta_2 + \theta_3)cos(\theta_2 + \theta_3) \\ sin\theta_3 cos\theta_3 + sin(\theta_2 + \theta_3)cos(\theta_2 + \theta_3) \end{bmatrix} \qquad (52)$$

Thus the necessary condition for maximizing manipulability given in (49) may be computed by taking the product of the expressions in (51) and (52) and setting the result equal to zero.

The performance of the configuration control algorithm with manipulability maximization capability is tested through a set of computer simulations. The control algorithm is implemented using the control structure shown in Figure 2, where the basic task is end-effector motion and the additional task is manipulability maximization. Thus the basic task is to track the desired end-effector trajectory $Y_d'(t) = [2.0 - 0.5\cos 0.5t \quad \sqrt{3}/2]$ and the additional task is to meet the necessary condition for manipulability maximization given in (49)–(52). The basic task and additional task are executed using the configuration control algorithm shown in Figure 2 and described in (11)-(19), where the initial controller gains, adaptation gains, weighting matrices, initial joint angles, and initial joint velocitites are all chosen exactly as in Case (i). The control law is applied to the dynamic model (32)-(33) through computer simulation on a SUN3 computer with a sampling period of one millisecond. It is noted that the manipulator is initially in the optimal configuration, which is a requirement of the proposed approach to kinematic optimization and is easily achieved [42]. The results of the simulation are given in Figure 5(vii) and indicate that tracking of the desired end-effector trajectory and the additional task requirement are achieved simultaneously.

3.2 Experiments with a PUMA 560 robot

In this section, we describe the experimental validation of the configuration control scheme on a PUMA 560 industrial robot.

The testbed facility at the JPL Robotics Research Laboratory consists of a six-jointed Unimation PUMA 560 robot/controller, and a DEC MicroVAX II computer, as shown in the functional diagram of Figure 6. A cylindrical aluminum "link" has been fabricated and attached to the PUMA wrist as shown in Figure 6, so that the end-effector can be mounted at the end of the link. The link acts as an extension for the PUMA robot. By activating the shoulder joint θ_2, the elbow joint θ_3, and the wrist joint θ_5, the upper-arm, forearm, and the extra link move in a vertical plane, and hence the PUMA robot can operate as a redundant 3-link planar arm.

The MicroVAX II computer hosts the RCCL (Robot Control "C" Library) software, which was originally developed at Purdue University [43] and subsequently modified and implemented at JPL. The RCCL creates a "C" programming environment in which the user can write his own software

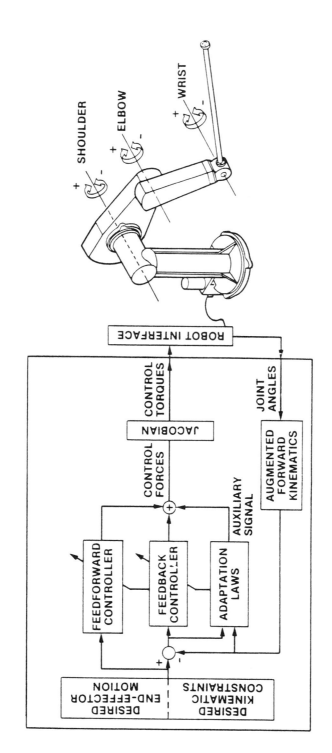

Figure 6. Functional Diagram of the Testbed Facility

for a trajectory generator and a control algorithm to directly control the robot motion. The MicroVAX communicates with the Unimation controller through a high-speed parallel link. During the operation of the robot, a hardware clock constantly interrupts the I/O program resident in the Unimation controller at the preselected sampling period of $T_s = 7$ milliseconds (i.e., $f_s = 143$ Hz), which is the lowest sampling period available in the present experimental set-up. At every interrupt, the I/O program gathers information about the state of the robot (such as joint encoder readings), and transmits this data to the control program in the MicroVAX. The I/O program then waits for the control program to issue a new set of control signals, which are then dispatched to the appropriate joint motors. Therefore, the MicroVAX acts as a digital controller for the PUMA robot and the Unimation controller is effectively by-passed and is utilized merely as an I/O device to interface the MicroVAX to the PUMA joint motors.

To test and evaluate the configuration control scheme of Section 2, the proposed controller is implemented on the three joints $[\theta_2, \theta_3, \theta_5]$ of the PUMA 560 robot; while the remaining three joints $[\theta_1, \theta_4, \theta_6]$ are held at their zero positions. For clarity of presentation, the three coplanar links of PUMA, namely upper-arm ($\ell_1 = 432$ mm), forearm ($\ell_2 = 433$ mm), and the extra link ($\ell_3 = 577$ mm) are redrawn in Figure 7 to form a 3-link planar arm, and the offset between upper-arm and forearm is ignored. The base coordinate frame (x, y) is then assigned to the planar arm as shown in Figure 7, with the origin coinciding with the shoulder joint. The joint angles $[\psi_1, \psi_2, \psi_3]$ are defined as "absolute" angles between the links and the positive x-direction; hence in terms of the PUMA relative angles $[\theta_2, \theta_3, \theta_5]$ measured from the PUMA zero position we have $\psi_1 = -\theta_2$, $\psi_2 = -\theta_2 - \theta_3 + 90°$, and $\psi_3 = -\theta_2 - \theta_3 - \theta_5 + 90°$. The problem is to control the Cartesian coordinates $[x, y]$ of the endpoint (tip of the extra link) in the base frame as the basic task, together with controlling a user-specified kinematic function ϕ which defines an appropriate additional task. The set $[x, y, \phi]$ then defines the configuration vector of the 3-link planar arm. The control scheme is implemented in a decentralized structure in task space, where each configuration variable is controlled *independently* by a simple feedback controller with adaptive gains.

Two different choices for the additional task variable ϕ will be considered. In the first case, ϕ is chosen in joint space as the shoulder angle; while, in the second case, ϕ is defined in Cartesian space as the wrist height.

Case (i) - Control of Shoulder Angle: In this case, we wish to use the redundancy to control the shoulder angle ψ_1. The shoulder joint requires the highest torque to cause motion, since the inertia of the whole arm is reflected back to the shoulder. Therefore, by controlling ψ_1 directly, we attempt to keep the largest joint torque under control. It is noted that in this case, the 3-link manipulator can be viewed as a 2-link arm (ℓ_2, ℓ_3) mounted on a

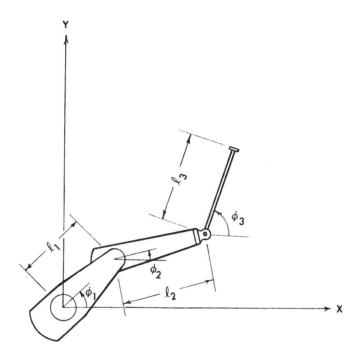

Figure 7. Three Coplanar Links of PUMA 560 Robot

moveable base. Therefore, by slow base motion and fast endpoint movement, we can achieve fast manipulation with low torques.

Referring to Figure 7, the configuration variables and the augmented Jacobian matrix are given by

$$x(t) = \ell_1 \cos \psi_1(t) + \ell_2 \cos \psi_2(t) + \ell_3 \cos \psi_3(t)$$
$$y(t) = \ell_1 \sin \psi_1(t) + \ell_2 \sin \psi_2(t) + \ell_3 \sin \psi_3(t) \qquad (53)$$
$$\phi(t) = \psi_1(t)$$

$$J = \begin{pmatrix} -\ell_1 \sin \psi_1 & -\ell_2 \sin \psi_2 & -\ell_3 \sin \psi_3 \\ \ell_1 \cos \psi_1 & \ell_2 \cos \psi_2 & \ell_3 \cos \psi_3 \\ 1 & 0 & 0 \end{pmatrix} \qquad (54)$$

The singular configurations of the arm are obtained from

$$\det[J] = -\ell_2 \ell_3 \sin(\psi_2 - \psi_3)$$

Hence, the arm is at a singularity when $\psi_2 = \psi_3$ or $\psi_2 = \psi_3 + 180°$, i.e. the links 2 and 3 are aligned. These can be recognized as the classic singularities of the two-link arm (ℓ_2, ℓ_3).

In the experiment, the initial configuration of the three-link arm is chosen as $\psi_{i1} = 0°$, $\psi_{i2} = 90°$, and $\psi_{i3} = 45°$. This results in the initial values of the configuration variables obtained from equation (53) as $(x_i, y_i, \phi_i) = (839$ mm, 842 mm, 0°$)$. The desired final values of the configuration variables are specified as $(x_f, y_f, \phi_f) = (1100$ mm, 600 mm, 60°$)$. The desired transition from the initial to the final values is described by the smooth cycloidal trajectories

$$\begin{aligned} x_d(t) &= 839 + \frac{1100 - 839}{2\pi} \left[\frac{2\pi t}{3} - \sin \frac{2\pi t}{3} \right] \ mm; & 0 \le t \le 3 \\ &= 1100 \ mm & ; \quad 3 < t \\ y_d(t) &= 842 + \frac{600 - 842}{2\pi} \left[\frac{2\pi t}{3} - \sin \frac{2\pi t}{3} \right] \ mm \ ; & 0 \le t \le 3 \\ &= 600 \ mm & ; \quad 3 < t \\ \phi_d(t) &= 0 + \frac{60 - 0}{2\pi} \left[\frac{2\pi t}{3} - \sin \frac{2\pi t}{3} \right] \ deg & ; \quad 0 \le t \le 3 \\ &= 60 \ deg & ; \quad 3 < t \end{aligned} \qquad (55)$$

and the transition time is 3 seconds. Note that the desired path for the endpoint is a straight line, since from the above equations we have $\frac{x_d - 839}{1100 - 839} = \frac{y_d - 842}{600 - 842}$.

The configuration variables $[x, y, \phi]$ are controlled by three *independent* adaptive feedback control laws of the general form

$$F_i(t) = d_i(t) + k_{pi}(t)e_i(t) + k_{vi}(t)\dot{e}_i(t) \quad ; \quad i = x, y, \phi \qquad (56)$$

as shown in the block diagram of Figure 8, where the feedforward terms are omitted to reduce the on-line computation time. In equation (56), F is the task-space control force, d is the auxiliary signal, (k_p, k_v) are the position and velocity feedback gains, and $(e, \dot{e})$ are the position and velocity errors; e.g. $e_x = x_d - x$ and $\dot{e}_x = \dot{x}_d - \dot{x}$. The current values of the configuration variables are computed from the joint angles using the augmented forward kinematic model (53). The terms in the control law (56) are updated on the basis of the weighted-error $q_i(t)$ as

$$q_i(t) = w_{pi}e_i(t) + w_{vi}\dot{e}_i(t) \qquad (57)$$

$$d_i(t) = d_i(0) + \delta_{1i}\int_0^t q_i(t)dt + \delta_{2i}q_i(t) \qquad (58)$$

$$k_{pi}(t) = k_{pi}(0) + \alpha_{1i}\int_0^t q_i(t)e_i(t)dt + \alpha_{2i}q_i(t)e_i(t) \qquad (59)$$

$$k_{vi}(t) = k_{vi}(0) + \beta_{1i}\int_0^t q_i(t)\dot{e}_i(t)dt + \beta_{2i}q_i(t)\dot{e}_i(t) \qquad (60)$$

It is noted that, from equations (57)-(58), the control law (56) can alternatively be rewritten in the PID form

$$F_i(t) = F_i(0) + \overline{k}_{pi}(t)e_i(t) + \overline{k}_{vi}(t)\dot{e}_i(t) + \overline{k}_{Ii}\int_0^t e_i(t)dt \qquad (61)$$

Once the task-space control forces $[F_x, F_y, F_\phi]$ are generated by the controllers in equation (56) or (61), the required joint control torques $[T_1, T_2, T_3]$ are obtained from

$$T(t) = J'(\psi)F(t) \qquad (62)$$

where J is the 3×3 augmented Jacobian matrix given in equation (54). It must be noted that the controller equations (56)-(60) are implemented in RCCL in discrete form, with the integrals computed by the trapezoidal rule. Furthermore, we ignore the dynamics of the joint motors by assuming that the demanded and the generated joint torques are always equal.

In the experiment, the adaptation gains in equations (57)-(60) are chosen

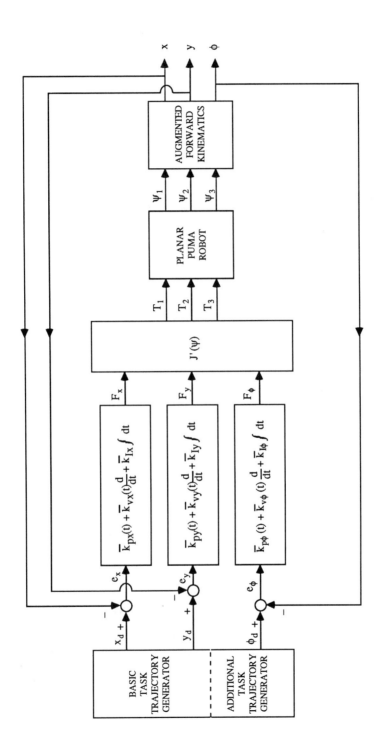

Figure 8. Configuration Control Scheme for Planar 3 dof PUMA Robot

after a few trial-and-errors as[3]

$$w_{px} = 50 \; ; \quad w_{vx} = 30 \; ; \quad w_{py} = 10 \; ; \quad w_{vy} = 10 \; ; \quad w_{p\phi} = 40 \; ; \quad w_{v\phi} = 20$$

$$\delta_{1x} = 50 \; ; \quad \delta_{1y} = 40 \; ; \quad \delta_{1\phi} = 50 \; ; \quad \delta_{2i} = 0 \quad \text{for } i = x, y, \phi$$

$$\alpha_{1x} = 100 \; ; \quad \alpha_{1y} = 100 \; ; \quad \alpha_{1\phi} = 100 \; ; \quad \alpha_{2i} = 0 \quad \text{for } i = x, y, \phi$$

$$\beta_{1x} = 200 \; ; \quad \beta_{1y} = 200 \; ; \quad \beta_{1\phi} = 800 \; ; \quad \beta_{2i} = 0 \quad \text{for } i = x, y, \phi$$

These values do not represent the "optimum" settings of the adaptation gains that can be chosen for the experiment. The initial values of the controller terms are chosen as follows

$$k_{pi}(0) = 0 \qquad\qquad i = x, y, \phi$$

$$k_{vx}(0) = 80 \; ; \quad k_{vy}(0) = 20 \; ; \quad k_{v\phi}(0) = 0$$

$$\begin{pmatrix} d_x(0) \\ d_y(0) \\ d_\phi(0) \end{pmatrix} = [J'(\psi_i)]^{-1} \begin{pmatrix} T_1(0) \\ T_2(0) \\ T_3(0) \end{pmatrix} + \begin{pmatrix} 3\text{sgn}\,[x_d(3) - x(0)] \\ 0 \\ 10\text{sgn}\,[\phi_d(3) - \phi(0)] \end{pmatrix}$$

where $[T_1(0), T_2(0), T_3(0)]$ are the initial joint torques used to approximately compensate for the initial gravity loading, and the sgn terms are chosen empirically to overcome the large stiction (static friction) present in the joints. The initial gravity torque for the third link is given by

$$T_3(0) = \frac{\ell_3}{2} m_3 g \cos \psi_3(0) = 0.8844 \cos \psi_3(0)$$

where $m_3 = 0.3125$kg; and the initial gravity torques for the first and second links are [44]

$$T_1(0) = 8.4 \cos \psi_2(0) + 37.2 \cos \psi_1(0) + 1.02 \sin \psi_1(0)$$

$$T_2(0) = 8.4 \cos \psi_2(0) - 0.25 \sin \psi_2(0)$$

It is important to note that although the arm is moving in the vertical plane with large joint frictions, , gravity and friction compensations are not used separately in addition to the control law (56), and are used merely as the initial auxiliary signal in order to improve the initial response of the arm.

In the experiment, the configuration variables $[x(t), y(t), \phi(t)]$, representing the endpoint coordinates and the shoulder angle, are commanded to change simultaneously from the initial to the final values in three seconds by tracking the desired cycloidal trajectories of equation (55). During the arm

---

[3] In the control program, the unit of length is "meter" and the unit of angle is "radian", and hence the numerical values of the adaptation gains are large.

motion, the joint encoder counts are recorded and transformed using equation (53) to obtain the current values of the configuration variables. Figures 9(i)-(iii) show the desired and actual trajectories of the configuration variables. It is seen that each variable $[x(t), y(t), \phi(t)]$ tracks the corresponding reference trajectory closely using the simple decentralized control scheme of equation (56), despite the coupled nonlinear robot dynamics. Some discrepancy is observed between $[x(t), y(t)]$ and $[x_d(t), y_d(t)]$ at low speed of endpoint motion which can be interpreted physically as follows. When the endpoint is moving at low speed, the joint angles are changing very slowly, and hence the stiction and Columb friction present in the joints become more dominant and oppose the motion, causing a slight tracking-error. Since the position and velocity tracking-errors $e(t)$ and $\dot{e}(t)$ are very small, the rate of adaptation of the controller terms $[d(t), k_p(t), k_v(t)]$ are also small. In this situation, it is primarily the output of the integral term $k_I \int edt$ that needs to build up sufficiently so as to overcome the large friction and cause proper joint motion to correct the error.

Case (ii) - Control of Wrist Height: In this case, the redundancy is used to control the vertical coordinate of the wrist. This is a suitable additional task in situations where we wish to avoid collision with a vertical obstacle, such as a wall, in the workspace. By controlling the wrist height, we ensure that the arm can go over the wall and the endpoint can reach a point behind the wall. This can provide a simple alternative approach to the more complicated obstacle avoidance schemes.

Referring to Figure 7, the configuration variables and the augmented Jacobian matrix in this case are

$$
\begin{aligned}
x(t) &= \ell_1 \cos \psi_1(t) + \ell_2 \cos \psi_2(t) + \ell_3 \cos \psi_3(t) \\
y(t) &= \ell_1 \sin \psi_1(t) + \ell_2 \sin \psi_2(t) + \ell_3 \sin \psi_3(t) \\
\phi(t) &= \ell_1 \sin \psi_1(t) + \ell_2 \sin \psi_2(t)
\end{aligned}
\tag{63}
$$

$$
J = \begin{pmatrix}
-\ell_1 \sin \psi_1 & -\ell_2 \sin \psi_2 & -\ell_3 \sin \psi_3 \\
\ell_1 \cos \psi_1 & \ell_2 \cos \psi_2 & \ell_3 \cos \psi_3 \\
\ell_1 \cos \psi_1 & \ell_2 \cos \psi_2 & 0
\end{pmatrix}
\tag{64}
$$

To find the singular configurations, we form

$$
\det[J] = \ell_1 \ell_2 \ell_3 \cos \psi_3 \sin(\psi_1 - \psi_2)
$$

Therefore, the singularities are defined by either $\psi_3 = \pm 90°$ or $\psi_1 - \psi_2 = 0, 180°$. In other words, the arm is at a singular configuration when the first and second links are aligned or when the third link is vertical. Note that when $\psi_3 = \pm 90°$, the wrist height and the endpoint vertical coordinate cannot be changed independently.

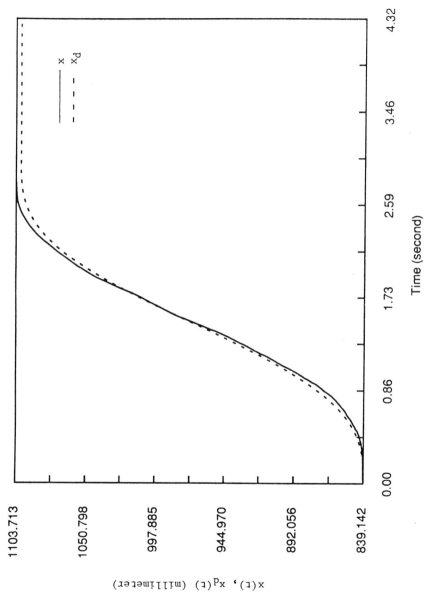

Figure 9(i). Response of Endpoint Horizontal Position x(t) in Case (i)

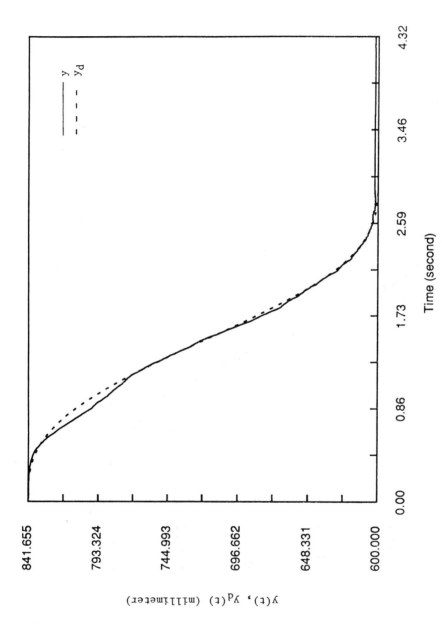

Figure 9(ii). Response of Endpoint Vertical Position y(t) in Case (i)

248

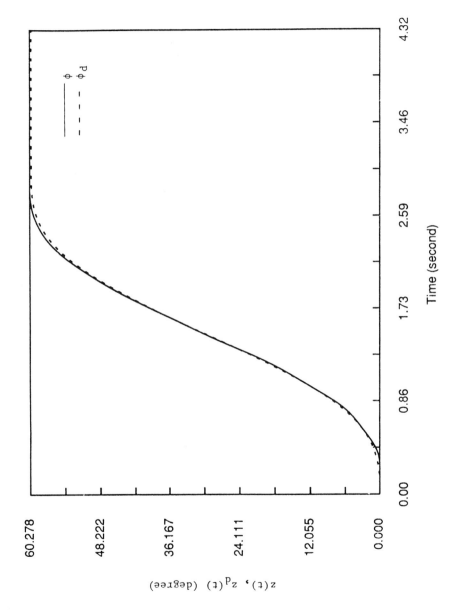

Figure 9(iii). Response of Shoulder Angle $\phi(t)$ in Case (i)

As in Case (i), the initial joint angles are $\psi_{i1} = 0°$, $\psi_{i2} = 90°$, $\psi_{i3} = 45°$; yielding the initial arm configuration variables as $(x_i, y_i, \phi_i) = (839$ mm, 842 mm, 433 mm$)$. It is desired to change the arm configuration variables in three seconds to the final values $(x_f, y_f, \phi_f) = (1000$ mm, 500 mm, 700 mm$)$ by tracking the following cycloidal reference trajectories

$$x_d(t) = 839 + \frac{1000 - 839}{2\pi}\left[\frac{2\pi t}{3} - \sin\frac{2\pi t}{3}\right] \quad \text{mm} \quad ; \quad 0 \le t \le 3$$

$$= 1000 \quad \text{mm} \qquad\qquad\qquad\qquad ; \quad 3 < t$$

$$y_d(t) = 842 + \frac{500 - 842}{2\pi}\left[\frac{2\pi t}{3} - \sin\frac{2\pi t}{3}\right] \quad \text{mm} \quad ; \quad 0 \le t \le 3 \qquad (65)$$

$$= 500 \quad \text{mm} \qquad\qquad\qquad\qquad ; \quad 3 < t$$

$$\phi_d(t) = 433 + \frac{700 - 433}{2\pi}\left[\frac{2\pi t}{3} - \sin\frac{2\pi t}{3}\right] \quad \text{mm} \quad ; \quad 0 \le t \le 3$$

$$= 700 \quad \text{mm} \qquad\qquad\qquad\qquad ; \quad 3 < t$$

Note that this yields a straight-line path for the endpoint, since we have $\frac{x_d - 839}{1000 - 839} = \frac{y_d - 842}{500 - 842}$.

The three configuration variables $[x, y, \phi]$ are controlled using three *independent* adaptive feedback controllers as described in Case (i) and shown in Figure 8. After a few trial-and-errors, the adaptation gains are chosen as

$$w_{px} = 90 ; \quad w_{vx} = 25 ; \quad w_{py} = 15 ; \quad w_{vy} = 20 ; \quad w_{p\phi} = 30 ; \quad w_{v\phi} = 60$$

$$\delta_{1x} = 120 ; \quad \delta_{1y} = 60 ; \quad \delta_{1\phi} = 200 ; \quad \delta_{2i} = 0 \qquad \text{for } i = x, y, \phi$$

$$\alpha_{1x} = 200 ; \quad \alpha_{1y} = 100 ; \quad \alpha_{1\phi} = 100 ; \quad \alpha_{2i} = 0 \qquad \text{for } i = x, y, \phi$$

$$\beta_{1x} = 200 ; \quad \beta_{1y} = 200 ; \quad \beta_{1\phi} = 200 ; \quad \beta_{2i} = 0 \qquad \text{for } i = x, y, \phi$$

The initial values of the controller gains are selected as

$$k_{pi} = 0 \qquad i = x, y, \phi$$

$$k_{vx} = 130 ; \quad k_{vy} = 30 ; \quad k_{v\phi} = 400$$

and the initial axiliary signals are chosen as in Case (i) with the friction coefficients $[3,0,40]$.

In the experiment, we command the configuration variables $[x(t), y(t), \phi(t)]$, representing the endpoint coordinates and the wrist height, to change from the initial to the final values in three seconds by tracking the desired cycloidal trajectories (65). Figures 10(i)-(iii) show the desired and actual trajectories of the configuration variables, and illustrate that each variable tracks the corresponding reference trajectory closely using a simple decentralized control scheme. There is a slight tracking-error at the terminal part

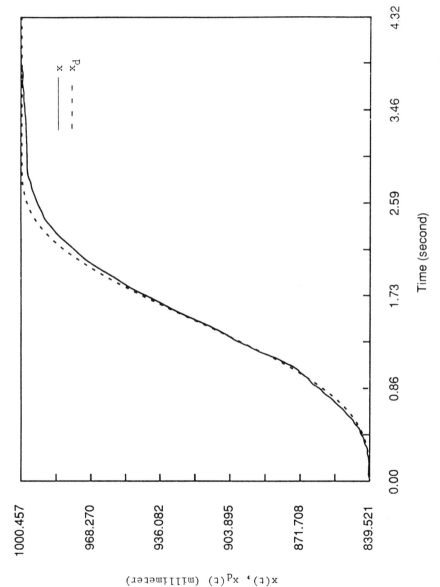

Figure 10(i). Response of Endpoint Horizontal Position x(t) in Case (ii)

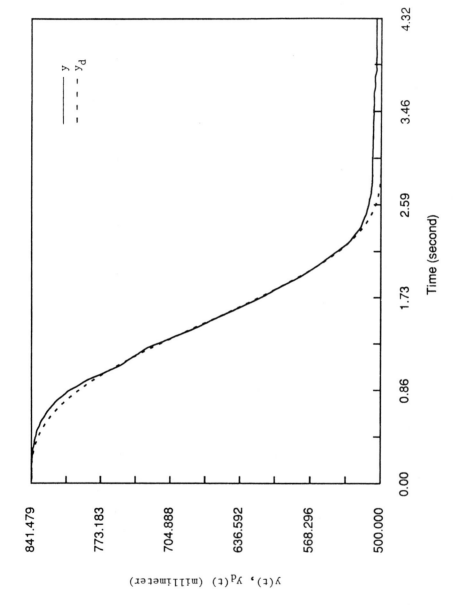

Figure 10(ii). Response of Endpoint Vertical Position y(t) in Case (ii)

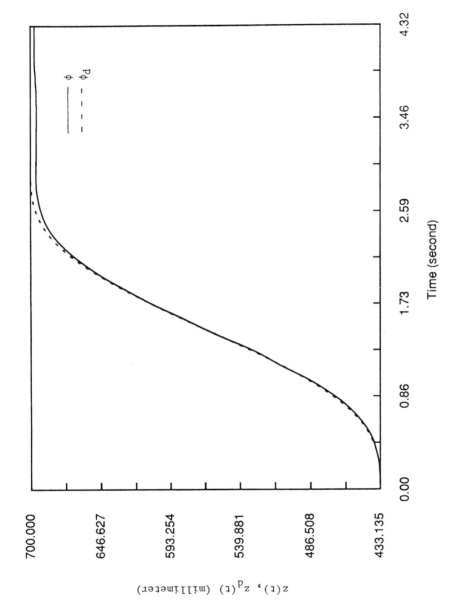

Figure 10(iii). Response of Wrist Vertical Position $\phi(t)$ in Case (ii)

of the trajectory due to the dominant effect of joint friction at low speed of motion. In other words, when the endpoint and wrist are moving slowly, the velocities of joint angles are so low that the friction effects dominate until the controller torques overcome the frictions and corrective actions are taken.

Finally, it is important to note that in both Cases (i) and (ii), by increasing the rate of sampling from $f_s = 143$ Hz to say $f_s = 1000$ Hz, considerable improvement is expected in the tracking performance of the controllers. However, the present experimental set-up cannot operate faster than $f_s = 143$ Hz. Furthermore, the PUMA robot has very large static and Columb frictions, and better results are expected with an inner-loop friction compensation. Nonetheless, given the limitations of the present experimental set-up, it is believed that the experimental results demonstrate the capabilities of the proposed configuration control scheme and validate the scheme in a realistic environment.

4. CONCLUSIONS

A simple formulation for configuration control of redundant manipulators has been developed in this paper. The controller achieves trajectory tracking for the end-effector directly in the Cartesian space to perform some desired basic task. In addition, the redundancy is utilized by imposing a set of kinematic constraints on the manipulator to accomplish an appropriate additional task. The proposed formulation incorporates the kinematic constraints (additional task) and the end-effector motion (basic task) in a conceptually simple and computationally efficient manner to resolve the redundancy. Furthermore, the adaptive controller has a very simple structure and the controller gains are updated in a simple manner to compensate for changing dynamic characteristics of the manipulator. The adaptation laws are based on the observed performance of the manipulator rather than on any knowledge of the manipulator dynamic model. Thus, the adaptive controller is capable of ensuring a satisfactory performance when the payload mass is unknown and time-varying. Any approach used to resolve redundancy should be implementable as a real-time algorithm, and therefore the speed of computation is a critical factor. The small amount of computations required by the proposed method offers the possibility of efficient real-time control of redundant manipulators.

It is important to appreciate the difference between the conventional pseudoinverse-based methods and the configuration control scheme. Pseudoinverse-based methods resolve the redundancy at the velocity level and yield *local* optimization results. In the configuration control scheme, the redundancy is resolved at the position (i.e., task) level to provide direct control over the *entire* motion. Furthermore, pseudoinverse-based methods do not generally produce cyclic motion of the manipulator, which is an essential requirement for repetitive operations; whereas the proposed scheme ensures cyclicity. It is also important to distinguish between the extended Jacobian method [8] and the proposed scheme. The extended Jacobian method is concerned with solving the inverse kinematic problem for redundant robots by extending the end-effector Jacobian to include an optimization subtask, assuming the initial robot configuration is optimal. The proposed approach, however, provides a dynamic control scheme for redundant manipulators based on any desired task augmentation, without requiring the inverse kinematic solution. It is also shown that the extended Jacobian method can be treated as a special case of the configuration control scheme. It is observed that when optimizing an objective function in joint space, the variation of the manipulator configuration in task space is not controlled directly, and this may lead to undesirable motions of the manipulator.

Current research is focussed on the development and implementation of configuration control schemes for the 7 dof Robotics Research arms [45]. Further research is also under way on the development of alternative methods for choosing the kinematic functions in order to diversify the utilization of redundancy.

5. ACKNOWLEDGEMENTS

The research described in this paper was performed at the Jet Propulsion Laboratory, California Institute of Technology, under contract with the National Aeronautics and Space Administration through the Goddard Space Flight Center in support of the Flight Telerobotic Servicer Project. The author is grateful to Professor Richard Colbaugh and Ms. Kristin Glass of New Mexico State University for the computer simulation results in Section 3.1. Thank are also due to Mr. Thomas Lee of JPL for developing the software on the MicroVAX computer for the experiments in Section 3.2.

6. REFERENCES

[1] D.E. Whitney: "Resolved motion rate control of manipulators and human prostheses," IEEE Trans. Man-Machine Systems, 1969, Vol. MMS-10, No. 2, pp. 47-53.

[2] A. Liégeois: "Automatic supervisory control of the configuration and behavior of multibody mechanisms," IEEE Trans. System, Man and Cybernetics, 1977, Vol. SMC-7, No. 12, pp. 868-871.

[3] H. Hanafusa, T. Yoshikawa and Y. Nakamura: "Analysis and control of articulated robot arms with redundancy," Proc. 8th IFAC Triennial World Congress, Kyoto, Japan, 1981, pp. 1927-1932.

[4] T. Yoshikawa: "Analysis and control of robot manipulators with redundancy," Proc. 1st Intern. Symp. on Robotics Research, New Hampshire, 1983, pp. 735-747.

[5] Y. Nakamura and H. Hanafusa: "Task priority based redundancy control of robot manipulators," Proc. 2nd Intern. Symp. on Robotics Research, Kyoto, August 1984.

[6] T. Yoshikawa: "Manipulability and redundancy control of robotic mechanisms," Proc. IEEE Intern. Conf. on Robotics and Automation, St. Louis, March 1985, pp. 1004-1009.

[7] J. Baillieul, J. Hollerbach and R. Brockett: "Programming and control of kinematically redundant manipulators," Proc. 23rd IEEE Conf. on Decision and Control, December 1984, pp. 768-774.

[8] J. Baillieul: "Kinematic programming alternatives for redundant manipulators," Proc. IEEE Intern. Conf. on Robotics and Automation, St. Louis, March 1985, pp. 722-728.

[9] J. Baillieul: "Avoiding obstacles and resolving kinematic redundancy," Proc. IEEE Intern. Conf. on Robotics and Automation, San Francisco, April 1986, pp. 1698-1704.

[10] J. Baillieul, R. Brockett, J. Hollerbach, D. Martin, R. Percy, and R. Thomas: "Kinematically redundant robot manipulators," Proc. NASA Workshop on Space Telerobotics, Pasadena, Vol. 2, pp. 245-255, January 1987.

[11] J. Baillieul: "Design of kinematically redundant mechanisms," Proc. 24th IEEE Conf. on Decision and Control, December 1985, Ft. Lauderdale, pp. 18-21.

[12] I.D. Walker and S.I. Marcus: "An approach to the control of kinematically redundant robot manipulators," Proc. American Control Conf., Seattle, June 1986, pp. 1907- 1912.

[13] C.A. Klein and C.H. Huang: "Review of pseudoinverse control for use with kinematically redundant manipulators," IEEE Trans. System, Man and Cybernetics, 1983, Vol. SMC-13, No. 3, pp. 245-250.

[14] S.Y. Oh, D. Orin, and M. Bach: "An inverse kinematic solution for kinematically redundant robot manipulators," Journal of Robotic Systems, 1984, Vol. 1, No. 3, pp. 235-249.

[15] S.Y. Oh: "Inverse kinematic control for redundant manipulators," Proc. IEEE Workshop on Intelligent Control, Troy, 1985, pp. 53-57.

[16] O. Khatib: "A unified approach for motion and force control of robot manipulators: The operational space formulation," IEEE Journal of Robotics and Automation, 1987, Vol. RA-3, No. 1, pp. 43-53.

[17] C.A. Klein: "Use of redundancy in the design of robotic systems," Proc. 2nd Intern. Symp. on Robotics Research, Kyoto, August 1984.

[18] A.A. Maciejewski and C.A. Klein: "Obstacle avoidance for kinematically redundant manipulators in dynamically varying environments," Intern. Journ. of Robotics Research, 1985, Vol. 4, No. 3, pp. 109-117.

[19] C.A. Klein and A.I. Chirco: "Dynamic simulation of a kinematically redundant manipulator system," Journal of Robotic Systems, 1987, Vol. 4, No. 1, pp. 5-23.

[20] D.R. Baker and C.W. Wampler: "Some facts concerning the inverse kinematics of redundant manipulators," Proc. IEEE Intern. Conf. on Robotics and Automation, Raleigh, March 1987, pp. 604-609.

[21] J.M. Hollerbach: "Optimum kinematic design for a seven degree of freedom manipulator," Proc. 2nd Intern. Symp. on Robotics Research, Kyoto, August 1984.

[22] J.M. Hollerbach and K.C. Suh: "Redundancy resolution of manipulators through torque optimization," Proc. IEEE Intern. Conf. on Robotics and Automation, St. Louis, March 1985, pp. 1016-1021.

[23] O. Egeland: "Cartesian control of a hydraulic redundant manipulator," Proc. IEEE Intern. Conf. on Robotics and Automation, Raleigh, April 1987, pp. 2081-2086.

[24] L. Sciavicco and B. Siciliano: "A dynamic solution to the inverse kinematic problem for redundant manipulators," Proc. IEEE Intern. Conf. on Robotics and Automation, Raleigh, April 1987, pp. 1081-1087.

[25] P. Hsu, J. Hauser, and S. Sastry: "Dynamic control of redundant manipulators," Proc. IEEE Intern. Conf. on Robotics and Automation, Philadelphia, April 1988, pp. 183-187.

[26] R.V. Dubey, J.A. Euler, and S.M. Babcock: "An efficient gradient projection optimization scheme for a 7 dof redundant robot with spherical wrist," Proc. IEEE Intern. Conf. on Robotics and Automation, Philadelphia, April 1988, p. 28-36.

[27] J.J. Craig: *Robotics - Mechanics and Control*, Addison Wesley Publishing Company, Reading, MA, 1986.

[28] R. Courant: *Differential and Integral Calculus*, Vol. II, Interscience Publishers Inc., New York, 1961.

[29] J. Burdick and H. Seraji: "Characterization and control of self-motions in redundant manipulators," Proc. NASA Conference on Space Telerobotics, Pasadena, 1989, Vol. 2, pp. 3-14.

[30] H. Seraji: "Direct adaptive control of manipulators in Cartesian space," Journal of Robotic Systems, 1987, Vol. 4, No. 1, pp. 157-178.

[31] H. Seraji: "A new approach to adaptive control of manipulators," ASME Journ. Dynamic Systems, Measurement and Control, 1987, Vol. 109, No. 3, pp. 193-202.

[32] H. Seraji: "Decentralized adaptive control of manipulators: Theory, simulation, and experimentation," IEEE Trans. Robotics and Automation, 1989, Vol. RA-5, No. 2, pp. 183-201.

[33] P.A. Ioannou: "Decentralized adaptive control of interconnected systems," IEEE Trans. Aut. Control, 1986, Vol. AC-31, No. 4, pp. 291-298.

[34] D.E. Kirk: *Optimal Control Theory: An Introduction*, Section 6.6, Prentice-Hall Publishing Company, 1970.

[35] R. Colbaugh, H. Seraji, and K. Glass: "Obstacle avoidance for redundant robots using configuration control," Journal of Robotic Systems, 1989, Vol. 6, No. 6, pp. 721-744.

[36] C.W. Wampler: "Inverse kinematic functions for redundant manipulators," Proc. IEEE Intern. Conf. on Robotics and Automation, Raleigh,

April 1987, pp. 610-617.

[37] T.W. Nye, D.J. LeBlanc and R.J. Cipra: "Design and Modeling of a Computer-Controlled Planar Manipulator," Intern. Journal of Robotics Research, 1987, Vol. 6, No. 1, pp. 85-95.

[38] A.A. Maciejewski and C.A. Klein: "Obstacle Avoidance for Kinematically Redundant Manipulators in Dynamically Varying Environments," Intern. Journal of Robotics Research, 1985, Vol. 4, No. 3, pp. 109-117.

[39] M.H. Raibert and J.J. Craig: "Hybrid Position/Force Control of Manipulators," ASME Journal of Dynamic Systems, Measurement, and Control, 1981, Vol. 102, No. 2, pp. 126-133.

[40] H. Seraji: "Adaptive Force and Position Control of Manipulators," Journal of Robotic Systems, 1987, Vol. 4, No. 4, pp. 551-578.

[41] C.A. Klein and B.E. Blaho: "Dexterity Measures for the Design and Control of Kinematically Redundant Manipulators," Intern. Journal of Robotics Research, 1987, Vol. 6, No. 2, pp. 72-83.

[42] R. Colbaugh: "Dynamic Performance Optimization of Redundant Robot Manipulators," Preprints 842nd Meeting of the American Mathematical Society, Las Cruces, NM, April 1988.

[43] V. Hayward and R. Paul: "Introduction to RCCL: A Robot Control 'C' Library," Proc. IEEE Intern. Conf. on Robotics, Atlanta, 1984, pp. 293-297.

[44] B. Armstrong, O. Khatib, and J. Burdick: "The explicit dynamic model and inertial parameters of the PUMA 560 arm," Proc. IEEE Intern. Conf. on Robotics and Automation, San Francisco, 1986, pp. 510-518.

[45] K. Kreutz, M. Long, and H. Seraji: "Kinematic functions for the 7 DOF Robotics Research Arm," Proc. NASA Conference on Space Telerobotics, Pasadena, 1989, Vol. 1, pp. 39-48.

NONLINEAR FEEDBACK
FOR FORCE CONTROL
OF ROBOT MANIPULATORS

XIAOPING YUN

Department of Computer and Information Science
University of Pennsylvania
Philadelphia, PA 19104

I. INTRODUCTION

Motions of robot manipulators are either constrained or unconstrained while they perform tasks. When manipulators perform pick-and-place tasks in the free space, their motions are unconstrained in the sense that there is no external geometric constraints imposed on the motion trajectories of manipulators. Constraints caused by the limited joint torques and the limited joint motion ranges are considered to be internal and are not topics of this discussion.

Many robotic applications such as assembly tasks require constrained motion of manipulators. In those applications, the motions of end effectors or of other part of manipulators are constrained by the presence of environments and interactions occur between manipulators and environments. In this chapter, the control of constrained manipulators will be discussed.

CONTROL AND DYNAMIC SYSTEMS, VOL. 40

Two cases of constrained motions are considered: (1) a single manipulator constrained by the environment, and; (2) multiple manipulators constrained with each other (as well as constrained by the environment). In both cases, degrees of freedom of the entire system is reduced by the presence of constraints. In terms of modeling, systems in both cases are characterized by their motion equations and constraint equations. In this chapter, these two classes of constrained systems will be treated in a unified framework using the state space representation.

Robot manipulators together with environments are mechanical systems. Their behaviors are described by physical laws in terms of motion equations and constraint equations. The motion description must be formulated into a system representation in order to utilize appropriate design techniques to design controllers. A dynamic system may be represented in various forms. However, when a nonlinear system is concerned, the state space representation becomes particularly important due to the existence of powerful design techniques such as Liapunov methods and linearization methods. When position control of robot manipulators is studied, the system representation is not a problem since the second order differential equations describing motion of manipulators are easily converted into a state space form by taking manipulator joint position and velocity variables as the state. However, behaviors of *constrained* manipulators are described by both differential motion equations and algebraic constraint equations. Those equations are not readily converted into the standard state space representation. In particular, certain forces must controlled in a constrained system. But which role should those forces play in a state space representation: state, input, output, or none of the three? Ironically, force control has been widely studied using state space techniques including Liapunov method without having a proper state space representation. It is therefore important to realize that the representation, priori to controller design, of force control is an issue of study. Towards this direction, McClamroch [1, 2] has presented a singular perturbation approach to modeling of constrained manipulators. Mills and Goldenberg [3] treated force control of manipulators using descriptor systems. Unseren and Koivo [4] used a reduced-order model. The representation approach taken in this chapter is in nature a form of reduced-order model.

Having mentioned issues on force control representation in the above, an equally important and yet confusing issue is force control stability. In comparison, position control is once again intrinsically simple and well formulated in this regard. Assuming perfect model parameters, any computed-torque-like control algorithms will result in stable controllers. Further, such controllers are robust with respect to model parameter variations. The robustness follows from the fact that a joint-wise constant PD feedback (without any other nonlinear compensations) is an overall stable controller, which can be shown using a proper Liapunov function. This may be somewhat surprising since dynamics of a typical manipulator is nonlinear and coupled. Unfortunately, force control does not share any "good" properties that position control entitles to. Force control stability side remains to be an active research issue. Nevertheless, there are numerous papers claiming the establishment of force control stability. The reason for having inconsistent results in the community stems from the *definition* of force control stability. Choosing an appropriate coordinates and applying a feedback, position and force control can be decoupled [5, pp. 415-418]. The force control subsystem is a static system, not a dynamic system. It becomes even clearer if a Cartesian manipulator is considered for performing tasks in which the manipulator is controlled to move along a flat frictionless surface parallel to x-y plane. In force controlled z direction, the input joint force is the output force to be controlled. What is then the stability definition of a static system whose input is its output? Adopting the bounded-input bounded-output stability for dynamic systems, force control of a constrained manipulator is always stable. The asymptotical stability is not applicable since the equilibrium is not defined in the first place.

In this chapter, we present an approach to force/position control in which force control is reformulated as a control problem of a *dynamic system*, instead of a static system. By doing so, all the stability concepts as well as design techniques can applied to force control. The reformulation can be done in a number of ways. Physically, we may include the dynamics of actuators in modeling force/position control of manipulators. For example, modeling the armature circuit inductance of a DC motor actuator introduces a first order dynamic system priori to the generation of joint torque/force. In general, we will introduce an integrator on each input channel. Force/position control will be properly represented in a state space form. We then apply a nonlinear state feedback to exactly linearize and decouple the nonlinear force/position control model. Asymptotical stability of force/position control will follow a final linear state feedback for pole assignment. The process of introducing integrators and applying static state feedback may also be viewed as application of a dynamic state feedback.

II.   The Dynamic Model of a Constrained Manipulator

We consider a robot manipulator with n joints. If the manipulator moves in the free space, its degrees of freedom is n, assuming that each joint (typically, rotary or prismatic joint) has one degree of freedom. If the manipulator is in contact with the environment, the degrees of freedom of the manipulator (the minimum number of independent variables to describe the motion of the manipulator) is in general less than n.

We consider geometric constraints imposed by the infinite stiff environment in the manipulator workspace which are characterized by smooth hypersurfaces

$$s(p) = \begin{bmatrix} s_1(p) \\ \vdots \\ s_r(p) \end{bmatrix} = 0 \tag{1}$$

defined in a world coordinate frame located in the workspace. The vector p represents the position and orientation coordinates. It is reasonable to assume that the number of constraints, r, is less than the number of the manipulator joints, n. An example of such a constraint is a table surface characterized by $z = $ constant. Let

$$q = [\, q_1 \quad \cdots \quad q_n \,]^T$$

denote the joint variables and let

$$p = p(q) \tag{2}$$

be the direct kinematics of the robot manipulator. Combining Equations (1) and (2), we obtain the constraint equations in the joint variables

$$s(p(q)) = 0. \tag{3}$$

To maintain contact with the environment, the joint variables have to satisfy Equation (3). This requirement is equivalent to that the joint velocities (or accelerations) satisfy the corresponding velocity (or accelerations) constraint equations and the initial conditions. To obtain the constraint equations of the joint velocities, we differentiate Equation (3) with respect to the time

$$\frac{ds}{dt} = \frac{\partial s}{\partial p}\frac{\partial p}{\partial q}\dot{q} \triangleq J_s J_p \dot{q} = 0. \tag{4}$$

While being in contact with the environment, the presence of the r constraints causes the manipulator to lose r degrees of freedom, leaving the manipulator with only $(n - r)$ degrees of freedom. In this case, $(n - r)$ linearly independent coordinates are sufficient to characterize the motion of the manipulator. We choose $n - r$ out of the n joint variables, denoted by

$$q^1 = \begin{bmatrix} q_1^1 & \cdots & q_{n-r}^1 \end{bmatrix}^T \tag{5}$$

to be the generalized coordinates to describe the motion of the constrained manipulator. It is noted that this is only one way among many to choose the generalized coordinates. We denote the remaining joint variables by

$$q^2 = \begin{bmatrix} q_1^2 & \cdots & q_r^2 \end{bmatrix}^T.$$ (6)

For linearly independent constraints in Equation (1), the velocity constraint in Equation (4) may be rewritten in the following form

$$\dot{q}^2 = M_1 \dot{q}^1$$ (7)

at nonsingular configurations of the manipulator. Differentiating both sides of Equation (7) with respect to the time, we obtain the acceleration constraints

$$\ddot{q}^2 = M_1 \ddot{q}^1 + M_2$$ (8)

where $M_2 = \dot{M}_1 \dot{q}^1$.

Having obtained the constraint equations, we now derive the motion equations of the constrained manipulator. The dynamic equations of an n-joint unconstrained manipulator are governed by the following second order differential equation [6]

$$D(q)\ddot{q} + E(q, \dot{q}) = \tau$$ (9)

where $D(q)$ is the inertia matrix, $E(q, \dot{q})$ consists of the Coriolis, centripetal, and gravity forces, and τ is the joint torque/force. When the manipulator is constrained, the reaction force applied to the manipulator by the environment has to be incorporated into the motion equations as follows

$$D(q)\ddot{q} + E(q, \dot{q}) = \tau + J_p^T J_s^T F_n$$ (10)

where F_n is the components of the contact force normal to the hypersurfaces. In writing Equation (10), we have assumed that the environment is frictionless. The matrix Equation (10) consists of n scalar equations, one for each joint. Without loss of generality, we assume that the elements of q^1 are chosen to be the first $n - r$ elements of q. If this is not the case, we can always reorder Equation (10) so that the first $n - r$ equations correspond to q^1, and the last r equations to q^2. Considering the acceleration constraint Equation (8), we can express Equation (10) in the following way

$$D(q) \begin{bmatrix} \ddot{q}^1 \\ M_1 \ddot{q}^1 + M_2 \end{bmatrix} + E(q, \dot{q}) = \tau + J_p^T J_s^T F_n.$$ (11)

The above equation can be further written in the form of

$$\begin{bmatrix} D(q) \begin{bmatrix} I \\ M_1 \end{bmatrix} & -J_p^T J_s^T \end{bmatrix} \begin{bmatrix} \ddot{q}^1 \\ F_n \end{bmatrix} + D(q) \begin{bmatrix} 0 \\ M_2 \end{bmatrix} + E(q, \dot{q}) = \tau.$$ (12)

Using a compact notation, it is convenient to write Equation (12) as

$$\bar{D}(q) \begin{bmatrix} \ddot{q}^1 \\ F_n \end{bmatrix} + \bar{E}(q, \dot{q}) = \tau \tag{13}$$

where $\bar{D}$ and $\bar{E}$ can be easily identified from Equation (12). We note that Equation (13) incorporates the dynamics of the manipulator and the constraints imposed by the environment. Solving Equation (13) for $\ddot{q}^1$ and F_n (see Appendix A), we obtain the reduced equations of motion and the equation of the contact force

$$\ddot{q}^1 = a + b\tau \tag{14}$$
$$F_n = F_1 + F_2\tau \tag{15}$$

where a, b, F_1, and F_2 are functions of the configuration and the velocity, which are given in Appendix A. Equation (14) is the reduced equations of motion of the constrained manipulator, which is free from the contact force. The joint torques/forces completely determine the projected motion of the manipulator onto the tangent subspace of the environment. Equation (15) is an explicit expression for the contact force in terms of the joint positions, the joint velocities, and the joint torques.

When designing a position controller for an unconstrained manipulator, we start from the equations of motion. Further, if the position control in the task space is concerned, the direct kinematics is available to relate the task space variables to the joint space variables. Position control of the unconstrained manipulator is a well-formulated problem, namely, the equations of motion are the state equation and the direct kinematics is the output equation of the robot dynamic system. It is the equations (14) and (15) that formulate the force control of the constrained manipulator as a standard nonlinear system: the reduced equations of motion are the state equation and the equation of the contact force is the output equation (or part of the output equation in case of the simultaneous motion and force control). This formulation makes it possible to study the force control of the constrained manipulator by using the advanced nonlinear system theory such as the differential geometric control theory for nonlinear systems [7].

III. Nonlinear Feedback for Force Control of a Constrained Manipulator

We have established the reduced equations of motion of the constrained manipulator and the equation of the contact force. These equations are suitable for controller design for three reasons. First, the equality constraint equations are embedded into the motion equations, resulting in an affine nonlinear system without constraints. Secondly, the recognition of the reduced degrees of freedom eliminates the redundancy in the modeling representation, which precludes the singularity from the system analysis. Finally, the explicit expression of the normal contact force F_n provides a unified framework for theoretically analyzing the stability of the force control and the position control.

To control the tangent motion and the normal contact force of the constrained manipulator, we take the following output equations

$$y = \left[\begin{array}{c} p_0(q) \\ F_n(q, \dot{q}, \tau) \end{array} \right] \tag{16}$$

where F_n is the normal contact force that is the same as in Equation (15), and $p_0(q)$ is a vector that describes the motion of the end effector tangent to the environment. The dimension of the vector $p_0(q)$ is $\min\{n - r, 6\}$, which reflects the fact that the manipulator has $n - r$ degrees of freedom and that the motion of the end effector is determined by at most six linearly independent quantities (three for position and three for orientation). By taking this form of the output equations, we place the same importance on controlling motion and force, and consequently we are able to analyze the properties of the motion control and the force control in the same framework.

The difference of including the contact force as part of the output equations is that the output equations now depend on the joint velocity and the joint torque as well as on the joint position, in contrast with the position control in which the output equations (the direct kinematics) depend only on the joint position. Various position control methods were developed based on this property, which may not be realized before. The simultaneous motion and force control of the constrained manipulator with the output equations (16) is inherently different from the pure position control of the unconstrained manipulator and is more challenging. In what follows, we employ the technique of feedback linearization from the differential geometric control theory for nonlinear systems [7]. Our purpose is to linearize and decouple the nonlinear dynamic system of the constrained manipulator by using a nonlinear feedback. However, the static state-feedback discussed by Isidori [7] and used for the position control [8, 9] can not be applied to the force control problem since the inputs (the joint torques τ) directly appear in the output equations. To resolve this problem, the idea of the dynamic extension [10] can be utilized to add integrators on the input channels so that τ becomes part of the state variables in the enlarged system. Viewing the enlarged system, the output equations depend only on the (enlarged) states.

From this point on, the technique of the static state-feedback [7, 8] can be used for the purpose of the state linearization and the output decoupling. The overall process is a special case of the dynamic state-feedback [10].

We introduce a new set of state variables x and simultaneously introduce an integrator on each input channel by letting

$$
\begin{aligned}
x_i &= q_i^1, & i &= 1, \ldots, n - r \\
x_i &= \dot{q}_{i-(n-r)}^1, & i &= n - r + 1, \ldots, 2(n - r) \\
x_i &= \tau_{i-2(n-r)}, & i &= 2(n - r) + 1, \ldots, 3n - 2r \\
\dot{\tau}_i &= u_i, & i &= 1, \ldots, n.
\end{aligned}
$$

For the notational convenience, we adopt the following compact block notation

$$
\begin{aligned}
x^1 &= q^1, & x^2 &= \dot{q}^1 \\
x^3 &= \tau, & \dot{\tau} &= u \\
x &= \left[\; (x^1)^T \quad (x^2)^T \quad (x^3)^T \; \right]^T
\end{aligned}
$$

where u is the inputs to the integrators. We note that the dimension of the state variables x is $2(n - r) + n = 3n - 2r$. Using the state variables x, Equations (14) and (16) become

$$
\dot{x} = \begin{bmatrix} \dot{x}^1 \\ \dot{x}^2 \\ \dot{x}^3 \end{bmatrix} = \begin{bmatrix} x^2 \\ a + bx^3 \\ 0 \end{bmatrix} + \begin{bmatrix} 0 \\ 0 \\ I \end{bmatrix} u \stackrel{\triangle}{=} f(x) + g(x)u \tag{17}
$$

$$
y = \begin{bmatrix} p_0(x^1) \\ F_n(x^1, x^2, x^3) \end{bmatrix} \stackrel{\triangle}{=} \begin{bmatrix} h^1 \\ h^2 \end{bmatrix} = h(x). \tag{18}
$$

Now the problem of the simultaneous motion and force control of the constrained manipulator becomes the control problem of a standard affine nonlinear system, *i.e.*, the system described by Equations (17) and (18). Our approach to the design problem of this nonlinear system is to find a nonlinear feedback

$$
u = \alpha(x) + \beta(x)v \tag{19}
$$

and a nonlinear state transformation $z = T(x)$ to exactly linearize the system. At the same time, we will decouple the output equations, that is, one input merely controls one output without affecting the rest of the output components.

The linearization and decoupling of affine nonlinear systems by means of nonlinear feedback have been studied extensively during the last decade. The necessary and sufficient conditions have been established so that a given nonlinear system can be checked to determine if it is linearizable by using a nonlinear feedback. In particular, an algorithm was developed to compute the required nonlinear feedback and the diffeomorphism that achieve the linearization and decoupling if the concerned system is linearizable [8]. At this point, we may work through the necessary and sufficient conditions to find if they are satisfied by the present system of the constrained manipulator. Nevertheless, for a system as complex as Equations (17) and (18), the verification of these condition is extremely tedious. Alternatively, we will straightforwardly compute the nonlinear feedback following the algorithm [8] and will then demonstrate that the computed nonlinear feedback indeed achieves the linearization and decoupling. Following the algorithm in [8], the $\alpha(x)$ and $\beta(x)$ in the nonlinear feedback (19) satisfy matrix equations

$$\Phi(x)\alpha(x) = -\begin{bmatrix} L_f^3 h^1 \\ L_f h^2 \end{bmatrix} \tag{20}$$

$$\Phi(x)\beta(x) = I \tag{21}$$

where $\Phi(x)$ is the decoupling matrix given by

$$\Phi(x) = \begin{bmatrix} L_g L_f^2 h^1 \\ L_g h^2 \end{bmatrix} = \begin{bmatrix} \frac{\partial h^1}{\partial x^i} & 0 \\ 0 & I \end{bmatrix} \bar{D}^{-1}. \tag{22}$$

Appendix B gives the computational details of the decoupling matrix. The existence of $\alpha(x)$ and $\beta(x)$ in Equations (20) and (21) is equivalent to the nonsingularity of the decoupling matrix $\Phi(x)$. By choosing linearly independent components for h^1 to describe the end effector motion, $\frac{\partial h^1}{\partial x^i}$ is generically nonsingular[1]. Consequently, the decoupling matrix $\Phi(x)$ is generically nonsingular by examining the most right hand side of Equation (22). Thus, we can solve for $\alpha(x)$ and $\beta(x)$ from Equations (20) and (21) by inverting the decoupling matrix $\Phi(x)$. The nonlinear transformation is given by [7]

$$z = T(x) = \begin{bmatrix} h_1^1 & L_f h_1^1 & L_f^2 h_1^1 & \cdots & h_{n-r}^1 & L_f h_{n-r}^1 & L_f^2 h_{n-r}^1 & h_1^2 & \cdots & h_r^2 \end{bmatrix}^T \tag{23}$$

whose dimension is $3(n - r) + r = 3n - 2r$, being exactly equal to the dimension of the state variables x. In the above, we have used the notation $h^1 = \begin{bmatrix} h_1^1 & h_2^1 & \cdots & h_{n-r}^1 \end{bmatrix}$ and $h^2 = \begin{bmatrix} h_1^2 & h_2^2 & \cdots & h_r^2 \end{bmatrix}$. After applying the nonlinear feedback (19) and the nonlinear transformation (23), we convert the system described by Equations (17) and (18) into a linear system [8]

$$\dot{z} = Az + Bv \tag{24}$$

$$y = Cz \tag{25}$$

---

[1]The set of the singular points forms a submanifold with dimensions strictly less than that of the embedded whole space.

where A, B and C are constant matrices with appropriate dimensions, and $z = T(x)$ is the transformed state variables. Furthermore, this linear system is decomposed into decoupled subsystems, each of which controls one component of the outputs. These subsystems possess the structure indicated below.

$$\dot{z}^i = \begin{bmatrix} \dot{z}^i_1 \\ \dot{z}^i_2 \\ \dot{z}^i_3 \end{bmatrix} = \begin{bmatrix} 0 & 1 & 0 \\ 0 & 0 & 1 \\ 0 & 0 & 0 \end{bmatrix} \begin{bmatrix} z^i_1 \\ z^i_2 \\ z^i_3 \end{bmatrix} + \begin{bmatrix} 0 \\ 0 \\ 1 \end{bmatrix} v_i \tag{26}$$

$$y_i = \begin{bmatrix} 1 & 0 & 0 \end{bmatrix} z^i \qquad i = 1, \ldots, n - r \tag{27}$$

and

$$\dot{z}^i = \begin{bmatrix} 0 \end{bmatrix} z^i + \begin{bmatrix} 1 \end{bmatrix} v_i \tag{28}$$

$$y_i = \begin{bmatrix} 1 \end{bmatrix} z^i, \qquad i = n - r + 1, \ldots, n. \tag{29}$$

Appendix C provides the verification that the application of the nonlinear feedback (19) and the transformation (23) does convert the nonlinear system of the constrained manipulator into the above linear subsystems described by Equations (26) through (29). The controller design of the constrained manipulators is simplified to the design problem of these decoupled linear subsystems. Since these linear subsystems are controllable, their characteristic poles can be placed at any desired location by using the constant feedback, $v_i = -K^i z^i + \bar{v}_i$, with an appropriate feedback gain K^i. In particular, the stability of these subsystems is guaranteed by placing the poles on the left half plane. The overall process of the controller design is depicted in Figure 1.

IV.   Control Equations of Two Cooperating Robot Manipulators

In this and the next sections, we will discuss the problem of coordinated control of two manipulators which grasp a common object. It has at least two properties in common with the problem treated in the preceding sections. First of all, when two manipulators hold a common object, their motions are constrained with each other. Secondly, force control is needed to coordinate motions of two manipulators. Due to these common properties, it turns out that the two problems can be treated in the same framework. This section addresses the modeling of two cooperating manipulators. The following section presents the nonlinear feedback for coordinated control of two manipulators.

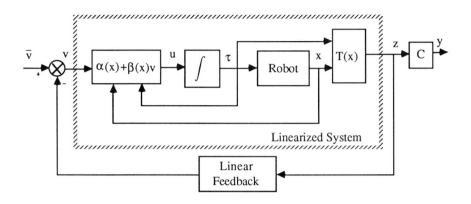

Figure 1: The Block Diagram of the Dynamic Nonlinear Feedback Control of the Constrained Manipulator

We consider tasks of transporting an object by rigidly grasping it with two manipulators. We consider gross motions of two manipulators and do not consider end effectors and grasping processes in this discussion. To perform tasks in a coordinated fashion, it is necessary to explicitly control the interaction force exerted by the grasped object as well as the motion trajectory of the object.

From the viewpoint of the Newtonian mechanics, the object held by two robot manipulators in the free space exerts three forces: the shared force applied by the first robot, that applied by the second robot, and its own gravity force. Specifying desired trajectory for the object determines the total force applied to the object at any instance of time. The gravity of the object can be either estimated from the knowledge of the task or obtained from measuring forces applied by two robots through force/torque sensors when robots and the object are at rest. By subtracting the gravity force from the total force, the sum of the two forces applied by two robots is known. We assign to each robot, in accordance with its load limit, the shared force that it must supply to perform the task. Now a question is how to ensure that the assigned forces are maintained during the execution of the task. We propose to control the motion of the object and the shared force applied to the object by the second robot. The first robot will provide the necessary interaction force since the total force is maintained by controlling the object's motion.

For convenience, we name the first robot as robot "a" and the second robot "b". We assume that robot "b" has a force/torque sensor installed at its end effector. Let F denote the vector of the force and moment applied to the object by robot "b", represented in the world coordinate frame. If we use subscript b to indicate arm "b", equations of motion for arm "b" are described by

$$D_b(q_b)\ddot{q}_b + E_b(q_b, \dot{q}_b) = \tau_b + J_b^T F \tag{30}$$

where D_b is the inertia matrix, F_b consists of Coriolis, centripetal, and gravity forces, τ_b is the joint torque/force, J_b is the Jacobian of the direct kinematics of arm "b", and J^T denotes the transpose of J. Likewise, equations of motion for arm "a" are governed by

$$D_a(q_a)\ddot{q}_a + E_a(q_a, \dot{q}_a) = \tau_a - J_a^T F. \tag{31}$$

We do not assume that arm "a" has a force sensor. The direction of force vector F is opposite as seen by arm "a". This fact is reflected by the minus sign appearing before J_a. The Jacobian matrix J_a, $D_a(q_a)$ and $E_a(q_a, \dot{q}_a)$ are to be computed by considering the object as part of the last link of arm "a".

Since two manipulators hold the same object, their motions are constrained. Let the constraints be given by

$$s(q_a, q_b) = 0. \tag{32}$$

Differentiating Equation (32) twice with respect to the time, we obtain the acceleration constraints

$$\ddot{q}_a = \phi_1(q_a, q_b)\ddot{q}_b + \phi_2(q_a, \dot{q}_a, q_b, \dot{q}_b) \tag{33}$$

in which ϕ_1 has full rank as long as robot "a" is out of its singular region. Equations (30), (31), and (33) are basic motion equations that include dynamics of two robot manipulators and constraints between them. To use state space control theory, these equations must be converted into a control system having state equations and output equations, which is suitable for designing controllers. Output equations define the relation between state variables and variables to be controlled. The position and orientation of the object are necessarily to be controlled, and thus are part of output equations. Further it would be easier to specify the position and orientation of the object in the world coordinate frame. We let p be a vector that represents the position and orientation of the object in the world coordinate frame. This vector p is related to the joint positions of manipulators through the direct kinematics. This relation can be denoted by

$$p = p(q_a, q_b). \tag{34}$$

As discussed early, we also need to control the force and moment F for coordinating two manipulators. We now derive an expression of the force F in terms of the manipulator joint variables and joint torques/forces. This expression will be another part of output equations. We solve for $\ddot{q}_a$ and $\ddot{q}_b$ from dynamic Equations (30) and (31) and substitute them into constraint Equation (33). We obtain the following

$$D_a^{-1}(q_a)[-E_a(q_a, \dot{q}_a) - J_a^T(q_a)F + \tau_a] =$$
$$\phi_1(q_a, q_b)D_b^{-1}(q_b)[-E_b(q_b, \dot{q}_b) + J_b^T(q_b)F + \tau_b] + \phi_2(q_a, \dot{q}_a, q_b, \dot{q}_b).$$

Solving for F from the above equation, we have

$$
\begin{aligned}
F &= R^{-1}[D_a^{-1}(\tau_a - E_a) - \phi_1 D_b^{-1}(\tau_b - E_b) - \phi_2] \\
&\triangleq F_1(q_a, \dot{q}_a, q_b, \dot{q}_b) + F_2(q_a, q_b)\tau
\end{aligned}
\tag{35}
$$

where

$$
\begin{aligned}
R &= D_a^{-1}J_a^T + \phi_1 D_b^{-1}J_b^T \\
F_1(q_a, \dot{q}_a, q_b, \dot{q}_b) &= (D_a^{-1}J_a^T + \phi_1 D_b^{-1}J_b^T)^{-1}(\phi_1 D_b^{-1}E_b - D_a^{-1}E_a - \phi_2) \\
F_2(q_a, q_b) &= (D_a^{-1}J_a^T + \phi_1 D_b^{-1}J_b^T)^{-1}\left[D_a^{-1} \quad -\phi_1 D_b^{-1}\right] \\
\tau &= \begin{bmatrix} \tau_a \\ \tau_b \end{bmatrix}.
\end{aligned}
$$

Equation (35) describes the dependence of the force and moment F on the joint positions q_a and q_b, the joint velocities $\dot{q}_a$ and $\dot{q}_b$, and the joint torques τ_a and τ_b. The nonsingularity of matrix R is showed in Appendix D. For clarity, we have omitted the dependent variables q_a, $\dot{q}_a$, q_b, and $\dot{q}_b$ from the above equation. To effectively control forces and moments, we have to find how they relate to states and inputs. After all, it is the joint torques (inputs), not forces and moments, that we can arbitrarily specify. The force and moment F, and the position and orientation p change as inputs change, following motion equations and output equations.

We now combine Equations (34) and (35) which constitute the entire output equations for coordinated control of two manipulators, and designate y to denote them

$$
y = \begin{bmatrix} p(q_a, q_b) \\ F(q_a, \dot{q}_a, q_b, \dot{q}_b, \tau_a, \tau_b) \end{bmatrix}.
\tag{36}
$$

We substitute the representation of force F, Equation (35), into robot dynamic Equations (30) and (31). Equations of motion of two robots are then purely governed by their position, velocity and input torques. For robot "a", we have

$$D_a(q_a)\ddot{q}_a + E_a(q_a, \dot{q}_a) = \tau_a - J_a^T(F_1 + F_2\tau)$$

which can be further written in the form

$$D_a(q_a)\ddot{q}_a + E_a(q_a, \dot{q}_a) + J_a^T F_1 = \tau_a - J_a^T F_2 \tau$$
$$= ([I_a \quad | \quad 0] - J_a^T F_2)\tau \qquad (37)$$

where I_a is the identity matrix of dimension equal to that of τ_a. Similarly, motion equations of robot "b" become

$$D_b(q_b)\ddot{q}_b + E_b(q_b, \dot{q}_b) - J_b^T F_1 = ([0 \quad | \quad I_b] + J_b^T F_2)\tau \qquad (38)$$

where I_b is the identity matrix of dimension equal to that of τ_b. We will use Equations (37) and (38) together with output equation (36) to design our control algorithm in the next section. We will use the exact linearization techniques to derive a nonlinear *dynamic* feedback to linearize the entire system. This makes it possible to design a coordinated controller for two robot manipulators by using linear system control theory.

V.   Nonlinear Feedback for Coordinated Control of Two Manipulators

Similar to Section III, in this section we will derive a nonlinear feedback for coordinated control of two manipulators. As a preliminary step, it is necessary to represent motion equations of two cooperative manipulators in the state space. As discussed early, since two manipulators are constrained with each other, the degrees of freedom of the system is reduced. It is desirable to choose as many variables as the number of system mobility to describe motions of two manipulators. However, this requires complex equation substitution and elimination. Alternatively, we will simply choose the original joint variables q_1 and q_2 to describe motions of two manipulators. As a result, the state space representation of motion equations is redundant. Consequently, a zero dynamics will be present after applying nonlinear feedback, which will be observed at the end of this section. Reference [11] presents a closed chain approach to modeling of two manipulators in which the minimal number of variables are used to describe motion equations.

For the sake of clarity, we proceed with two steps to arrive at a standard affine nonlinear state space representation. First of all, Let

$$x^1 = \begin{bmatrix} q_a \\ q_b \end{bmatrix}, \qquad x^2 = \begin{bmatrix} \dot{q}_a \\ \dot{q}_b \end{bmatrix}, \qquad x = \begin{bmatrix} x^1 \\ x^2 \end{bmatrix}.$$

Equations (37) and (38) can then be written together in terms of x as follows

$$\dot{x} = \begin{bmatrix} \dot{x}^1 \\ \dot{x}^2 \end{bmatrix} = \begin{bmatrix} x^2 \\ \tilde{E} \end{bmatrix} + \begin{bmatrix} 0 \\ \tilde{D}^{-1} \end{bmatrix} \tau \qquad (39)$$

where

$$\tilde{E} = \left[\begin{array}{c} -D_a^{-1}(E_a + J_a^T F_1) \\ -D_b^{-1}(E_b - J_b^T F_1) \end{array} \right]$$

and

$$\tilde{D}^{-1} = \left[\begin{array}{c} D_a^{-1}([I_a \ \mid \ 0] - J_a^T F_2) \\ D_b^{-1}([0 \ \mid \ I_b] + J_b^T F_2) \end{array} \right].$$

Comparing with the matrix form of single arm dynamic equations, we notice that $\tilde{D}^{-1}$ plays the role of the inverse of the inertia matrix. Equation (39) can be further written in a compact form as

$$\dot{x} = f(x) + g(x)\tau \qquad (40)$$

where $f(x)$ and $g(x)$ can be easily identified from Equation (39). Using the variable x, output equation (36) can be expressed as

$$y = h(x, \tau) = \left[\begin{array}{c} h^1(x^1) \\ h^2(x, \tau) \end{array} \right] = \left[\begin{array}{c} p(x^1) \\ F_1(x) + F_2(x^1)\tau \end{array} \right]. \qquad (41)$$

We note that the output equation is directly dependent on input τ. For the same reason as for force control of a constrained manipulator, we will eliminate this direct dependence by introducing an integrator on each input channel. Specifically, we let

$$\dot{\tau} = w.$$

The enlarged state variables are

$$\xi = \left[\begin{array}{c} x \\ \tau \end{array} \right] = \left[\begin{array}{c} x^1 \\ x^2 \\ \tau \end{array} \right].$$

The extended system under the state variable ξ is then

$$\dot{\xi} = \bar{f}(\xi) + \bar{g}(\xi)w \qquad (42)$$
$$y = h(\xi) \qquad (43)$$

where

$$\bar{f}(\xi) = \left[\begin{array}{c} x^2 \\ \tilde{E} + \tilde{D}^{-1}\tau \\ 0 \end{array} \right], \qquad \bar{g}(\xi) = \left[\begin{array}{c} 0 \\ 0 \\ I \end{array} \right].$$

The output function h is the same as Equation (41) except renaming its arguments. w is the new reference input. The system described by Equations (42) and (43) is in the standard form of affine nonlinear systems for

which a static state feedback may be derived for the purpose of linearization
and output decoupling.

We may now apply the same algorithm used in Section III to compute
the nonlinear feedback for coordinated control of two manipulators. The key
part of the algorithm is the computation of the decoupling matrix. For the
present system, the decoupling matrix is found to be (see Appendix E for
detailed computations)

$$M(\xi) = \begin{bmatrix} L_{\bar{g}} L_{\bar{f}}^2 h^1 \\ L_{\bar{g}} h^2 \end{bmatrix} = \begin{bmatrix} M_{11} & M_{12} \\ M_{21} & M_{22} \end{bmatrix} \tag{44}$$

with

$$
\begin{aligned}
M_{11} &= \frac{\partial h^1}{\partial x^1} D_a^{-1}([I_a \quad | \quad 0] - J_a^T F_2) \\
M_{12} &= \frac{\partial h^1}{\partial x^2} D_b^{-1}([0 \quad | \quad I_b] + J_b^T F_2) \\
M_{21} &= (D_a^{-1} J_a^T + \phi_1 D_b^{-1} J_b^T)^{-1} D_a^{-1} \\
M_{22} &= -(D_a^{-1} J_a^T + \phi_1 D_b^{-1} J_b^T)^{-1} \phi_1 D_b^{-1}.
\end{aligned}
$$

$\alpha(\xi)$ and $\beta(\xi)$ in the nonlinear feedback satisfy the following matrix equations
[12]

$$M(\xi)\alpha(\xi) = -\begin{bmatrix} L_{\bar{f}}^3 h^1 \\ L_{\bar{f}} h^2 \end{bmatrix} \tag{45}$$

$$M(\xi)\beta(\xi) = I. \tag{46}$$

It is shown in Appendix E that $M(\xi)$ is generically nonsingular. Therefore,
$\alpha(\xi)$ and $\beta(\xi)$ can be solved from the above equations as to construct the
nonlinear feedback. The required nonlinear diffeomorphic state space transformation is given by [7]

$$T(\xi) = \begin{bmatrix} T^1(\xi) \\ T^2(\xi) \end{bmatrix} \triangleq \begin{bmatrix} z \\ \eta \end{bmatrix}. \tag{47}$$

The first part of the transformation, $z = T^1(\xi)$, is defined in terms of the
components of output functions and their Lie derivatives along the vector
field $\bar{f}(\xi)$. Specifically,

$$T^1(\xi) = \begin{bmatrix} h_1^1 & L_{\bar{f}} h_1^1 & L_{\bar{f}}^2 h_1^1 & \cdots & h_{n_1}^1 & L_{\bar{f}} h_{n_1}^1 & L_{\bar{f}}^2 h_{n_1}^1 & h_1^2 & \cdots & h_{n_2}^2 \end{bmatrix}^T.$$

The remaining part of the transformation, $\eta = T^2(\xi)$, is so defined to make
$T(\xi)$ a diffeomorphism. In the above, we have used the notation

$$
\begin{aligned}
h^1 &= \begin{bmatrix} h_1^1 & h_2^1 & \cdots & h_{n_1}^1 \end{bmatrix}^T \\
h^2 &= \begin{bmatrix} h_1^2 & h_2^2 & \cdots & h_{n_2}^2 \end{bmatrix}^T.
\end{aligned}
$$

After applying the nonlinear feedback and the diffeomorphic transformation, Equation (47), we convert the system described by Equations (42) and (43) into a system of the following structure

$$\dot{z} = Az + Bv \tag{48}$$
$$\dot{\eta} = \tilde{f}(\eta, z) \tag{49}$$
$$y = Cz \tag{50}$$

where A, B and C are constant matrices with appropriate dimensions. We note that we have decomposed the original system into two subsystems. The first subsystem, Equation (48), is linear and its state, z, linearly affects the output y as indicated by Equation (50). The second subsystem is the zero dynamics of the system whose structure is determined by the choice of $T^2(\xi)$. Furthermore, linear state Equation (48) with output Equation (50) is decomposed into decoupled linear subsystems, each of which controls one component of outputs [12]. The z space is partitioned into $n_1 + n_2$ subspaces

$$z = \begin{bmatrix} z^1 & \cdots & z^{n_1} & z^{n_1+1} & \cdots & z^{n_1+n_2} \end{bmatrix}^T$$

with $z^i, i = 1, \ldots, n_1$, having three components and $z^i, i = n_1 + 1, \ldots, n_1 + n_2$, having only one component. These decoupled subsystems possess the structure indicated below.

$$\dot{z}^i = \begin{bmatrix} \dot{z}^i_1 \\ \dot{z}^i_2 \\ \dot{z}^i_3 \end{bmatrix} = \begin{bmatrix} 0 & 1 & 0 \\ 0 & 0 & 1 \\ 0 & 0 & 0 \end{bmatrix} \begin{bmatrix} z^i_1 \\ z^i_2 \\ z^i_3 \end{bmatrix} + \begin{bmatrix} 0 \\ 0 \\ 1 \end{bmatrix} v_i \tag{51}$$
$$y_i = \begin{bmatrix} 1 & 0 & 0 \end{bmatrix} z^i, \qquad i = 1, \ldots, n_1 \tag{52}$$

and

$$\dot{z}^i = \begin{bmatrix} 0 \end{bmatrix} z^i + \begin{bmatrix} 1 \end{bmatrix} v_i \tag{53}$$
$$y_i = \begin{bmatrix} 1 \end{bmatrix} z^i, \qquad i = n_1 + 1, \ldots, n_1 + n_2. \tag{54}$$

To demonstrate the validity of the nonlinear feedback and the state transformation, in what follows we show that they do linearize and decouple the nonlinear dynamics of the two robots. That is, we do have the linear subsystems described by Equations (51) through (54). Under the state coordinate z, we compute $\dot{z}^i$ for $i = 1, \ldots, n_1$,

$$\dot{z}^i = \frac{dz^i}{dt} = \frac{d}{dt} \begin{bmatrix} h^1_i \\ L_{\tilde{f}} h^1_i \\ L^2_{\tilde{f}} h^1_i \end{bmatrix} = \frac{\partial}{\partial \xi} \begin{bmatrix} h^1_i \\ L_{\tilde{f}} h^1_i \\ L^2_{\tilde{f}} h^1_i \end{bmatrix} \frac{d\xi}{dt}$$

$$
= \begin{bmatrix} \frac{\partial h_i^1}{\partial \xi} \\ \frac{\partial L_{\bar f} h_i^1}{\partial \xi} \\ \frac{\partial L_{\bar f}^2 h_i^1}{\partial \xi} \end{bmatrix} (\bar f(\xi) + \bar g(\xi) w) = \begin{bmatrix} L_{\bar f} h_i^1 \\ L_{\bar f}^2 h_i^1 \\ L_{\bar f}^3 h_i^1 \end{bmatrix} + \begin{bmatrix} L_{\bar g} h_i^1 \\ L_{\bar g} L_{\bar f} h_i^1 \\ L_{\bar g} L_{\bar f}^2 h_i^1 \end{bmatrix} w
$$

$$
= \begin{bmatrix} z_2^i \\ z_3^i \\ 0 \end{bmatrix} + \begin{bmatrix} 0 \\ 0 \\ L_{\bar f}^3 h_i^1 \end{bmatrix} + \begin{bmatrix} 0 \\ 0 \\ L_{\bar g} L_{\bar f}^2 h_i^1 \end{bmatrix} w.
$$

In the above, we used equalities, $L_{\bar g} h_i^1 = 0$ and $L_{\bar g} L_{\bar f} h_i^1 = 0$, which are shown in Appendix E. We note that the first two rows, defining $\dot z_1^i$ and $\dot z_2^i$, are linear and also match with the corresponding rows in Equation (51). It is the third row, defining $\dot z_3^i$, that needs to be linearized by using nonlinear feedback. Likewise, for $i = n_1 + 1, \ldots, n_1 + n_2$, we have

$$
\dot z^i = \frac{dz^i}{dt} = \frac{dh_{i-n_1}^2}{dt} = \frac{\partial h_{i-n_1}^2}{\partial \xi} \frac{d\xi}{dt}
$$

$$
= \frac{\partial h_{i-n_1}^2}{\partial \xi} (\bar f(\xi) + \bar g(\xi) w) = L_{\bar f} h_{i-n_1}^2 + L_{\bar g} h_{i-n_1}^2 w.
$$

We write together the third rows in the first part $(i = 1, \ldots, n_1)$ and all the second part $(i = n_1 + 1, \ldots, n_1 + n_2)$

$$
\begin{bmatrix} \dot z_3^1 \\ \vdots \\ \dot z_3^{n_1} \\ \dot z^{n_1+1} \\ \vdots \\ \dot z^{n_1+n_2} \end{bmatrix} = \begin{bmatrix} L_{\bar f}^3 h_1^1 \\ \vdots \\ L_{\bar f}^3 h_{n_1}^1 \\ L_{\bar f} h_1^2 \\ \vdots \\ L_{\bar f} h_{n_2}^2 \end{bmatrix} + \begin{bmatrix} L_{\bar g} L_{\bar f}^2 h_1^1 \\ \vdots \\ L_{\bar g} L_{\bar f}^2 h_{n_1}^1 \\ L_{\bar g} h_1^2 \\ \vdots \\ L_{\bar g} h_{n_2}^2 \end{bmatrix} w
$$

$$
= \begin{bmatrix} L_{\bar f}^3 h^1 \\ L_{\bar f} h^2 \end{bmatrix} + \begin{bmatrix} L_{\bar g} L_{\bar f}^2 h^1 \\ L_{\bar g} h^2 \end{bmatrix} w = \begin{bmatrix} L_{\bar f}^3 h^1 \\ L_{\bar f} h^2 \end{bmatrix} + M(\xi) w
$$

$$
= \begin{bmatrix} L_{\bar f}^3 h^1 \\ L_{\bar f} h^2 \end{bmatrix} + M(\xi)(\alpha(\xi) + \beta(\xi) v).
$$

If $\alpha(\xi)$ and $\beta(\xi)$ satisfy Equations (45) and (46), the most right hand side of the above equation evaluates to be v. Therefore, under the state coordinate z, the system is linear and decoupled. Equations (51) through (54) are the representation of the linearized system.

Up to this point, the control design of two robot manipulators is made equivalent to a design problem of decoupled linear subsystems described by Equations (51), (52), (53), and (54). Figure 2 depicts the block diagram of the dynamic nonlinear feedback as applied to two robot manipulators. In

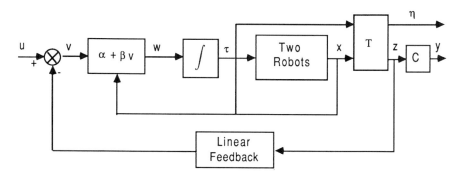

Figure 2: The Block Diagram of the Dynamic Nonlinear Feedback Control of Two Cooperative Manipulators

this figure, the relation from v to z (and hence to y since C is a constant matrix) is linear. A linear feedback may be further designed to satisfy the performance requirement by using standard design methods for linear systems. Since these linear subsystems are controllable, their eigenvalues can be placed anywhere by using a proper constant feedback. In particular, stability of these subsystems, and consequently input-output stability of the two robot system, is guaranteed by placing the eigenvalues on the left half plane. Stability of Equation (49) does not affect the input-output behavior but does affect the internal behavior of the system, which is to be further studied.

VI. Conclusions

In this chapter, nonlinear feedback for force control of a constrained manipulator and coordinated control of two manipulators are developed. The two problems are linked together by the fact that in both cases motions of manipulators are constrained and force control is needed for achieving the required tasks. By applying a nonlinear feedback, it is shown that simultaneous motion and force control of a constrained manipulator or two cooperative manipulators is converted into the design problem of decoupled linear subsystems.

It is noted that in both problems output equations depend directly on inputs. Because of this, dynamic state feedback has been used in place of static state feedback for pure position control of manipulators. It is assumed in both problems that dynamic models of manipulators are available. In control of a constrained manipulator, it is further assumed that a model of the environment is available and that the environment is smooth, frictionless, and stiff.

VII.   Appendices

A.   The Derivation of the Reduced Equations of Motion

In this appendix we derive the closed-form expression of a, b, F_1 and F_2 in Equations (14) and (15). Let us differentiate Equation (4) once more with respect to the time

$$J_s J_p \ddot{q} + \dot{J}_s J_p \dot{q} + J_s \dot{J}_p \dot{q} = 0. \tag{55}$$

From Equation (10), we have

$$\ddot{q} = D^{-1}(q)(\tau - E + J_p^T J_s^T F_n). \tag{56}$$

Now multiplying both sides of Equation (56) by $J_s J_p$ and making use of Equation (55), we obtain

$$J_s J_p D^{-1}(q)(\tau - E + J_p^T J_s^T F_n) = -\dot{J}_s J_p \dot{q} - J_s \dot{J}_p \dot{q}.$$

From the above equation, we obtain F_n as follows:

$$
\begin{aligned}
F_n &= -(J_s J_p D^{-1} J_p^T J_s^T)^{-1}(J_s J_p D^{-1}(\tau - E) + \dot{J}_s J_p \dot{q} + J_s \dot{J}_p \dot{q}) \\
&= -(J_s J_p D^{-1} J_p^T J_s^T)^{-1}(\dot{J}_s J_p \dot{q} + J_s \dot{J}_p \dot{q} - J_s J_p D^{-1} E) \\
&\quad -(J_s J_p D^{-1} J_p^T J_s^T)^{-1} J_s J_p D^{-1} \tau \\
&\triangleq F_1 + F_2 \tau.
\end{aligned}
\tag{57}
$$

Equation (11) can be rewritten as

$$N(q)\ddot{q}^1 + \bar{E}(q, \dot{q}) = \tau + J_p^T J_s^T F_n \tag{58}$$

where

$$N(q) = D(q) \begin{bmatrix} I \\ M_1 \end{bmatrix}.$$

Multiplying both sides of Equation (58) by $N^T(q)$

$$N^T(q)N(q)\ddot{q}^1 + N^T(q)\bar{E}(q, \dot{q}) = N^T(q)(\tau + J_p^T J_s^T F_n) \tag{59}$$

we get

$$\ddot{q}^1 = (N^T N)^{-1} N^T (\tau - \bar{E} + J_p^T J_s^T F_n). \tag{60}$$

Substituting Equation (57) into Equation (60), we have

$$
\begin{aligned}
\ddot{q}^1 &= (N^T N)^{-1} N^T [-\bar{E} - J_p^T J_s^T (J_s J_p D^{-1} J_p^T J_s^T)^{-1}(\dot{J}_s J_p \dot{q} + J_s \dot{J}_p \dot{q} \\
&\quad - J_s J_p D^{-1} E)] + (N^T N)^{-1} N^T (I - J_p^T J_s^T [(J_s J_p D^{-1} J_p^T J_s^T)^{-1} J_s J_p D^{-1}]\tau \\
&\triangleq a + b\tau
\end{aligned}
\tag{61}
$$

Since we solved $\ddot{q}^1$ and F_n from Equation (13) for arbitrary nonsingular configurations of the manipulator and arbitrary input torques, it follows that the matrix $\bar{D}(q)$ is nonsingular if the manipulator is at nonsingular configurations.

B.   The Relative Degrees and the Decoupling Matrix for the Force Controlled Manipulator

Here we compute the relative degrees of the system described by Equations (17) and (18). The structure of f, g and h is important in the following computations. The relative degrees are defined by [8]

$$\rho_i = \min\{s \,|\, L_g L_f^s h_i \neq 0\} \tag{62}$$

where h_i is the ith component of the outputs h. For the first part of the outputs, h^1, the following computation is straightforward.

$$L_g h^1 = 0 \tag{63}$$

$$L_f h^1 = \frac{\partial h^1}{\partial x^1} x^2 \tag{64}$$

$$L_g L_f h^1 = 0 \tag{65}$$

$$L_f^2 h^1 = \frac{\partial}{\partial x^1}\left(\frac{\partial h^1}{\partial x^1} x^2\right)x^2 + \frac{\partial h^1}{\partial x^1}(a + bx^3) \tag{66}$$

$$L_g L_f^2 h^1 = \frac{\partial h^1}{\partial x^1} b \neq 0. \tag{67}$$

Therefore, the relative degrees of h^1 are all equal to two. Those of h^2 are equal to zero since

$$L_g h^2 = F_2 \neq 0.$$

It follows that the decoupling matrix then is [8]

$$\Phi(x) = \begin{bmatrix} L_g L_f^2 h^1 \\ L_g h^2 \end{bmatrix} = \begin{bmatrix} \frac{\partial h^1}{\partial x^1} b \\ F_2 \end{bmatrix}$$

$$= \begin{bmatrix} \frac{\partial h^1}{\partial x^1} & 0 \\ 0 & I \end{bmatrix} \begin{bmatrix} b \\ F_2 \end{bmatrix} = \begin{bmatrix} \frac{\partial h^1}{\partial x^1} & 0 \\ 0 & I \end{bmatrix} \bar{D}^{-1}.$$

C.   The Verification of Linearization and Output Decoupling

To demonstrate the validity of the nonlinear feedback and the state transformation, in this appendix we show that they do linearize and decouple the nonlinear dynamics of the constrained manipulator. Under the transformed state coordinate z, we compute $\dot{z}^i$ for $i = 1, \ldots, n - r$,

$$
\dot{z}^i = \frac{dz^i}{dl} = \frac{d}{dl} \begin{bmatrix} h_i^1 \\ L_f h_i^1 \\ L_f^2 h_i^1 \end{bmatrix} = \frac{\partial}{\partial x} \begin{bmatrix} h_i^1 \\ L_f h_i^1 \\ L_f^2 h_i^1 \end{bmatrix} \frac{dx}{dt}
$$

$$
= \begin{bmatrix} \frac{\partial h_i^1}{\partial x} \\ \frac{\partial L_f h_i^1}{\partial x} \\ \frac{\partial L_f^2 h_i^1}{\partial x} \end{bmatrix} (f(x) + g(x)u) = \begin{bmatrix} L_f h_i^1 \\ L_f^2 h_i^1 \\ L_f^3 h_i^1 \end{bmatrix} + \begin{bmatrix} L_g h_i^1 \\ L_g L_f h_i^1 \\ L_g L_f^2 h_i^1 \end{bmatrix} u
$$

$$
= \begin{bmatrix} z_2^i \\ z_3^i \\ 0 \end{bmatrix} + \begin{bmatrix} 0 \\ 0 \\ L_f^3 h_i^1 \end{bmatrix} + \begin{bmatrix} 0 \\ 0 \\ L_g L_f^2 h_i^1 \end{bmatrix} u.
$$

In the above, we used equalities, $L_g h_i^1 = 0$ and $L_g L_f h_i^1 = 0$, which are from Equations (63) and (65) in Appendix B. We note that the first two rows, defining $\dot{z}_1^i$ and $\dot{z}_2^i$, are linear and also match with the corresponding rows in Equation (26). It is the third row, defining $\dot{z}_3^i$, that needs to be linearized by using the nonlinear feedback. Likewise, for $i = n - r + 1, \ldots, n$, we have

$$
\dot{z}^i = \frac{dz^i}{dt} = \frac{dh_{i-n+r}^2}{dt} = \frac{\partial h_{i-n+r}^2}{\partial x} \frac{dx}{dt}
$$

$$
= \frac{\partial h_{i-n+r}^2}{\partial x} (f(x) + g(x)u) = L_f h_{i-n+r}^2 + L_g h_{i-n+r}^2 u.
$$

We write together the third rows in the first part $(i = 1, \ldots, n - r)$ and all the second part $(i = n - r + 1, \ldots, n)$

$$
\begin{bmatrix} \dot{z}_3^1 \\ \vdots \\ \dot{z}_3^{n-r} \\ \dot{z}^{n-r+1} \\ \vdots \\ \dot{z}^n \end{bmatrix} = \begin{bmatrix} L_f^3 h_1^1 \\ \vdots \\ L_f^3 h_{n-r}^1 \\ L_f h_1^2 \\ \vdots \\ L_f h_r^2 \end{bmatrix} + \begin{bmatrix} L_g L_f^2 h_1^1 \\ \vdots \\ L_g L_f^2 h_{n-r}^1 \\ L_g h_1^2 \\ \vdots \\ L_g h_r^2 \end{bmatrix} u
$$

$$
= \begin{bmatrix} L_f^3 h^1 \\ L_f h^2 \end{bmatrix} + \begin{bmatrix} L_g L_f^2 h^1 \\ L_g h^2 \end{bmatrix} u = \begin{bmatrix} L_f^3 h^1 \\ L_f h^2 \end{bmatrix} + \Phi(x)u
$$

$$
= \begin{bmatrix} L_f^3 h^1 \\ L_f h^2 \end{bmatrix} + \Phi(x)(\alpha(x) + \beta(x)v).
$$

If $\alpha(x)$ and $\beta(x)$ satisfy Equations (20) and (21), the most right hand side of the above equation evaluates to be v. Therefore, under the transformed state coordinate z, the system of the constrained manipulator is linearized and decoupled. Equations (26) through (29) are the representation of the linearized system.

D.   The Nonsingularity of Matrix R

Assuming that both manipulators have 6 joints and are not in singular configuration, we will show that the matrix R appeared in Equation (35) is invertible. Let us rewrite the definition of R below

$$R = D_a^{-1} J_a^T + \phi_1 D_b^{-1} J_b^T. \tag{68}$$

Premultiplying Equation (68) by $J_a^{-T} D_a$ and postmultiplying by $J_b^{-T} D_b J_b^{-1}$, we obtain

$$J_a^{-T} D_a R J_b^{-T} D_b J_b^{-1} = J_b^{-T} D_b J_b^{-1} + J_a^{-T} D_a \phi_1 J_b^{-1}. \tag{69}$$

The first term on the right hand side of Equation (69) is the Cartesian mass matrix of arm "b" [13, p.212], and the second term is that of arm "a" by observing $\phi_1 J_b^{-1} = J_a^{-1}$. Since both terms on the right hand side of Equation (69) is positive definite, so is the sum. Therefore, the product of six matrices on the left hand side is positive definite. It follows that matrix R is nonsingular.

E.   The Decoupling Matrix of the Two-Arm System

We here show the detailed computation of the decoupling matrix and prove its nonsingularity. We first compute relative degrees of the system which are defined in [7, 12]

$$\rho_i = \min\{s \left| L_{\bar{g}} L_{\bar{f}}^{s-1} h_i \neq 0\right.\} \tag{70}$$

where h_i is the ith component of output h. The decoupling matrix for the present system with $n_1 + n_2$ outputs is defined as

$$M(\xi) = \begin{bmatrix} L_{\bar{g}} L_{\bar{f}}^{\rho_1-1} h_1 \\ \vdots \\ L_{\bar{g}} L_{\bar{f}}^{\rho_{n_1+n_2}-1} h_{n_1+n_2} \end{bmatrix}. \tag{71}$$

For the first part of outputs, h^1, the following computation is straightforward.

$$L_{\bar{g}} h^1 = \frac{\partial h^1}{\partial \xi} \bar{g} = 0$$

$$L_{\bar{f}} h^1 = \frac{\partial h^1}{\partial \xi} \bar{f} = \frac{\partial h^1}{\partial x^1} x^2$$

$$L_{\bar{g}} L_{\bar{f}} h^1 = \frac{\partial (L_{\bar{f}} h^1)}{\partial \xi} \bar{g} = 0$$

$$L_f^2 h^1 = \frac{\partial(L_f h^1)}{\partial \xi}\bar{f} = \frac{\partial(\frac{\partial h^1}{\partial x^1}x^2)}{\partial x^1}x^2 + \frac{\partial h^1}{\partial x^1}(\tilde{E}(x) + \tilde{D}^{-1}(x^1)\tau)$$

$$L_{\bar{g}}L_f^2 h^1 = \frac{\partial(L_f^2 h^1)}{\partial \xi}\bar{g} = \frac{\partial h^1}{\partial x^1}\tilde{D}^{-1}(x^1) \neq 0.$$

Therefore, the relative degrees of h^1 are all equal to three. Those of h^2 are equal to one since

$$L_{\bar{g}}h^2 = \frac{\partial h^2}{\partial \xi}\bar{g} = F_2 \neq 0.$$

It follows that the decoupling matrix is

$$M(\xi) = \begin{bmatrix} L_{\bar{g}}L_f^2 h^1 \\ L_{\bar{g}}h^2 \end{bmatrix} = \begin{bmatrix} \frac{\partial h^1}{\partial x^1}\tilde{D}^{-1}(x^1) \\ F_2 \end{bmatrix}. \tag{72}$$

Next we show that the decoupling matrix $M(\xi)$ is generically nonsingular. By generic nonsingularity, we mean that the set composed of the singular points is of Lebesgue measure zero. For example, the Jacobian matrix of a nonredundant manipulator is generically nonsingular. Without loss of generality, we assume that h^1 (position and orientation vector of the grasped object) is a function only of q_a. The upper part of $M(\xi)$ can be futher written in the following form

$$\frac{\partial h^1}{\partial x^1}\tilde{D}^{-1}(x^1) = \frac{\partial h^1}{\partial q_a}D_a^{-1}([I_a \mid 0] - J_a^T F_2).$$

Replacing F_2 by its contents and using a shorthand notation $R = D_a^{-1}J_a^T + \phi_1 D_b^{-1}J_b^T$, we obtain an expression of the decoupling matrix.

$$M(\xi) =$$
$$\begin{bmatrix} \frac{\partial h^1}{\partial q_a}D_a^{-1}(I_a - J_a^T R^{-1}D_a^{-1}) & \frac{\partial h^1}{\partial q_a}D_a^{-1}J_a^T R^{-1}\phi_1 D_b^{-1} \\ R^{-1}D_a^{-1} & -R^{-1}\phi_1 D_b^{-1} \end{bmatrix} =$$
$$\begin{bmatrix} \frac{\partial h^1}{\partial q_a}D_a^{-1}(D_a R - J_a^T) & \frac{\partial h^1}{\partial q_a}D_a^{-1}J_a^T \\ I & -I \end{bmatrix} \begin{bmatrix} R^{-1}D_a^{-1} & 0 \\ 0 & R^{-1}\phi_1 D_b^{-1} \end{bmatrix}$$

The second matrix above is generically nonsingular. The nonsingularity of the first matrix above is examined easily by adding the second "column" to the first "column", resulting in an upper triangular matrix

$$\begin{bmatrix} \frac{\partial h^1}{\partial q_a}R & \frac{\partial h^1}{\partial q_a}D_a^{-1}J_a^T \\ 0 & -I \end{bmatrix}$$

which is nonsingular provided that $\frac{\partial h^1}{\partial q_a}$ is. In practice, we always choose independent (actually orthogonal) quantities to represent the position and orientation of the object in the task space. It follows that the Jacobian matrix of h^1, $\frac{\partial h^1}{\partial q_a}$, will be generically nonsingular. Therefore, the decoupling matrix of the two arm system is generically nonsingular.

References

[1] N. Harris McClamroch. Singular systems of differential equations as dynamic models for constrained robot systems. In *Proceedings of 1988 International Conference on Robotics and Automation*, pages 21–28, San Francisco, CA, 1986.

[2] N. Harris McClamroch. A singular perturtation approach to modeling and control of manipulators constrained by a stiff environment. In *The 28th IEEE Conference on Decision and Control*, pages 2407–2411, Tampa, Florida, December 1989.

[3] James K. Mills and Andrew A. Goldenberg. Force and position control of manipulators during constrained motion tasks. *IEEE Transactions on Robotics and Automation*, 5(1):30–46, February 1989.

[4] M. A. Unseren and A. J. Koivo. Reduced order model and decoupled control architecture for two manipulators holding an object. In *Proceedings of 1989 International Conference on Robotics and Automation*, pages 1240–1245, Scottsdale, Arizona, May 1989.

[5] A. J. Koivo. *Fundamentals for Control of Robotic Manipulators*. John Wiley & Sons, Inc., 1989.

[6] A. K. Bejczy. *Robot Arm Dynamics and Control*. Technical Report 33-669, Jet Propulsion Laboratory, 1974.

[7] A. Isidori. *Nonlinear Control Systems: An Introduction*. Springer-Verlag, Berlin, New York, 1985.

[8] T. J. Tarn, A. K. Bejczy, A. Isidori, and Y. Chen. Nonlinear feedback in robot arm control. In *Proceedings of 23rd IEEE Conference on Decision and Control*, pages 736–751, Las Vegas, Nevada, December 1984.

[9] E. Freund. Fast nonlinear control with arbitrary pole-placement for industrial robots and manipulators. *International Journal of Robotic Research*, 1(1):65–78, 1982.

[10] A. Isidori, C. H. Moog, and A. De Luca. A sufficient condition for full linearization via dynamic state feedback. In *Proceedings of 25th IEEE Conference on Decision and Control*, pages 203–208, Athens, Greece, December 1986.

[11] T. J. Tarn, A. K. Bejczy, and X. Yun. Design of dynamic control of two cooperating robot arms: closed chain formulation. In *Proceedings of 1987 International Conference on Robotics and Automation*, pages 7–13, Raleigh, North Carolina, March 1987.

[12] Y. L. Chen. *Nonlinear Feedback and Computer Control of Robot Arms*. PhD thesis, Washington University, St. Louis, Missouri, December 1984.

[13] John J. Craig. *Introduction to Robotics: Mechanics and Control*. Addison Wesley Publishing Company, 2nd edition, 1989.

SYSTOLIC ARCHITECTURES FOR
DYNAMIC CONTROL OF MANIPULATORS

Masoud Amin-Javaheri

GMFanuc Robotics Corporation
2000 South Adams Road
Auburn Hills, Michigan 48057-2090

I. Introduction

A major challenge in effectively realizing advanced control schemes for robotic systems is the difficulty of implementing the kinematic and dynamic equations required for coordination and control in real time. While the total amount of computation appears to be somewhat less than that of many scientific computations, implementation in real-time implies that these same computations must be repeated at rates of hundreds or perhaps thousands of times per second. This, then, results in an important computational problem in robotics control, and consequently, the computational aspects of dynamic control techniques are the main thrust of this chapter.

While the dynamics of manipulators were not considered in earlier control schemes, the more recent development of model-based control [1, 2] includes the dynamics in an integral way. Typically, model-based control schemes are based on an Inverse Dynamics

CONTROL AND DYNAMIC SYSTEMS, VOL. 40

computation that determines the required actuator torques for a
desired system trajectory. In model-based control, accurate modeling
of the dynamic parameters of the plant is very crucial for dynamic
stability. Equally important is the rate at which the dynamics may
be computed.

The general dynamic equations of motion for a single N-degree-of-
freedom open-chain manipulator, in joint space, with physical
constraint forces and moments between the links eliminated, may be
represented in the following form:

$$\mathcal{T} = \mathbf{H}(\mathbf{q})\ddot{\mathbf{q}} + \mathbf{C}(\mathbf{q}, \dot{\mathbf{q}})\dot{\mathbf{q}} + \mathbf{G}(\mathbf{q}) + \mathbf{J}^T(\mathbf{q})\mathbf{f} \qquad (1)$$

where

$\mathcal{T}$	$N \times 1$	vector of joint torques (forces),
$\mathbf{q}, \dot{\mathbf{q}}, \ddot{\mathbf{q}}$	$N \times 1$	vectors of joint positions, rates, and accelerations,
$\mathbf{C}(\mathbf{q}, \dot{\mathbf{q}})$	$N \times N$	coriolis and centrifugal force matrix,
$\mathbf{G}(\mathbf{q})$	$N \times 1$	vector of gravitational forces,
$\mathbf{J}(\mathbf{q})$	$6 \times N$	Jacobian matrix,
$\mathbf{f}$	6×1	vector of external forces and moments exerted by link N,
$\mathbf{H}(\mathbf{q})$	$N \times N$	symmetric, positive-definite inertia matrix

Inverse Dynamics problem has generally been implemented using
either a recursive Newton-Euler or a Lagrange-Euler formulation, since
explicit determination of the terms of the dynamic equations of
motion, which includes determination of the inertia matrix and the
Jacobian matrix, is not computationally efficient. However, if systolic
architectures in conjunction with multiple rates for the computation
of the individual terms are considered, it may be that the latter approach, which involves explicit computation of the inertia matrix, the
Jacobian matrix, the coriolis and centrifugal matrix, and the gravitational vector, will be the most efficient.

Systolic architectures for computation of the Jacobian matrix are
fully explored in [3]. Multiple rate concept and its application
toward the computation of the manipulator inertia matrix are presented in [4]. Thus, this chapter focuses on the development of several

systolic architectures to compute the inertia matrix, which is the most computationally intensive term in Eq. (1), in real time. Although the development of the systolic architectures presented here is specific to the computation of the inertia matrix, the concept may be applied to a wide range of robotics computations.

In addition to computation of Inverse Dynamics, implementation of a number of dynamic control schemes is based upon computation of the inertia matrix so as to decouple the dynamics along the several axes of the manipulator [5, 6, 7]. This allows either linear or nonlinear control schemes to be more effectively applied. Other applications in which the inertia matrix has been explicitly used include surface tracking and object identification using force control [8], and computation of the collision effects between a manipulator and its environment or between two manipulators in a shared work space [9].

This chapter presents the development of several systolic architectures used to compute the inertia matrix to reduce the total computational delay.

Separate processors may be assigned to compute individual rows, columns, etc. Such mapping is used to partition the processes among several processors connected in a systolic configuration. Thus, the total computational delay is expected to be reduced significantly. A VLSI-based (Very Large Scale Integrated) Robotics Processor [3] is used as the basic processing element in the systolic array for computing the inertia matrix. It contains a pipelined 32-bit floating-point multiplier and a 32-bit floating-point adder/subtractor functioning in parallel to achieve maximum throughput. A detailed description of this processing element, which is used primarily to exploit fine-grain parallelism, is given in a later section.

The algorithm implemented is based upon the determination of the mass, center of mass, and inertia of a series of composite rigid bodies for the manipulator [10]. These composite rigid bodies are made up of sets of links at the end of the manipulator that are assumed to be fixed with respect to each other (no relative joint movement). The equations for the algorithm are developed and summarized in a form that shows much of the parallelism inherent in the algorithm. The computation required is reduced from $O(N^2)$ for a single processor to $O(N)$ for N processors.

In the next section, the $O(N)$ parallel algorithm is given. In the section following, the architecture of the VLSI-based Robotics Processor is described. Then the development of various systolic architectures, using 1, N, and $N(N + 1)/2$ processors, for computing the inertia matrix, is presented. The various configurations are evaluated in terms of the compute time delay, which is of $O(N)$ for both the N and $N(N+1)/2$ processor cases. In addition to this basic delay parameter, other important criteria are used to evaluate the various configurations. These include the initiation rate (sampling rate), CPU utilization, speedup, and on-chip memory required. In all cases, the I/O communication time and idle time, required to synchronize the processors, are fully considered in the evaluation. Finally, the work is summarized and conclusions are made.

II. $O(N)$ Parallel Algorithm for the Inertia Matrix

In this section, the development of a linear recursive algorithm used for computing the manipulator inertia matrix is fully explained. The algorithm is derived from [10] but expressed in a form that explicitly shows the reduction of the computation from $O(N^2)$ to $O(N)$ when N processors are used. Details of the development are given to provide additional insight into the algorithm so that subsequent selection of appropriate systolic architectures may be facilitated.

The algorithm may be development by considering the general dynamic equations of motion given in Eq. (1). For a given set of joint position, velocity, and acceleration vectors for a specified trajectory, and for given external forces and moments, the required torques to be applied by the actuators may be computed from Eq. (1). This is the problem of Inverse Dynamics that computes the actuating signals for a given velocity and acceleration profile.

The most efficient algorithm for Inverse Dynamics is generally considered to be based on the Newton-Euler formulation [11, 12] that does not require computation of the individual terms of Eq. (1). Briefly, the formulation employs a two-step process, a forward recursion and a backward recursion, to give a computation that is linearly related to N, the number of degrees of freedom.

The forward recursion ($i = 1$ to N) begins at the base of the manipulator. The velocity and acceleration of a link are computed based on the values of these variables for the previous link and the relative values at the connecting joint. The resultant forces and moments exerted on each link are first computed in the backward recursion $(i = N$ to 1) by applying Newton's and Euler's equations. Then, beginning with the external forces and moments applied to the end-effector, the forces and moments applied to a link (at the near end) are computed based on the resultant values for the link and those applied by it to the previous link.

A straightforward approach in solving for $\mathbf{H(q)}$ is to apply Inverse Dynamics to Eq. (1) $N + 1$ times. To do so, Eq. (1) may be rewritten in the following form:

$$\mathcal{T} = \mathbf{H(q)\ddot{q}} + \mathbf{b} \tag{2}$$

where

$$\mathbf{b} = \mathbf{C(q, \dot{q})\dot{q}} + \mathbf{G(q)} + \mathbf{J}^T\mathbf{(q)f}. \tag{3}$$

The vector $\mathbf{b}$ may be solved for by simply setting $\mathbf{\ddot{q}}$ equal to $\mathbf{0}$ when applying Inverse Dynamics. That is,

$$\mathbf{b} = \mathcal{T}\big|_{\mathbf{\ddot{q}}=[00...0]^T} . \tag{4}$$

Now, the columns of $\mathbf{H(q)}$ may be solved for in Eq. (2) by applying a unit vector of acceleration to the joints. That is,

$$\mathbf{H}_1 = (\mathcal{T} - \mathbf{b})\big|_{\mathbf{\ddot{q}}=[100...]^T} \tag{5}$$

where $\mathbf{H}_1$ is the first column of the $\mathbf{H(q)}$ matrix and in general,

$$\mathbf{H}_i = (\mathcal{T} - \mathbf{b})\big|_{\mathbf{\ddot{q}}=[00...1...0]^T} . \tag{6}$$

By repeating the above process N times (not necessarily recursively), all of the elements of $\mathbf{H(q)}$ may be extracted.

Further simplifications may be made to decrease the amount of computation needed. One obvious simplification is to set $\mathbf{\dot{q}}$, $\mathbf{G(q)}$, and $\mathbf{f}$ to zero when applying Inverse Dynamics. This itself eliminates the need to compute $\mathbf{b}$ and all other subsequent subtractions of $\mathbf{b}$ from $\mathcal{T}$. Also, it should be noted that the inertia matrix is a symmetric matrix. That is, the diagonal and the upper off-diagonal elements of $\mathbf{H(q)}$ are all that are needed to fully specify it. This indicates that

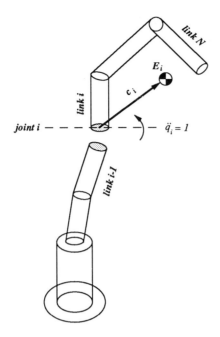

Figure 1: Two Composite Rigid Body Modeling of an Open-Chain Mechanism.

all of the elements of the $\mathcal{T}$ vector need not be obtained in each step; this further simplifies the computation.

One additional concept that is of computational significance is that the application of a unit acceleration to a joint (for instance $\ddot{q}_i = 1$ at joint i), with all joint velocities and other joint accelerations equal to zero, divides the manipulator chain into two sets of composite rigid bodies with one degree of freedom between them, as shown in Fig. 1. Links i through N constitute one of these composite rigid bodies and may be modeled as a single rigid body with a composite mass (M_i), composite center of mass ($\mathbf{c}_i$), and composite moment of inertia ($\mathbf{E}_i$) as shown in Fig. 2. For this single composite rigid body, the forces and moments at joint i, due to a unit acceleration there, may be rather simply computed by applying the Newton-Euler equations to the composite body. While the torques within the composite body

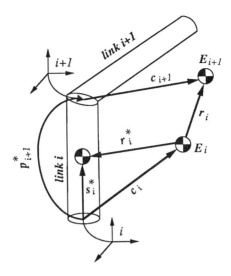

Figure 2: Composite Center of Mass (c_i) and Composite Moment of Inertia ($\mathbf{E}_i$) for links i through N.

are not computed, this is not a problem since they are related to the off-diagonal elements of the lower half of the inertia matrix.

Since the computation of M_i, $\mathbf{c}_i$, and $\mathbf{E}_i$ is an important part of the determination of the inertia matrix, a linear recursive technique has been derived for each and is given as follows. From basic physics principles,

$$M_i = M_{i+1} + m_i \tag{7}$$

where m_i is the mass of link i. Also, $\mathbf{c}_i$, the composite center of mass with respect to the origin of link i coordinates is given as (Fig. 2)

$$\mathbf{c}_i = \frac{1}{M_i} \left\{ m_i \left(\mathbf{s}_i^* \right) + M_{i+1} \left(\mathbf{c}_{i+1} + \mathbf{p}_{i+1}^* \right) \right\} \tag{8}$$

where $\mathbf{p}_{i+1}^*$ denotes the position of the origin of link $i+1$ coordinates with respect to the origin of link i coordinates, and $\mathbf{s}_i^*$ denotes the position of the center of mass of link i with respect to the origin of link i coordinates. Finally, $\mathbf{E}_i$ may be determined by using the three-dimensional version of the parallel axis theorem, which states:

$$\mathbf{I}_O = \mathbf{I}_G + M\left[\left(\mathbf{r}^T\mathbf{r}\right)\mathbf{1} - \left(\mathbf{r}\mathbf{r}^T\right)\right] \tag{9}$$

where $\mathbf{1}$ is the 3×3 identity matrix, $\mathbf{I}_O$ is the inertia matrix of a rigid body about an arbitrary origin O, $\mathbf{I}_G$ is the inertia matrix relative to the center of mass, and $\mathbf{r}$ is the position vector between the two points. Thus, noting Fig. 2:

$$\begin{aligned}\mathbf{E}_i &= \mathbf{E}_{i+1} + M_{i+1}\left[\left(\mathbf{r}_i^T\mathbf{r}_i\right)\mathbf{1} - \left(\mathbf{r}_i\mathbf{r}_i^T\right)\right] + \\ &\quad \mathbf{I}_i + m_i\left[\left(\mathbf{r}_i^{*T}\mathbf{r}_i^*\right)\mathbf{1} - \left(\mathbf{r}_i^*\mathbf{r}_i^{*T}\right)\right]\end{aligned} \tag{10}$$

where $\mathbf{I}_i$ is the inertia matrix of link i at its center of mass.

Links $i - 1$ through 1 constitute the second composite rigid body. For these links ($j \leq i-1$), particular components of the corresponding forces and moments at joint j due to a unit acceleration at joint i ($\mathbf{f}_{j,i}$ and $\mathbf{n}_{j,i}$ respectively), that are needed to keep the links fixed with respect to each other, are simply the elements of the inertia matrix. The Newton-Euler approach may be used to determine the necessary force and moment components (along or about the joint axes). The appropriate equations are developed below for revolute joints while the final equations for both revolute and prismatic joints are given in Table I.

For the upper composite rigid body represented by the three quantities: M_i, $\mathbf{c}_i$, and $\mathbf{E}_i$, the total resultant force ($\mathbf{F}_i$) and moment ($\mathbf{N}_i$) exerted on link i may be computed by applying the Newton-Euler equations:

$$\mathbf{F}_i = M_i\ddot{\mathbf{c}}_i \tag{11}$$

and

$$\mathbf{N}_i = \mathbf{E}_i\dot{\omega}_i + \omega_i \times (\mathbf{E}_i \cdot \omega_i). \tag{12}$$

The vector ω_i is the angular velocity of the ith link (and composite body) and is zero, since all velocities are zero in the analysis. Also, for a unit acceleration at joint i,

$$\dot{\omega}_i = \ddot{\mathbf{q}}_i\mathbf{z}_i = \mathbf{z}_i. \tag{13}$$

Thus,

Table I. $O(N)$ Parallel Algorithm for Computing the Inertia Matrix.

CONST

$$M_{N+1} \;=\; 0$$
$$\mathbf{c}_{N+1} \;=\; \mathbf{0}$$
$$\mathbf{E}_{N+1} \;=\; \mathbf{0}$$
$$\mathbf{z}_0 \;=\; [0 \; 0 \; 1]^T$$
$$\sigma_i \;=\; \begin{cases} 1 & \text{revolute joint} \\ 0 & \text{prismatic joint} \end{cases}$$

BEGIN

{* Computation of composite rigid body parameters and diagonal elements of *}
{*the inertia matrix *}

FOR i := N TO 1 DO
   BEGIN1

$$M_i \quad := \quad M_{i+1} + m_i$$

$$\mathbf{c}_i \quad := \quad \left\{ \frac{1}{M_i} \left[m_i \left(\mathbf{s}_i^* \right) + M_{i+1} \left({}^i U_{i+1}\, \mathbf{c}_{i+1} + {}^i \mathbf{p}_{i+1}^* \right) \right] \right\}$$

$$\mathbf{E}_i \quad := \quad {}^i U_{i+1} \mathbf{E}_{i+1}\, {}^{i+1} U_i +$$
$$M_{i+1} \left[\left({}^i U_{i+1}\, \mathbf{c}_{i+1} + {}^i \mathbf{p}_{i+1}^* - \mathbf{c}_i \right)^T \left({}^i U_{i+1}\, \mathbf{c}_{i+1} + {}^i \mathbf{p}_{i+1}^* - \mathbf{c}_i \right) \mathbf{1} \right.$$
$$\left. - \left({}^i U_{i+1}\, \mathbf{c}_{i+1} + {}^i \mathbf{p}_{i+1}^* - \mathbf{c}_i \right) \left({}^i U_{i+1}\, \mathbf{c}_{i+1} + {}^i \mathbf{p}_{i+1}^* - \mathbf{c}_i \right)^T \right]$$
$$+ \mathbf{I}_i + m_i \left[\left(\mathbf{s}_i^* - \mathbf{c}_i \right)^T \left(\mathbf{s}_i^* - \mathbf{c}_i \right) \mathbf{1} - \left(\mathbf{s}_i^* - \mathbf{c}_i \right) \left(\mathbf{s}_i^* - \mathbf{c}_i \right)^T \right]$$

$$\mathbf{F}_i \quad := \quad \sigma_i \left(\mathbf{z}_0 \times M_i\, \mathbf{c}_i \right) + \bar{\sigma}_i \left(M_i\, \mathbf{z}_0 \right)$$

$$\mathbf{N}_i \quad := \quad \sigma_i \left(\mathbf{E}_i\, \mathbf{z}_0 \right)$$

$$\mathbf{f}_{i,i} \quad := \quad \mathbf{F}_i$$

$$\mathbf{n}_{i,i} \quad := \quad \mathbf{N}_i + \mathbf{c}_i \times \mathbf{F}_i$$

$$H_{i,i} \quad := \quad \sigma_i \left(\mathbf{n}_{i,i} \cdot \mathbf{z}_0 \right) + \bar{\sigma}_i \left(\mathbf{f}_{i,i} \cdot \mathbf{z}_0 \right)$$

   END1

{* Computation of off-diagonal elements of the inertia matrix *}

FOR ALL i := N TO 1 DO
   FOR j := i-1 TO 1 DO
      BEGIN2

$$\mathbf{f}_{j,i} \quad := \quad {}^j U_{j+1}\, \mathbf{f}_{j+1,i}$$

$$\mathbf{n}_{j,i} \quad := \quad {}^j U_{j+1} \mathbf{n}_{j+1,i} + {}^j \mathbf{p}_{j+1}^* \times \mathbf{f}_{j+1,i}$$

$$H_{j,i} \quad := \quad \sigma_j \left(\mathbf{n}_{j,i} \cdot \mathbf{z}_0 \right) + \bar{\sigma}_j \left(\mathbf{f}_{j,i} \cdot \mathbf{z}_0 \right)$$

      END2
END

$$\ddot{\mathbf{c}}_i = \dot{\boldsymbol{\omega}}_i \times \mathbf{c}_i = \mathbf{z}_i \times \mathbf{c}_i. \tag{14}$$

Substituting Eqs. (13) and (14) into Eqs. (11) and (12) gives:

$$\mathbf{F}_i = M_i \left(\mathbf{z}_i \times \mathbf{c}_i \right) \tag{15}$$

and

$$\mathbf{N}_i = \mathbf{E}_i \mathbf{z}_i. \tag{16}$$

Consequently, $\mathbf{f}_{i,i}$ and $\mathbf{n}_{i,i}$ may be determined by simply resolving these forces and moments to the origin of link i coordinates with the following results:

$$\mathbf{f}_{i,i} = \mathbf{F}_i \tag{17}$$

and

$$\mathbf{n}_{i,i} = \mathbf{N}_i + \mathbf{c}_i \times \mathbf{F}_i. \tag{18}$$

Having obtained $\mathbf{f}_{i,i}$ and $\mathbf{n}_{i,i}$, Newton-Euler backward recursion may be applied to links $i - 1$ through 1 to compute the necessary inter-link constraint forces and moments needed to keep the joints stationary. Since the composite rigid body nearest the base is completely stationary, the resulting equations involve resolution of the same force and moment to the various origins of the associated link coordinate systems. This gives:

$$\mathbf{f}_{j,i} = \mathbf{f}_{j+1,i} \tag{19}$$

and

$$\mathbf{n}_{j,i} = \mathbf{n}_{j+1,i} + \mathbf{p}_{j+1}^* \times \mathbf{f}_{j+1,i}. \tag{20}$$

The joint torque is simply the $\mathbf{z}$ component of the moment vector for rotational joints. That is,

$$\mathcal{T}_{j,i} = \mathbf{n}_{j,i} \cdot \mathbf{z}_j \qquad j \leq i. \tag{21}$$

Finally, the elements of the inertia matrix may be determined by using the following equality:

$$H_{j,i} = \mathcal{T}_{j,i} \qquad j \leq i. \tag{22}$$

The equations in the foregoing are completely summarized in Table I and are generalized to cover the case of prismatic joints. Note that $^iU_{i+1}$ is the 3×3 transformation between link i coordinates and link $i+1$ coordinates. It is used to keep most quantities in local coordinates to reduce the overall computation in a way similar to the most efficient Newton-Euler method for Inverse Dynamics [11]. Quantities are expressed in the local coordinate system, indicated by their leading superscript (in the case of $\mathbf{p}^*$), or by their first trailing subscript (all other cases).

It should be emphasized that the algorithm listed in Table I is a task which is composed of two processes. The two processes are signified by BEGIN1 and BEGIN2 blocks. The BEGIN1 process generates the composite rigid body parameters and the diagonal elements of the inertia matrix. The BEGIN2 process generates the off-diagonal elements of the inertia matrix. Once the BEGIN1 process is completed, the BEGIN2 process may begin; this involves backward recursions that may all be performed in parallel. This fact is indicated in the algorithm listing by a FOR ALL statement. Thus, it can be seen that the total computation is simply of $O(N)$.

III. VLSI-Based Robotics Processor

To facilitate implementation of robotics computations such as that just presented for the inertia matrix, the design of a VLSI-based Processing Element (PE) using $3\,\mu m$ n-channel Metal Oxide Semiconductor (nMOS) technology has been completed [3]. A block diagram showing the data paths of this Robotics Processor is shown in Fig. 3. The architecture, which is general enough to handle most robotics algorithms, is used here as the basic processing element within the systolic architectures. Below, the major blocks within the processor are explained, and its most important features are discussed. Further details of the Robotics Processor are given in [3].

The processor is composed of five major elements: a 32-bit floating-point multiplier (FPM), a 32-bit floating-point adder/subtractor (FPA/S), a 32-bit triple-port register file (RF), four 16-bit unidirectional I/O ports, and four 16-bit to 32-bit format converters (FC). Both the floating-point multiplier and adder have three stages of internal pipelining which provides increased throughput. The triple-

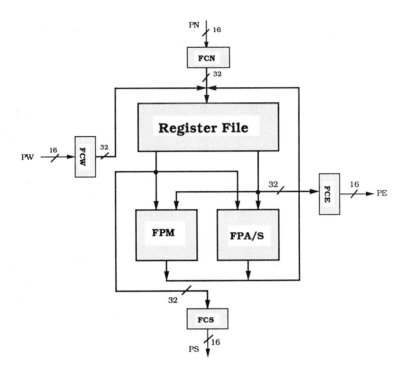

Figure 3: Block Diagram of the Robotics Processor [3].

port memory is used to hold the intermediate results; and six different locations may be accessed in each clock cycle to load/store all FPM and FPA/S operands/results during an arithmetic operation. Four 16-bit unidirectional I/O ports provide local and/or global broadcasting of data to their adjacent processors and/or to a high-level controller. The I/O ports are 16-bit, as opposed to 32-bit, to satisfy the I/O pin constraint on an integrated circuit chip. The north and west ports (PN and PW) are used for input while the south and east ports (PS and PE) are used for output. Four 32-bit words may be simultaneously transferred through all ports in two clock cycles through use of a single I/O operation.

The Robotics Processor is a three-bus horizontal microprogrammed machine that uses a 16 MHz external clock frequency and a 1 MHz internal clock frequency (clock period = $1\,\mu s$). In addition to the major data path elements, a microprogram sequencer, a control RAM, a microprogram register, and a boot-strap unit are incorporated into the chip for control (not shown in Fig. 3). The microcode instruction width is 40–46 bits depending on the size of the register file (64–128 words) [3].

Three major instructions are available to perform either arithmetic, branch, or I/O operations. No means for handshaking is furnished through the port interfaces so that communication among processors must be synchronized through careful microcoding. That is, delays in program execution are necessary and result in processor idle periods during which a NO-OP instruction is executed.

The use of a systolic array of Robotics Processors, as opposed to a more traditional attached processor, not only eliminates a potential I/O bottleneck to a great extent, but also makes a VLSI-based implementation, which emphasizes design simplicity and regularity (modularity), feasible. Also, the architecture must facilitate locality of communication so that it is usually compute-bound as opposed to I/O-bound. Taking all these criteria into account, several systolic architectures were designed using this Robotics Processor to implement the inertia matrix and are discussed in the next section.

IV. Implementation of the Inertia Matrix with the Robotics Processor

This section presents several systolic architectures used to implement the inertia matrix using the Robotics Processor. Various systolic architectures are explored to implement the $O(N)$ parallel algorithm presented in Table I. In each case, the I/O (communications overhead) and idle times are fully accounted for. In addition, in writing the microcode, all features of the processor including its inherent pipelining and parallelism are fully explored. Various hardware configurations have been considered with the objectives of minimizing the idle time and communications overhead and the resulting execution time delay, and of maximizing throughput and CPU utilization. However, unlike many of the previously proposed architectures for robotics computations, attention has also been focused on feasibility and cost effectiveness based on current or near-term technology.

From Table I, it may be observed that most of the operations are either 3×1 vector operations (vector addition (V+V), dot product (V·V), cross product (V×V), and scalar-vector product (SV)) or 3×3 matrix operations (matrix addition (M+M); matrix-matrix (MM), matrix-vector (MV), and scalar-matrix (SM) multiplication). Reservation tables have been developed to schedule these operations on the Robotics Processor.

A task graph (Fig. 4) has been developed for the $O(N)$ algorithm using matrix-vector operations as its primitives. This task graph incorporates information on the precedence of operations, the degree of parallelism, and the relative complexity of the different operations. The complexity is explicitly indicated by the height of a computation node. All primitives are normalized with respect to the V+V operation. For example, with one unit of computation for V+V, MV would be 4 units of computation, MM would be 6 units of computation, and so on.

Based on the task graph, various systolic configurations are proposed and will be evaluated. Among them are N, and $N(N+1)/2$-processor architectures. With the aid of the task graph and with full knowledge of the architecture, efficient code has been written to take complete advantage of parallelism and pipelining. An evaluation of the microcode steps needed to compute the inertia matrix has been used to calculate a number of parameters including the compute time, I/O time, initiation rate, CPU utilization, speedup, and total memory required.

The following points need to be considered in implementing algorithms on the Robotics Processor. An I/O operation takes 2 clock cycles (since the I/O ports are only 16 bits wide); whereas, any other operation takes only one clock cycle ($1\,\mu s$). As it was mentioned previously, NO-OP operations are used as required to synchronize input/output operations among processors. In addition, the Robotics Processor in its present configuration is incapable of performing sine, cosine, and inverse operations, $\sin \theta_i$, $\cos \theta_i$, and $\frac{1}{M_i}$ must be furnished by the host processor (involves relatively small amounts of computation). Also, $\mathbf{s}_i^*$, $\mathbf{p}_{i+1}^*$, m_i, and $\mathbf{I}_i$, for a particular manipulator, should be initially downloaded to the Robotics Processors before real-time

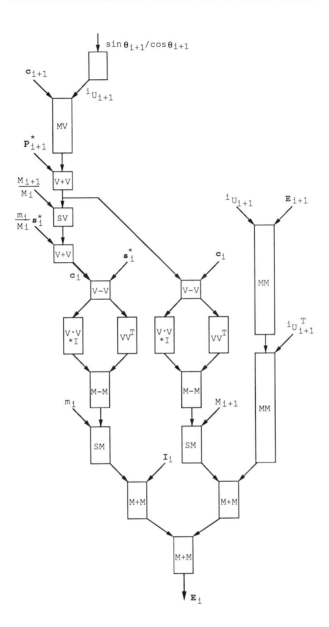

Figure 4: Task Graph for the $O(N)$ Algorithm.

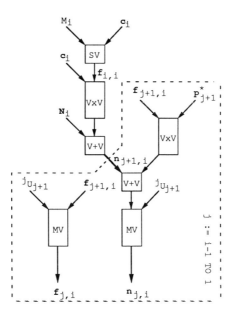

Figure 4 (continued)

execution begins.

In addition, in writing the microcode for all the proposed architectures, the following important observations have been made in order to make the code more efficient.

- The inertia matrix is a symmetric matrix. Thus, the resultant matrix produced by the transformation ${}^{j}U_{j+1}$ $\mathbf{E}_{j+1}$ ${}^{j}U_{j+1}^{T}$ is also symmetric.

- For a given vector v, $(v \cdot v)\mathbf{1}$ is a diagonal matrix.

- For a given vector v, (vv^{T}) is a symmetric matrix.

A description of the various architectures which have been explored to execute the microcode required to compute the inertia matrix follows.

One-processor, several N-processor (computation of the inertia matrix in column, row, and diagonal form) and $N(N+1)/2$-processor have been considered and will be compared and contrasted.

For an N-processor architecture, in column form, it is meant that the computation is partitioned in such a way that a particular processing element generates the corresponding column elements of the inertia matrix. That is, processor N generates all the Nth column, processor $N-1$ generates all the $N-1^{st}$ column, and in general, processor i generates all the ith column of the inertia matrix as shown in Fig. 5. However, since the inertia matrix is symmetric, only the diagonal and upper off-diagonal elements are computed.

The timing diagram for the column form is shown in Fig. 6. The computation of various subtasks is indicated, to scale, by letters on the timing diagram. In particular, note that time slot "a" corresponds to the time period during which the host computer outputs the required data to the corresponding PE's, "b" corresponds to computation of c_i, etc. Also, the timing diagram clearly shows the critical path of the computation (determination of E_i) from which values for compute time, I/O time, and execution time are derived.

The same approach is used to configure N processors in row or diagonal form, as shown in Figs. 7 and 9, respectively. Their corresponding timing diagrams are shown in Figs. 8 and 10.

Finally, the interconnections for the $N(N+1)/2$-processor architecture are shown in Fig. 11. This figure indicates that the critical computational path is dictated by the processing elements located on the diagonal. It may be noticed that the diagonal processing elements by themselves have a very close resemblance to the column configuration. The timing diagram for this critical path is very similar to the column form and therefore is not shown here.

The performance characteristics of the N-processor configurations are summarized in Table II along with 1-processor and $N(N+1)/2$-processor configurations. A number of parameters are used to evaluate the various configurations of Robotics Processors.

> Compute time is the processing delay time needed for the entire array to compute the inertia matrix.

> I/O time is the delay time spent doing local and global communications for computing the inertia matrix.

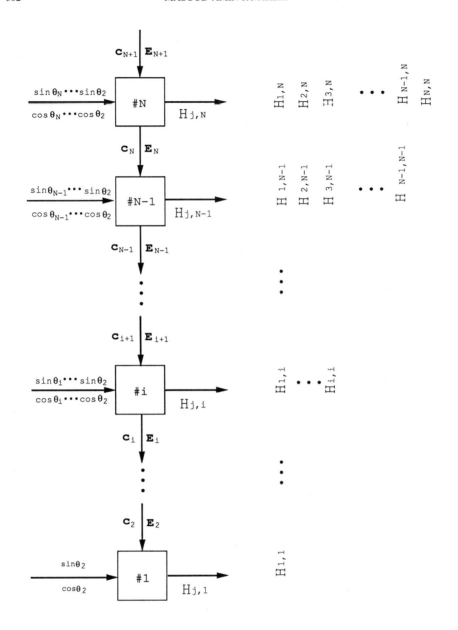

Figure 5: N-Processor Systolic Architecture in Column Form.

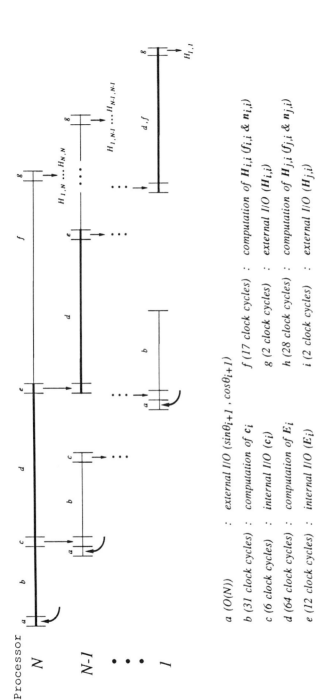

Figure 6: Timing Diagram for N Processors in Column Form.

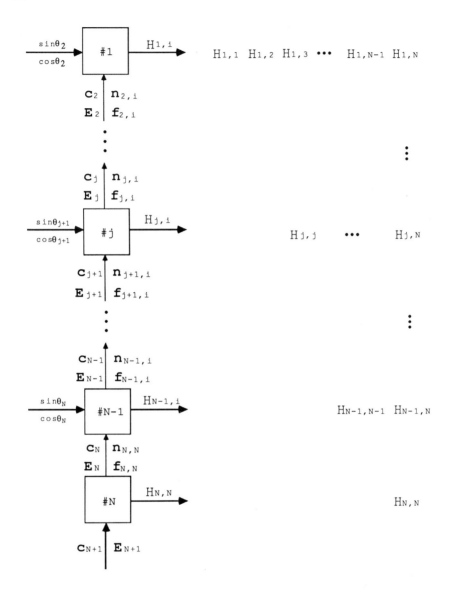

Figure 7: N-Processor Systolic Architecture in Row Form.

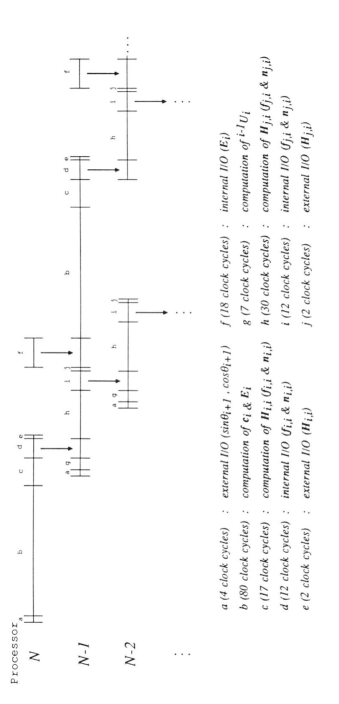

Figure 8: Timing Diagram for N Processors in Row Form.

$H_{1,1}$ $H_{1,2}$ $\cdots$ $H_{1,N-j+1}$ $\cdots$ $H_{1,N}$

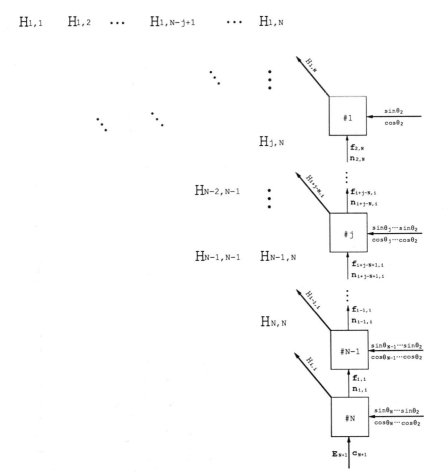

Figure 9: N-Processor Systolic Architecture in Diagonal Form.

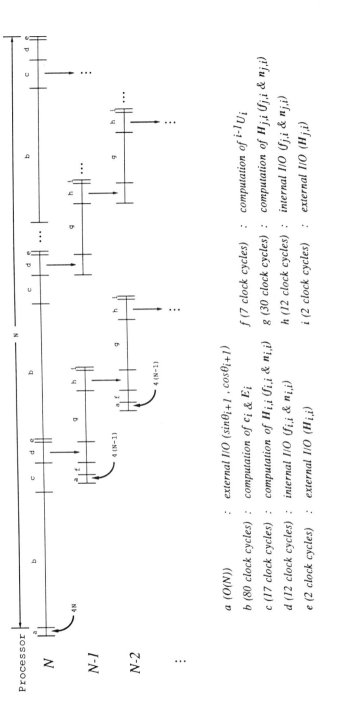

Figure 10: Timing Diagram for N Processors in Diagonal Form.

a (O(N)) : external I/O ($sin\theta_{i+1}$, $cos\theta_{i+1}$)

b (80 clock cycles) : compuation of c_i & E_i

c (17 clock cycles) : computation of $H_{i,i}$ ($f_{i,i}$ & $n_{i,i}$)

d (12 clock cycles) : internal I/O ($f_{j,i}$ & $n_{j,i}$)

e (2 clock cycles) : external I/O ($H_{i,i}$)

f (7 clock cycles) : computation of i-$1U_i$

g (30 clock cycles) : computation of $H_{j,i}$ ($f_{j,i}$ & $n_{j,i}$)

h (12 clock cycles) : internal I/O ($f_{j,i}$ & $n_{j,i}$)

i (2 clock cycles) : external I/O ($H_{j,i}$)

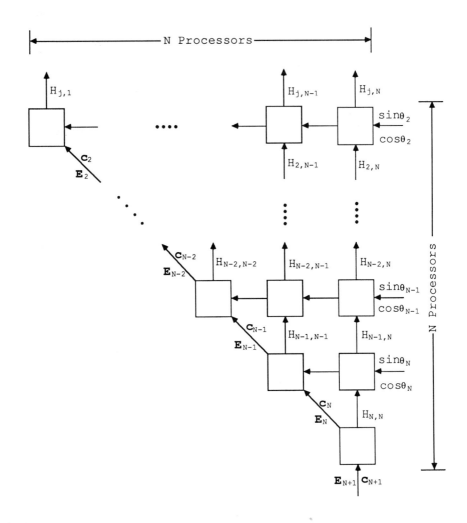

Figure 11: $N(N+1)/2$-Processor Systolic Architecture.

Table II. Architecture Parameters as a Function of the Number of Processors Used. (Numbers correspond to $N = 8$)

	Number of Processors				
	1	N Column	N Row	N Diagonal	$\frac{N(N+1)}{2}$
Compute Time	1728	560	986	776	560
T_c (μs)	$17N^2 + 80N$	$64N + 48$	$127N - 30$	$97N$	$64N + 48$
I/O Time	84	120	302	128	92
T_{io} (μs)	$N^2 + 3N - 4$	$16N - 8$	$44N - 50$	$18N - 16$	$12N - 4$
Execution Time	1812	680	1288	904	652
T_e (μs)	$18N^2 + 83N - 4$	$80N + 40$	$171N - 80$	$115N - 16$	$76N + 44$
Initiation Rate	1/1812	1/444	1/1006	1/904	1/408
IR $(\mu s)^{-1}$	$\frac{1}{18N^2 + 83N - 4}$	$\frac{1}{44N + 92}$	$\frac{1}{129N - 26}$	$\frac{1}{115N - 16}$	$\frac{1}{39N + 97}$
Average CPU Utilization %	100.0	65.0	35.6	37.0	21.4
Speedup (η)	1.0	2.66	1.41	2.0	2.78
Number of 32-bit Registers	245	119	63	119	63
	$26N + 37$	$8N + 55$	63	$8N + 55$	63
Control Memory Size (K bits)	11.55	18.8	39.6	38.6	15.6
Total Memory (K bits)	19.3	22.6	41.6	42.45	17.6

Execution time is the summation of compute time and I/O time and is the total delay in producing the inertia matrix.

Initiation rate is the rate at which successive computations of the inertia matrix may be initiated.

For the 1-processor configuration, the initiation rate is just the inverse of the execution time. The initiation rates for all other systolic configurations are determined so as to ensure all PE's, in a given configuration, are tightly synchronized under lock-step control. This simply implies the existence of some idle time which does not effect the execution time.

The speedup, η, is defined as the ratio of the execution time for the 1-processor case to the execution time for a particular case considered. RN is the number of 32-bit registers required in the register file to store all of the known parameters and computed variables needed in the computation. The Control Memory Size (CMS) is defined as follows:

$$CMS = MC \left(4 + 6 \lceil \log_2 RN \rceil \right) \tag{23}$$

where MC is the number of microinstruction steps required for the entire operation. The value of 6 is associated with the number of register file address fields and 4 is the number of additional control/opcode bits within a microinstruction. Finally, the total memory is just the sum of RN (times 32) and CMS.

The results tabulated in Table II convey several important facts. First of all, as expected, as the number of processors increases the compute time and execution time decrease. These do not decrease substantially with more than N processors though, because very little additional overlap can be obtained along the critical path (computing the composite rigid body parameters). Correspondingly, the initiation rate increases since a specific computation is more widely shared among the processors. The speedup is about 2.7 for the N-column and $N(N + 1)/2$ cases. Among the N-processor configurations, the column form gives the best results with $\eta = 2.66$, $T_e = 680 \, \mu s$, and approximately 22.6 K bits of memory for $N = 8$. Also, on the average, the processors are more fully utilized for this configuration.

Another very important figure in Table II is the amount of memory, which is a major factor in determining the required silicon area for the chip. The $N(N + 1)/2$-processor configuration has by far the best results in this case. This configuration not only requires the smallest memory, but also has the smallest execution time. The major drawback for this case is its cost, which is related to the large number of processors required $(O(N^2))$.

Overall, the implementation of the $O(N)$ algorithm using N processors in column form appears to give the best results for computing the manipulator inertia matrix. The $N(N+1)/2$-processor case had better performance in certain aspects, but the cost of the large number of processors $(O(N^2))$ may be prohibitive.

V. Summary and Conclusions

This chapter has presented the development of an $O(N)$ parallel algorithm plus several systolic architecture configurations for computing the $N \times N$ inertia matrix for a robot manipulator. The algorithm used is generally based on employment of the computationally efficient Newton-Euler method for Inverse Dynamics (a total of N times). However, considerable simplification is possible since in the use of Inverse Dynamics, the joint velocities and joint accelerations (all but one) are equal to zero. This results in sets of composite rigid bodies for which the composite mass, center of mass, and inertia are computed. The equations are summarized in Table I, which explicitly shows much of the parallelism that is inherent in the algorithm.

One- and two-dimensional systolic architectures consisting of 1, N, and $N(N+1)/2$ processors were presented for implementing the $O(N)$ parallel algorithm. The architectures presented were based upon a VLSI Robotics Processor [3] used to exploit fine-grain parallelism and to facilitate execution of matrix-vector operations. I/O time and idle time due to processor synchronization, as well as CPU utilization and on-chip memory size, were fully included in the evaluation of all proposed architectures. These performance parameters, along with speedup and latency, indicate the feasibility and effectiveness of the designs.

Good results are obtained for N-processor and $N(N + 1)/2$-processor configurations which give a compute time delay of $O(N)$. Overall, the column configuration of N processors exhibited the best performance when cost-effectiveness and interconnection cost, which are both related to the number of processors used, are considered.

While, in general, increasing the number of processors will decrease the total execution time delay, the results indicate that increased communications overhead and idle time (due to increased problems of synchronization) negate many of the advantages of going to more processors. In addition, the average CPU utilization decreases, thus decreasing the overall cost-effectiveness. However, the initiation rate increases significantly with increasing numbers of processors. A higher initiation rate may have a positive impact on control system stability.

Although the development of parallel algorithms and systolic architectures presented here are specific to the computation of the inertia matrix, the concept, approach, and findings are general enough to be applied to a wide range of robotics computations. In specific, implementation of dynamic control schemes proposed for robotics applications may certainly benefit from the conclusions reached in this chapter.

Acknowledgments

The author wishes to thank, Emily S. Bopp for carefully reviewing and proof reading the manuscript.

REFERENCES

1. C. H. An, C. G. Atkeson, and J. M. Hollerbach, *Model-Based Control of a Robot Manipulator.* Cambridge, MA: The MIT Press, 1988.

2. M. B. Leahy and G. N. Saridis, "Compensation of unmodeled PUMA manipulator dynamics," in *Proc. of IEEE International Conference on Robotics and Automation*, pp. 151–156, Raleigh, North Carolina, 1987.

3. D. E. Orin, K. W. Olson, and H. H. Chao, "Systolic architectures for computation of the Jacobian for robot manipulators," in *Computer Architectures for Robotics and Automation*, J. H. Graham, Ed., pp. 39–67, New York: Gordon and Breach Science Publishers, 1987.

4. M. Amin-Javaheri, "Parallel algorithms and architectures for the manipulator inertia matrix," Ph.D. Dissertation, The Ohio State University, March 1989.

5. J. R. Hewit and J. Padovan, "Decoupled feedback control of a robot and manipulator arms," in *Proc. of the 3rd CISM-IFToMM Symposium on the Theory and Practice of Robots and Manipulators*, pp. 251–266, New York: Elsevier, 1979.

6. M. Leborgne, R. Dumas, J. J. Borrelly, C. Samson, and B. Espiau, "Nonlinear control of robot manipulators, Part 2: simulation and implementation of a robust control method," *IRISA/INRIA Report*, Rennes, France, 1986.

7. O. Khatib, "A unified approach for motion and force control of robot manipulators: the operational space formulation," *IEEE Journal of Robotics and Automation*, vol. RA-3, no. 1, pp. 43–53, February 1987.

8. J. Bay, "Constrained motion of a 3-D manipulator over unknown constraints: the robotic groping problem," Ph.D. Dissertation, The Ohio State University, September 1988.

9. Y. Zheng and H. Hemami, "Mathematical modeling of a robot collision with its environment," *Journal of Robotics Systems*, vol. 2, no. 3, pp. 289–307, 1985.

10. M. W. Walker and D. E. Orin, "Efficient dynamic computer simulation of robotic mechanisms," *Journal of Dynamic Systems, Measurement, and Control*, vol. 104, pp. 205–211, September 1982.

11. J. Y. S. Luh, M. W. Walker, and R. P. C. Paul, "On-line computational scheme for mechanical manipulators," *Journal of Dynamic Systems, Measurement, and Control*, vol. 102, pp. 69–76, June 1980.

12. J. M. Hollerbach, "A recursive Lagrangian formulation of manipulator dynamics and a comparative study of dynamics formulation complexity," *IEEE Transactions on Systems, Man, and Cybernetics*, vol. SMC-10, no. 11, pp. 730–736, November 1980.

TECHNIQUES FOR PARALLEL COMPUTATION OF MECHANICAL MANIPULATOR DYNAMICS PART I: INVERSE DYNAMICS

Amir Fijany and Antal K. Bejczy

Jet Propulsion Laboratory
California Institute of Technology
Pasadena, CA 91109

I. INTRODUCTION

Several model-based control schemes have been developed for linearization and decoupling of the equations of motion of the manipulator. The linearization and decoupling can be achieved in joint space [1,2] or in task space [3]-[6]. The task space control schemes, taking into account the kinematic and dynamic properties of the manipulator, are expected to achieve significant improvement over the conventional independent joint schemes. However, this improvement is achieved at the expense of a significant increase in the computational requirements of the control scheme, leading to inefficiency for real-time implementation.

To be more precise, the difficulty for real-time implementation results from the lower bound on the sampling rate since the convergence of the control algorithm requires a sampling rate much higher than the structural resonant frequency of the manipulator [3]. Furthermore, at each sampling interval, besides the kinematic transformations, the inverse dynamic problem should be solved, which alone represents a heavy computational load.

CONTROL AND DYNAMIC SYSTEMS, VOL. 40

For the past two decades, efficient formulation of manipulator dynamics has been an active subject of research in the field of robotics. At present the Newton-Euler (N-E) formulation [7] represents the most efficient algorithm for the computation of the inverse dynamics problem. In fact, it seems unlikely that any but minor improvements in the efficiency of the inverse dynamics solution can be achieved [8]. However, despite this relative efficiency, the amount of computation involved in the N- E formulation is still a major obstacle for its real-time implementation.

The advent of VLSI technology, providing the possibility of integrating many processor/memory modules into one architecture, has stimulated the application of parallel processing to the computation-intensive problems in different scientific fields. Recently, researchers in the field of robotics are showing more interest in application of parallel processing to different computational problems in general, and to the N-E formulation in particular. However, despite the rich literature, few attempts have been made at the practical implementation of the proposed algorithms. Furthermore, usually one specific problem e.g., Jacobian or inverse dynamics, is considered, while the successful implementation of the advanced control schemes required efficient computation of a class of problems, i.e., forward kinematics, Jacobian and its time derivative, and inverse dynamics.

Recognizing the necessity of a highly parallel architecture for real-time control and simulation in robotics, we have investigated parallel algorithms for the class of kinematic and dynamic problems as a key step toward developing a Universal Real-time Robotic Controller and Simulator (URRCS) architecture [9]. The main criteria in developing the algorithms and architecture have been high performance and generality with respect to the class of problems and arms. That is, the algorithms and architecture should be capable of fast computation of different problems for different arms with different Degree-of-Feedeom (DOF). Particularly, a good performance for redundant arms has been of prime interest since it can lead to the removal of one of the obstacles in their application. This investigation has led to the development of a set of efficient parallel algorithms for the class of problems. A highly parallel architecture has been designed and implemented which is capable of exploiting the common properties of the set of algorithms and efficiently computing different problems.

In this paper we present the parallel algorithm for computation of the N-E formulation and the developed architecture. A hierarchical graph-based mapping approach is devised to analyze the inherent parallelism in the Newton-Euler formulation at several computational levels and to derive the features of an abstract architecture for exploitation of parallelism. At each computational level a parallel algorithm represents the application of a parallel model of computation which transforms the computation into a graph whose structure defines the features of an abstract architecture, i.e., processors and communication structure. Data flow analysis is then employed to determine the time lower bound in the computation as well as the sequencing of the abstract architecture. The features of the target architecture are defined by optimization of the abstract architecture to exploit maximum parallelism while minimizing various overheads and architectural complexity. An algorithmically specialized, highly parallel, MIMD-SIMD architecture is designed and implemented which is capable of efficient exploitation of parallelism at several computational levels. The computation time of the Newton-Euler formulation for a six DOF general manipulator is measured as $187\mu s$. The increase time for each additional DOF is $23\mu s$, which leads to a computation time of less than $500\mu s$, even for a 12 DOF redundant arm. The architecture also achieves a good performance for other computational problems. For a six DOF general manipulator, the computation time of the forward kinematics is measured as $60\mu s$, and the forward kinematics and Jacobian as $75\mu s$ [9].

This paper is organized as follows. In Sec. II, the previous proposals for parallel computation of the N-E formulation are briefly reviewed. In Sec. III, the mechanisms for mapping parallel algorithms onto parallel architectures are discussed; this provides a framework for presenting the mapping methodology adopted in URRCS. In Sec. IV, parallelism in the N-E formulation is studied and the time lower bound in computation is determined. The architectural optimization is discussed in Sec. V. The developed architecture is presented in Sec. VI. In Sec. VII, a comparative study of different proposals along with a discussion about several aspects of our approach are presented. Finally, some concluding remarks are made in the last section.

II. BACKGROUND

In the literature, a wide variety of MIMD (Multiple Instruction- Multiple Data) and SIMD (Single Instruction-Multiple Data) algorithms (and architectures) for computation of the inverse dynamics problem are proposed. Several pipeline algorithms are also reported.

In their pioneering work, Luh and Lin [13] developed an MIMD algorithm by decomposing the N-E formulation for the Stanford arm into a task graph. They considered a linear array of processors, i.e., an MIMD architecture with local shared memory, and proposed a branch-and-band technique for optimal mapping of the task graph on the architecture. Kasahara and Narita [14] also considered the Stanford arm and used a different scheduling scheme for a bus connected architecture, i.e., an MIMD architecture with global shared memory. Barhen [15,16] considered a hypercube architecture (NCUBE), i.e., an MIMD message passing architecture, and developed a load-balancing scheme to map the task graph of the Stanford arm onto the hypercubespace. SIMD algorithms are reported by Lathrop [17], and Lee and Chang [18]. Pipeline algorithms are proposed by Lathrop [17], Lee and Chang [18], and Orin *et al.* [19].

Some researchers have argued that the N-E formulation, due to its recursive form, is inherently serial and have proposed different formulations or modified versions of the N-E formulation to achieve a higher degree of parallelism in the computation [20]-[23]. Zheng and Hemami [20] presented an algorithm based on the N-E state-space formulation developed by Hemami [21]. Binder and Herzog [22] proposed an algorithm in which parallelism in the N-E formulation was increased by replacing the propagating variables with predicted ones. It is important to note that in both approaches an MIMD architecture with local shared memory was considered.

However, practical implementation of the N-E formulation has only been reported in [14], in which a limited number of processors could have been employed leading to a small speedup factor. Simulation and analysis of other MIMD algorithms have also shown that only a limited speedup can be achieved [13]-[16]. The SIMD algorithm, proposed in [18], also achieves a small speedup factor. The SIMD algorithms developed in [17] are capable of achieving a greater speedup and hence reduction in the computation time. However, the proposed architectures requires an excessive amount of hardware and more importantly, from the implementation point of view, a complex communication structure and excessive buffering. The pipeline algorithms in [17,18] achieve a very high throughput. However, in addition to demanding an excessive amount of hardware, this increase in the throughput is not mainly achieved by reducing the computation time (or latency) which is of prime importance in real-time control.

The review of the various proposals suggests that an analysis is required to define the nature and the amount of inherent parallelism in the N-E formulation. Such an analysis can determine the suitability of the N-E formulation for parallel computation and the architectural features which are best suited for its parallel implementation.

III. ALGORITHM-TO-ARCHITECTURE MAPPING PROBLEM

"It is easy to design computers, but it is hard to know what kind of computer to design ...." [24].

A. The Mapping Problem

Solving a problem on an architecture is the process of mapping the problem onto the architecture through several stages. The mapping process in serial processing is systematic and straightforward due to the abstraction at each stage. Parallel processing is considered as an *ad hoc* approach which lacks both theoretical foundations and engineering methodologies for the mapping process [25]. Due to the lack of abstraction, the mapping process is more difficult and less straightforward than the serial case and a poor mapping between different stages can lead to a drastic loss of efficiency.

The analysis of the serial algorithms is based on their behavioral characteristics, which include input/output specifications, and computational steps and complexity. This provides an abstraction since the analysis can be performed independently of the features of any target architecture for the implementation of the algorithm. Particularly, the efficiency of the serial algorithms can be defined based on the computational complexity alone. For parallel algorithms, however, the analysis should rather be based on their structural characteristics, i.e., granularity, communication and synchronization structures, data dependency, type and regularity of the operations, etc. In fact, these are the structural characteristics which are determining factors in mapping parallel algorithms onto the parallel architectures. That is, given a parallel algorithm and a target parallel architecture, the efficiency of the algorithm when mapped onto the architecture is a strong function of the degree to which the architectural features match the algorithm's structural characteristics. The lack of abstraction in the mapping process for parallel processing mainly results from the complex relationships between the features of the parallel algorithms and architectures [26]. A better understanding of these relationships is necessary for a better design of parallel algorithms and architectures. It can also provide a framework for developing an analytical model for measuring the efficiency of parallel algorithms and the computational model supported by parallel architectures.

B. The Mapping Approaches

The proposals in the previous section represent two common mapping approaches. In the first approach, given a problem, a target architecture is considered whose features define the parallel model of computation in advance. An algorithm is then developed to exploit parallelism in the problem, taking into account, to some degree, the given model of parallel computation. The MIMD proposals and the SIMD proposal in [18] represent such an approach. The drawback of this approach is that the architectural features can limit the algorithm's performance. The limitations result from the cardinality variation (i.e., the difference between the desired algorithm's processes and the actual architecture's processors) and the topological variation (i.e., the difference between the communication structure inherent to the algorithm and that of the architecture). Also, the architectures considered are usually capable of exploiting only one type of parallelism at one computational level, which may not be the most significant exploitable parallelism in the problem. Note that in any problem, there is parallelism at several computational levels [27,28]. However, the exploitation of parallelism, while at the higher levels is an algorithmic issue, at the lower levels is rather an architectural one.

When performance is the main criteria, a second approach is chosen in which, given a problem, an algorithm is developed to efficiently exploit the inherent (or increased) parallelism in the problem. An algorithmically specialized architecture is then designed to fully support the algorithm. The SIMD proposal in [17] as well as the pipeline proposals represent this second approach. The mapping process in the approach is more difficult than in the first one since, given a problem, different algorithms can be developed. However, the real efficiency of any algorithm depends on its suitability for architectural optimization. Hence, the problem is not so much developing an algorithm to exploit parallelism, as is designing an architecture to exploit algorithmic properties. That is, in this approach, the challenge is not to develop any algorithm but the algorithms with structural properties amenable to architectural optimization [29]. Here, the careful analysis of the structural properties of the algorithms is even more crucial than for the first approach since it provides a better insight into the design of the appropriate architecture. Furthermore, for different problems from an application domain, it may be possible to develop a class of parallel algorithms with similar properties. In such a case, based on the common properties of the class of algorithms, a unique architecture can be designed to efficiently implement the different problems of the application domain. For URRCS, this has been the key issue in the design of parallel algorithms for the class of kinematic and dynamic problems.

Algorithmically specialized architectures represent an algorithmic/architectural approach to the mapping problem which requires knowledge of the algorithmic and architectural issues and their relationships. The difficulty resides in the design procedure which incorporates both top down and bottom-up strategies, wherein conflicting objectives must be optimized, e.g., the balance between the computation complexity and the communication and synchronization complexity (maximizing the exploitable parallelism by reducing the grain size of the algorithm, while minimizing the complexity of the communication and synchronization structures of the algorithm). The key difficulty in achieving satisfactory balances, as stated before, resides in the complex relationships between algorithmic and architectural features.

C. The Mapping Approach in URRCS

The above discussion provides the required framework for presenting the mapping strategy in URRCS. A hierarchical approach is adopted for the mapping process which allows the study of the algorithmic and architectural features, their relationships, and efficient tradeoffs at several levels of abstraction.

A computation problem can be presented at several levels of abstraction. At each level, an algorithm can be thought of as the application of a parallel model of the computation which transforms the computation into a graph, wherein each node represents some operation on a data object and the directed arcs the data dependency among the nodes. The computation graph is mapped onto an abstract architecture, with no cardinality and topological variations, which defines the features of an abstract architecture, i.e., processors and communication structure. This procedure is continued through more detailed levels, which further refines the features of the abstract architecture at each level. The behavior of the algorithm is described by the flow of the data through the computation graph, which defines the computation time and sequencing of the abstract architecture.

The mapping process is mainly determined by the selection of the type of operation and data of the parallel model of the computation, since they define the structure of the computation graph and hence the abstract architecture. At each level, the selection can be made based on different tradeoffs. This hierarchical approach, by creating abstraction, provides additional degrees of freedom in the mapping process, since conflicting objectives can be optimized at several abstraction levels.

The exact features of the target architecture are defined by optimizing the abstract architecture for maximum exploitation of the algorithmic properties, while minimizing the architectural complexity. Note that this approach represents a multilevel approach to the exploitation of parallelism since parallelism and architectural means for its exploitation are studies at several computational levels. Particularly, the analysis of the relative significance of parallelism at different computational levels provides useful insights into the architectural optimization. This hierarchical approach also allows the architectural optimization to be performed based on the common properties of a class of algorithms (and abstract architectures). The architecture of URRCS represents such an optimization based on the class of parallel algorithms for the kinematic and dynamic problems. We further discuss different aspects of this approach through its application to the parallel computation of the N-E formulation.

IV. PARALLELISM IN THE NEWTON-EULER FORMULATION

A. The Newton-Euler Formulation

The N-E formulation, as developed in [7], is reported in Table II where Eqs. (1)-(6) represent the backward recursion and Eqs. (7)-(9) represent the forward recursion. For the sake of simplicity, an all-rotary-joint arm is considered. However, the same results can be obtained for the arms with sliding joint(s) [9]. A modified version of the forward recursion procedure, given by Eqs. (10)-(15), is used which is slightly more efficient, since Eq. (15) does not require any computation. Note that f_{n+1} and n_{n+1} are usually described with respect to the End-Effector (EE) coordinated frame. The purpose of this modification, however, is not to increase the efficiency in the serial sense but to make the computation more suitable for parallel processing. In the original formulation, both A_i and A_{i+1} are needed for the computation of link i, while by this modification, only A_i is needed. This modification, by enforcing the locality in the computation, reduces the amount of communication and

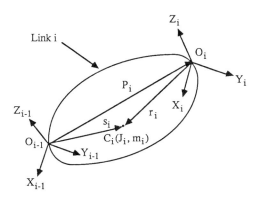

Figure 1. Link, Frames, and Position Vectors

a_i, d_i, and α_i are Denavit- Hartenberg parameters of link i.

$\theta_i, \dot{\theta}_i$, and $\ddot{\theta}_i$ are joint i position, velocity, and acceleration, respectively.

A_i 3x3 matrix describing orientation of frame i with respect to frame $i-1$.

m_i Mass of link i.

J_i Inertia tensor of link i about its center of mass.

ω_i Angular velocity of link i.

$\dot{\omega}_i$ Angular acceleration of link i.

$\ddot{P}_i$ Linear acceleration of point O_i.

$\ddot{r}_i$ Linear acceleration of link i center of mass.

F_i Total force exerted on link i.

N_i Total moment exerted at the center of mass of link i.

f_i Force exerted on link i by link $i-1$.

n_i Moment exerted on link i by link $i-1$

τ_i Torque applied at joint i.

Link Kinematic and Dynamic Parameters

A Addition.

M Multiplication.

MV Matrix-Vector multiplication.

VC Vector Cross product.

VA Vector Addition.

SV Scalar Vector multiplication.

Notation used in Computation Complexity Analysis

Table I. Summary of the Notation

Backward Recursion

Initial Conditions: $^o w_o$, $^o \dot{w}_o$, and $^o \ddot{P}_o$ are known. $^o \ddot{P}_o = [g_x \ g_y \ g_z]^t$ is the base gravitional acceleration.

For $i = 1, 2, \ldots, n$

$$^i \omega_i = A_i^t \left(^{i-1}\omega_{i-1} + Z_o \dot{\theta}_i\right) \qquad \text{with } Z_o = \,^{i-1}Z_{i-1} = [0 \ 0 \ 1]^t \tag{1}$$

$$^i \dot{\omega}_i = A_i^t \left(^{i-1}\dot{\omega}_{i-1} + \,^{i-1}\omega_{i-1} \times Z_o \dot{\theta}_i + Z_o \ddot{\theta}_i\right) \tag{2}$$

$$^i \ddot{P}_i = A_i^t \,^{i-1}\ddot{P}_{i-1} + \,^i \dot{\omega}_i \times P_i + \,^i \omega_i \times \left(^i \omega_i \times P_i\right) \tag{3}$$

$$^i \ddot{r}_i = \,^i \ddot{P}_i + \,^i \dot{\omega}_i \times r_i + \,^i \omega_i \times \left(^i \omega_i \times r_i\right) \tag{4}$$

$$^i F_i = m_i \,^i \ddot{r}_i \tag{5}$$

$$^i N_i = \,^i J_i \,^i \dot{\omega}_i + \,^i \omega_i \times \left(^i J_i \,^i \omega_i\right) \tag{6}$$

Forward Recursion

Initial Conditions: f_{n+1} and n_{n+1} are known force and moment exerted on the EE by the environment.

For $i = n, n-1, \ldots, 1$

$$^i f_i = \,^i F_i + A_{i+1} \,^{i+1} f_{i+1} \tag{7}$$

$$^i n_i = \,^i N_i + A_{i+1} \,^{i+1} n_{i+1} + s_i \times \,^i F_i + P_i \times \left(A_{i+1} \,^{i+1} f_{i+1}\right) \tag{8}$$

$$\tau_i = \,^i n_i \cdot \left(A_i^t Z_o\right) = \,^i n_i \cdot \,^i z_{i-1} \qquad \text{with } \,^i Z_{i-1} = [0 \ S\alpha_i \ C\alpha_i]^t \tag{9}$$

Modified Forward Recursion

$$^i f_{i+1} = A_{i+1} \,^{i+1} f_{i+1} = f'_{i+1} \text{ and } \,^i n_{i+1} = A_{i+1} \,^{i+1} n_{i+1} = n'_{i+1} \tag{10}$$

For $i = n, n-1, \ldots, 1$

$$^i f_i = \,^i F_i + \,^i f_{i+1} \tag{11}$$

$$^i n_i = \,^i N_i + \,^i n_{i+1} + s_i \times \,^i F_i + P_i \times \,^i f_{i+1} \tag{12}$$

$$^{i-1} f_i = A_i \,^i f_i \tag{13}$$

$$^{i-1} n_i = A_i \,^i n_i \tag{14}$$

$$\tau_i = \,^{i-1} n_i \cdot Z_o \tag{15}$$

$$A_i = \begin{bmatrix} a_{11} & a_{12} & a_{13} \\ a_{21} & a_{22} & a_{23} \\ a_{13} & a_{32} & a_{33} \end{bmatrix} = \begin{bmatrix} C\theta_i & -S\theta_i C\alpha_i & S\theta_i S\alpha_i \\ S\theta_i & C\theta_i C\alpha_i & -C\theta_i S\alpha_i \\ 0 & S\alpha_i & C\alpha_i \end{bmatrix} \tag{16}$$

Table II. N-E Formulation for an All-Rotary- Joint Arm

simplifies parallel computation. It also increases the uniformity of the operations by completely eliminating the need to perform the vector dot product. Note that, also, compared to other matrix-vector operations, there is less parallelism in the vector dot product.

B. Parallelism at the Problem Level

The computation cost of the N-E formulation is $(50n - 48)M + (131N - 48)A$ [8]. The computation cost for each link is about $150M + 131A$. The input data for the computation of link i in backward recursion are $\omega_{i-1}, \dot{\omega}_{i-1}$, and $\ddot{P}_{i-1}$, and in forward recursion are f'_{i+1} and n'_{i+1}. The output data are $\omega_i, \dot{\omega}_i$, and $\ddot{P}_i$ in backward recursion, and f'_i and n'_i in forward recursion. Other data, $A_i, \bar{r}_i, F_i, N_i, F_i, N_i$, and τ_i, are generated and consumed within the computation of link i. Therefore, the computation of each link exhibits a high degree of locality since the amount of computation is much greater than that of input and output. There is parallelism between the computation of adjacent links since, in backward recursion, the computation of link $i + 1$ can be started by the availability of ω_i and, in forward recursion, the computation of link i can be started by the availability of f'_{i+1} or n'_{i+1}. Hence, the computation of link $i+1(i)$ and the rest of the computation for link $i(i+1)$ can be performed in parallel and, for the most part, asynchronously.

At this level (called the problem level), we can choose between exploitation of linear or logarithmic parallelism [17,18]. We exploit linear parallelism, which leads to a simple interconnection structure. Other reasons for such a selection are the high degree of locality in link computation, asynchronous parallelism between adjacent link computation, and the poor speedup of the logarithmic parallelism for all practical ranges of the size of the problem, the number of DOF (see Sec. VII). Note that this choice represents a computation/communication tradeoff. In fact, while it results in a bounded parallelism, i.e., a constant speedup factor, at the problem level, as will be shown later, it leads to the communication and synchronization overhead minimization as well as the simplicity and flexibility of the architecture.

At this level, the computation for each link (in backward and forward recursion) and its input and output data are taken to be the operation and data of the parallel model of computation. Based on the above model, the computation at the problem level can be presented by the graph of Fig. 2. A perfect mapping, i.e., a mapping with no cardinality or topological variation of this graph defines the features of the abstract architecture at

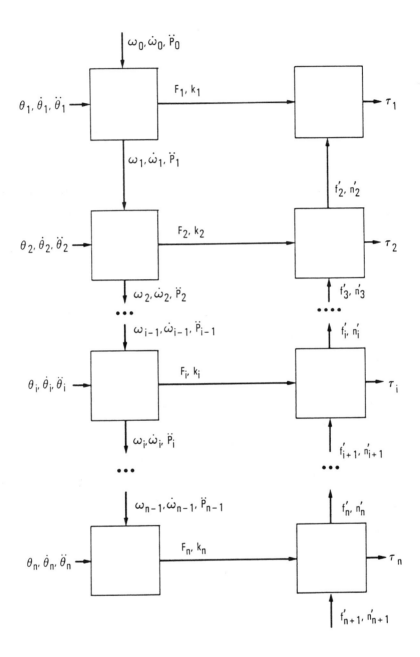

Figure 2. Computation graph at problem level

the problem level. Since, for any link, there is no parallelism between its backward and forward recursion computations, only one processor (the exact definition of the processor at this level will be given later) is assigned for each link and the backward and forward recursion computations are performed sequentially. The abstract architecture at this level is an MIMD local shared memory with n processors and a linear interconnection structure.

C. Parallelism at the Link Level

At the second computation level, the link level, matrix-vector operations and their corresponding vectors are selected as the basic operation and data of the parallel model of computation. There is a high degree of inherent parallelism in matrix-vector operations. However, the exploitation of this parallelism is rather a pure architectural issue than an algorithmic one. In fact, the matrix-vector operations represent a suitable structure which can be exploited by architectural optimization. This selection also simplifies the algorithmic analysis since it can be done based on the higher level (matrix-vector) operations instead of scalar multiplication and addition. In analyzing the algorithm behavior, a minimum number of assumptions about the performance of the architecture in matrix-vector operations is made. In what follows, the matrix-vector operations are considered as the basic operations and the scalar multiplication and addition as the primitive operations.

The decomposition of the link computation leads to a computation graph which is presented in Fig. 3. Again, a perfect mapping of this graph defines the abstract architecture at link level. The generated subtasks and their computational costs are given as follows:

1. Backward Recursion

$$\omega_i' = \left[\omega_{i-1(x)} \ \omega_{i-1(y)} \ \left(\omega_{i-1(z)} + \dot{\theta}_i\right)\right]^t \text{ and}$$

$$\omega_i = A_i^t \omega_i' = 1MV + 1VA \tag{17}$$

$$\dot{\omega}_i' = \left[\left(\dot{\omega}_{i-1(x)} + \omega_{i-1(x)}\dot{\theta}_i\right) \ \left(\dot{\omega}_{i-1(y)} - \omega_{i-1(y)}\dot{\theta}_i\right) \ \left(\dot{\omega}_{i-1(z)} + \ddot{\theta}_i\right)\right]^t \text{ and}$$

$$\dot{\omega}_i = A_i^t \dot{\omega}_i' = 1MV + 1SV + 1VA \tag{18}$$

$$a_{1i} = \omega_i \times \left(\omega_i \times P_i\right) = 2VC \text{ and } b_{1i} = \dot{\omega}_i \times P_i = 1VC \tag{19}$$

$$a_{2i} = \omega_i \times \left(w_i \times r_i\right) = 2VC \text{ and } b_{2i} = \dot{\omega}_i \times r_i = 1VC \tag{20}$$

$$a_{3i} = \omega_i \times \left(J_i w_i\right) = 1MV + 1VC \text{ and } b_{3i} = J_i \dot{\omega}_i = 1MV \tag{21}$$

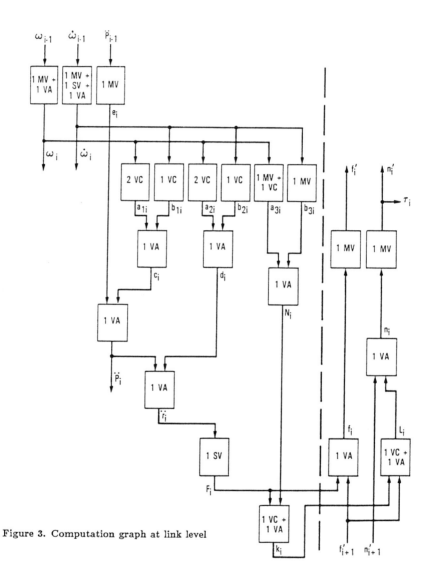

Figure 3. Computation graph at link level

$$e_i = A_i^t \ddot{P}_{i-1} = 1MV, \quad c_i = a_{1i} + b_{1i} = 1VA, \quad \text{and} \quad \ddot{P}_i = e_i + c_i = 1VA \qquad (22)$$

$$d_i = a_{2i} + b_{2i} = 1VA \quad \text{and} \quad \ddot{r}_i = \ddot{P}_i + d_i = 1VA \qquad (23)$$

$$N_i = a_{3i} + b_{3i} = 1VA \qquad (24)$$

$$F_i = m_i \ddot{r}_i = 1SV \quad \text{and} \quad k_i = N_i + s_i \times F_i = 1VC + 1VA \qquad (25)$$

2. Forward Recursion

$$f_i = F_i + f'_{i+1} = 1VA \quad \text{and} \quad f'_i = A_i f_i = 1MV \qquad (26)$$

$$L_i = K_i + P_i \times f'_{i+1} = 1VC + 1VA \qquad (27)$$

$$n_i = n'_{i+1} + L_i = 1VA, \quad \text{and} \quad n'_i = A_i n_i = 1MV \qquad (28)$$

The computational costs given in Eqs. (17)-(18) are based on the assumption that the additions and multiplications in VA and SV operations are performed in parallel. Hence, the cost of one addition in Eq. (17) is the same as $1VA$ and the cost of two multiplications and one addition in Eq. (18) is the same as $1SV + 1VA$. Note that the subtasks represented by a_{1i}, a_{2i}, and a_{3i} are ω_i dependent while those represented by b_{1i}, b_{2i}, and b_{3i} are $\dot{\omega}_i$ dependent.

D. Data Flow Analysis

The degree of exploitable parallelism at the problem level and link level depends on the relative delay in propagation of the angular velocities and the angular and linear accelerations in backward recursion, and the forces and moments of constraint in forward recursion. In order to define the impact of the delay on the parallelism and derive the time lower bound, a data flow analysis is required.

In backward recursion and at link 1, $w_o, \dot{w}_o$, and $\ddot{P}_o$ are available at the same time since they are known initial conditions. With zero angular velocity and acceleration of base, the computation cost of ω_1 and $\dot{\omega}_1$ is $1SV$. Hence, ω_1 and $\dot{\omega}_1$ are available at the same time for the computation of link 1 (and link 2). With nonzero initial conditions or for any other link, there is a delay between the availability of ω_i and $\dot{\omega}_i$, due to the difference in their computation costs, which is rather small. Therefore, we assume the same computation cost for ω_i and $\dot{\omega}_i$, i.e., $1MV + 1SV + 1VA$. With this assumption, ω_i and $\dot{\omega}_i$ will be available simultaneously. This assumption not only simplified the data flow analysis but also reduces the overhead due to the communication, since, for a small amount of data, the overhead is largely dominated by that of the communication initiation.

From the graph in Fig. 3, the relative delay in the availability (Av) of different variables can be derived as:

$$Av(c_i) = Av(\omega_i) + 2VC + 1VA \tag{29}$$

$$Av(N_i) = Av(\omega_i) + 1MV + 1VC + 1VA \tag{30}$$

$$Av(\ddot{P}_i) = \text{Max}\left((Av(\ddot{P}_{i-1}) + 1MV),\ (Av(\omega_{i-1}) + 1MV + 1SV + 2VC + 2VA)\right) + 1VA \text{ and}$$

$$Av(\ddot{P}_i) = \text{Max}\left((Av(\ddot{P}_{i-1} + 1MV),\ (Av(\omega_i) + 2VC + 1VA)\right) + 1VA \tag{31}$$

$$Av(\tilde{r}_i) = \text{Max}\left((Av(\ddot{P}_i),\ (Av(\omega_i) + 2VC + 1VA)\right) + 1VA \tag{32}$$

$$Av(F_i) = Av(\tilde{r}_i) + 1SV = \text{Max}\left((Av(\ddot{P}_i),\ (Av(\omega_i) + 2VC + 1VA)\right) + 1VA + 1SV \tag{33}$$

$$Av(K_i) = \text{Max}\left(Av(F_i),\ Av(N_i)\right) + 1VC + 1VA \tag{34}$$

The computation for backward recursion is completed by the availability of K_i and the computation for forward recursion can not be started unless f'_{i+1} or n'_{i+1} becomes available.

In forward recursion and at link f'_{n+1} and n'_{i+1} are available at the same time since they are known initial conditions. If the architecture at the problem level, as it is assumed, contains n processors, then the term $n'_{n+1} + P_n \times f'_{n+1}$ can be computed before the backward recursion reaches the link n. Hence, forward recursion can be started as soon as F_n or N_n becomes available. The relative delays of f_n and N_n are

$$Av(f_n) = Av(F_n) + 1VA \tag{35}$$

$$Av(n_n) = \text{Max}\left((Av(F_n) + 1VC),\ Av(N_n)\right) + 1VA \tag{35}$$

As can be seen, f_n is made available before n_n. Also, f'_n is made available before n'_n which leads to the delay in propagation of forces and moments of constraint among the links. For any other link:

$$Av(f_i) = Av(f'_{i+1}) + 1VA \text{ and } Av(f'_i) = Av(f_i) + 1MV \tag{37}$$

$$Av(n_i) = \text{Max}\left(\left(Av(f'_{i+1} + 1VC + 1VA), \left(Av(n'_{i+1} + 1VA)\right)\right) + 1VA \text{ and}\right.$$

$$Av(n'_i) = Av(n_i) + 1MV = Av(\tau_i) \tag{38}$$

As will be shown later, less than n processors at the problem level are required. Hence in the following, the forward recursion computation of link n is assumed to be performed similarly to the other links. Various time intervals in the computation of each link can be recognized as (Fig. 4):

1) T_o is the initialization time, i.e., the computation time of A_i. If four multiplications in Eq. (16) are performed in parallel, then $T_o = 1SV$. Except for link 1, A_i can be computed before backward recursion reaches the link i.

2) T_{w1i} is the idle time before the start of backward recursion computation, i.e., the delay between the availability of A_i and ω_{i-1}. For link 1, $T_{w11} = 0$.

3) T_{1i} is the delay between the availability of ω_{i-1} and ω_i. With zero initial condition, $T_{11} = 1SV$ and $T_{1i} = 1MV + 1SV + 1VA = T_1$ for $i \geq 2$. We consider nonzero initial conditions which implies that $T_{1i} = T_1$ for $i \geq 1$.

4) T_{2i} is the delay between the availability of $\ddot{P}_i$ and ω_i or $\ddot{P}_{i-1}$, whichever is made available later.

5) T_{3i} is the delay between the availability of $\ddot{P}_i$ and K_i.

6) T_{w2i} is the idle time between the completion of the backward recursion and the start of the forward recursion computation, $T_{w2n} = 0$.

7) T_{4i} is the delay between the availability of f'_{i+1} and f'_1, $T_{4i} = 1VA + 1MV = T_4$ for $i \geq 1$.

8) T_{5i} is the delay between the availability of n'_i and f'_i.

9) T_{6i} is the delay between the availability of n'_{i+1} and n'_i (or τ_i).

For link i, the computation time of the backward recursion, TB_i, is (Fig. 4):

$$TB_i = T_1 + T_{2i} + T_{3i} = 1MV + 1SV + 1VA + T_{2i} + T_{3i} \tag{39}$$

where T_{2i} and T_{3i} depend on the relative delay in the availability of ω_{i-1} and $\ddot{P}_{i-1}$. The computation time of the forward recursion, TF_i, is (Fig. 4):

$$TF_i = T_4 + T_{5i} = 1MV + 1VA + T_{5i} \tag{40}$$

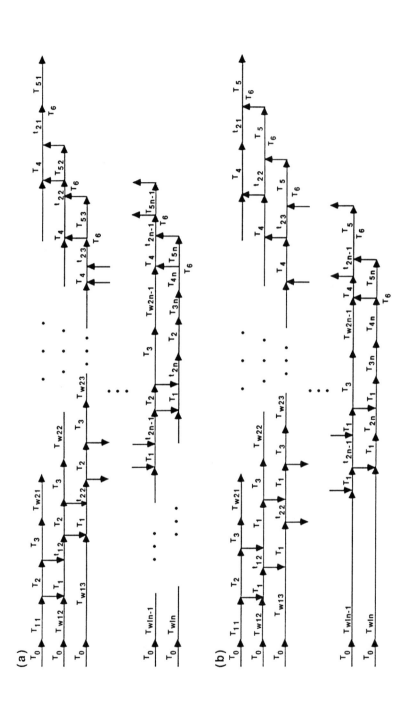

Figure 4. Timing sequence of link computation: a) $T_1 > T_2$ b) $T_1 < T_2$

where T_{5i} depends on the relative delay in the availability of f'_{i+1} and n'_{i+1}. In order to define TB_i and TF_i, and derive the time lower bound in the computation while exploiting parallelism at the problem and link computation levels, T_{2i}, T_{3i}, and T_{5i} need to be defined.

At link 1, $\omega_o, \dot{\omega}_0$, and $\ddot{P}_0$ are available at the same time. Hence, e_1 can be computed in parallel with ω_1 and $\dot{\omega}_1$. From Eqs. (18) and (31), it follows that

$$T_{21} = 2VC + 2VA = T_2 \tag{41}$$

Three different cases can be distinguished as:

$$T_1 > T_2, \ T_1 = T_2, \text{ and } T_1 < T_2 \tag{42}$$

The detailed data flow analysis for the above three cases is presented in Appendix 1. This analysis shows the regularity of the flow of data and the fact that no additional delay in the link computation is introduced while exploiting maximum parallelism and imposing different constraints. It also shows that the above three cases lead to the same results as

$$TB_i = T_1 + T_2 + T_3 = 1MV + 3VC + 2SV + 5VA \tag{43}$$
$$TF_i = T_4 + T_5 = 1MV + 1VC + 2VA \tag{44}$$

E. Time Lower Bound in the Computation

Let TPB and TPF denote the computation time of the backward and forward recursion wile exploiting parallelism at the problem and link level. From Eqs. (43)-(44) and Fig. 4, it follows that

$$TPB = T_0 + nT_1 + T_2 + T_3 = n(1MV + 1SV + 1VA) + 3VC$$
$$+ 2SV + 4VA \tag{45}$$
$$TPF = nT_4 + T_5 = n(1MV + 1VA) + 1VC + 1VA \tag{46}$$

Also, let $TP = TPB + TPF$, denote the computation time of the N-E formulation while exploiting parallelism at the problem and link level. It follows that

$$TP = T_0 + n(T_1 + T_4) + T_2 + T_3 + T_5 = n(2MV + 2VA + 1SV)$$
$$+ 4VC + 2SV + 5VA = k'_1 n + k'_2 \tag{47}$$

The third level of parallelism exists in the matrix-vector operations. If this parallelism is fully exploited, then the cost of parallel computation of different matrix-vector operations will be:

$$1MV = 1M + 2A, \ 1VC = 1M + 1A, \ 1SV = 1M, \text{ and } 1VA = 1A. \tag{48}$$

The number of Processor Element (PE) required to achieve the above computation time for SV and VA is 3, for VC is 6, and for MV is 9. However, in addition to the excessive number of PEs, the data alignment for VC and MV can become a major architectural bottleneck. A better scheme is to assign 3 PEs for each matrix- vector operation. In this case,

$$1MV = 3M + 2A, \ VC = 2M + 1A, \ 1SV = 1M, \text{ and } 1VA = 1A. \tag{49}$$

The SV or VA operations are computed by performing 3 multiplications or additions in parallel. The MV and VC operations are computed by performing 3 sets of inner-product (multiply and add/subtract) operations in parallel. Such a scheme leads to a uniformity in the exploitation of parallelism in matrix-vector operations (which is important for architectural optimization) and a constant speedup factor of 3 in computing different matrix-vector operations.

The lower bound in the computation, TFP, while exploiting parallelism at three computational levels, is then obtained:

$$TFP = n(7M + 6A) + 10M + 9A = k_1 n + k_2 \tag{50}$$

This computation time represents the lower bound while exploiting parallelism at three computation levels subject to different computation/communication and algorithmic/ architectural tradeoffs rather than the theoretical lower bound. This result shows the high degree of parallelism in the N-E formulation which can be exploited with minimum overhead. Although the computation time is still $O(n)$, representing a bounded parallelism in the computation, the coefficient of n is significantly reduced, which is also suitable for redundant arms since the increase in computation time for each additional DOF is small. If the time of multiplication and addition is taken to be the same, then, compared with the serial computation of the N-E formulation, the result in Eq. (50) represents an asymptotic speedup factor of about 21.

F. Relative Significance of Parallelism

For the purpose of architectural optimization, it is constructive to analyze the relative significance of parallelism at different computational levels. In order to define the relative significance of parallelism at the problem level and link levels, it is assumed that there is only one processor in the architecture at problem level, i.e., the nodes of the graph in Fig. 2 are processed sequentially, but all the nodes of the graph in Fig. 3 are mapped onto the available processors, i.e., parallelism at link level is fully exploited. In this case, $\omega_{i-1}, \dot{\omega}_{i-1}$, and $\ddot{P}_{i-1}$ in backward recursion, and f'_{i+1} and n'_{i+1} in forward recursion, are available at the same time at link i. Therefore, the computation time of each link, in forward recursion, is equal to that of link n, computation derived in the last section. Hence for $i \geq 1$

$$TB_i = TB_1 = T_0 + T_1 + T_2 + T_3 = 1MV + 3VC + 3SV + 4VA \tag{51}$$

$$TF_i = TF_n = T_4 + T_5 = 1MV + 1VC + 2VA \tag{52}$$

Let TSG denote the computation time of the N-E formulation while exploiting parallelism at the link computation level only. It follows that

$$TSG = n(TB_1 + TF_n) = n(2MV + 4VC + 3SV + 6VA) \tag{53}$$

In Table III, the speedup factors resulting from exploitation of parallelism at different computational levels are compared where the time of addition and multiplication is taken to be the same. $TS, TSG, TP,$ and TFP denote the computation time of the N-E formulation exploiting no parallelism, parallelism at the link level only, parallelism at both the problem and link levels, and parallelism at the problem, link, and matrix-vector operations levels, respectively. The comparison shows that the parallelisms at the link level and at the matrix-vector operations level have equal significance and that they are more significant than the parallelism at the problem level.

V. ARCHITECTURAL OPTIMIZATION

A. Nonoptimized Architecture

In order to appreciate the significant reduction in the hardware complexity which can be achieved by architectural optimization, let us first consider a nonoptimized architecture. The computation time in Eq. (50) can be achieved by a perfect mapping of the graphs of

Comp. Time	Mul.	Add.	Speedup
TS	150n-48	131n-56	1
TSG	51n	42n	3
TP	21n	18n	7
TFP	7n	6n	22

Table III. Speedup Factors by Exploiting Different Parallelism

Figs. (2) and (3). For each node of the graph of Fig. 3, an SIMD processor can be designed
with 3 PEs, to perform the matrix-vector operations in parallel, along with the required data
paths, a small local memory (a multiport register file) for storing the link parameters, and a
simple local control unit. Since the mapping is performed with no topological variation, the
interconnection structure among the SIMD processors is the same as the graph of Fig. 3.
The two nodes computing $\ddot{r}_i$ and F_i can be combined since their computations are strictly
serial which simplifies the interconnection structure and reduces the number of required
PEs.

    If all the SIMD processors are run with the same clock, then the synchronization among
the processors can be simplified. Based on the data flow analysis in Sec. IV.D., a data-driven
mechanism can be employed. The nodes with a single input are activated by the availability
of their single input. The nodes with two inputs are activated by the availability of the
input which is made available later. For the nodes with two inputs, additional registers
are required for temporary storage of the input which is made available first. Simultaneous
availability of ω_i and $\dot{\omega}_i$ can be achieved by adding idle (no operation) cycles to the SIMD
processor of the node computing ω_i.

    With the perfect mapping, the architecture at the problem level is not an MIMD.
The nonoptimized architecture combines the synchronization mechanisms of systolic arrays
[30], i.e., global clock-driven mechanisms, and wave front arrays [31], i.e., local data-driven
mechanisms, wherein each cell, instead of being a simple inner-product processor, is a pro-

grammable SIMD processor capable of performing matrix-vector operations in parallel. The number of required PEs for backward recursion computation is 45, and for forward recursion is 15, which leads to a total of 60 PEs per link. For a six DOF arm, the required number of PEs is 360 which is rather excessive and represents a major obstacle in the implementation of the architecture, even considering VLSI and WSI possibilities. There are also other drawbacks in this architecture. The first is the poor efficiency in hardware utilization since the SIMD processors are idle most of the time. The efficiency can be increased by a multilevel pipelining scheme [9]. However, this pipelining scheme, while increasing the throughput and efficiency, does not reduce the computation time. The second drawback is the lack of flexibility, since due to its fixed interconnection structure, it can not be used for other problems.

As will be shown in the next section, the optimization of the abstract architecture allows one to achieve (or closely approach) the computation time in Eq. (50) while minimizing the architectural complexity and providing flexibility.

B. Architectural Optimization at the Problem Level

From Fig. 3, if

$$mT_1 \geq T_2 + T_3 \tag{54}$$

then before the backward recursion reaches the link $m + 1$, the computation at link 1 is completed. If $m < n$, then the backward recursion computation for link i and link $m + i$ can be performed by the same processor. Thus, only m processors with a ring structure are needed to fully exploit parallelism at the problem level. With m processors, the forward recursion computation for link $m+i$ and link i can also be performed by the same processor. In fact, it can be shown that if Eq. (54) holds for a given m, then with m processors and for different cases in Eq. (41) no additional delay is introduced in the forward recursion computation [9]. The value of m does not depend on the degree to which the parallelism in matrix-vector operations is exploited. If maximum parallelism at the link level is exploited, then, assuming the same time for multiplication and addition, $m \geq 2$, which implies that at most 3 processors are required to fully exploit parallelism at the problem level. This results from the fact that due to the data dependency in the links computation, there is a limited parallelism at the problem level which is also consistent with the results of Table III. The exact value of m depends on the performance of the architecture at the link level, i.e., the value of $T_1, T_2,$ and T_3.

Eq. (54), though trivial, is of fundamental importance. As will be shown, if maximum parallelism at link level is not exploited, then T_2 and T_3 will increase faster than T_1 which means that m also increases. However, insofar as $m \geq n$, maximum parallelism at problem level is exploited and the computation time is given by Eq. (47). Note that the computation time in Eq. (47) is based on the assumption that there is no additional delay at the problem level, i.e., that maximum parallelism at the problem level is exploited. Hence, Eq. (54) represents a tradeoff between the complexity of the architecture (number of processors) at the problem level and that at the link level. It implies that a given achievable speedup factor, resulting from the exploitation of parallelism at the problem and link levels, can be obtained by increasing the number of processors at link level and reducing the number of processors at the problem level and vice versa. This provides an additional DOF in architectural design since the architectural complexity can be shifted to a level where it can be more easily handled.

C. Architectural Optimization at the Link Level

The optimization of the architecture at the link level can be performed by analyzing the sequencing of the abstract architecture in the link computation, based on the data flow analysis in Sec. IV.D. From Fig. 4, four major computational cycles for each link can be distinguished:

1) In the first cycle, T_1, ω_i and $\dot{\omega}_i$ are computed in parallel. Due to the assumption of the simultaneous availability of w_{i-1} and $\dot{w}_{i-1}$, computation of w_i' and $\dot{w}_1'$ can be done by performing $1SV$ and then $2VA$ in parallel. ω_i and $\dot{\omega}_i$ are then computed by performing $2MV$ in parallel. If $T_1 > T_2$, then e_i is partly computed in parallel with ω_i and $\dot{\omega}_i$. Therefore, in order to exploit maximum parallelism in this cycle, $2VA$ and then $2MV$ ($3MV$ for $T_1 > T_2$) need to be performed in parallel.

2) In the second cycle, T_2, the a_i and b_i terms are computed in parallel. If $T_2 > T_1$, then e_i is also computed in parallel with these terms. Hence, exploitation of maximum parallelism requires that $2MV$ ($3MV$ for $T_2 > T_1$) and $4VC$ be performed in parallel.

3) In the third cycle, T_3, with d_i being already computed, N_i is partly computed in parallel with $\ddot{r}_i$ and F_i, and then K_i is computed. Hence parallel computation of $1VC$ (or $1MV$) and $1SV$ is required.

4) In the fourth cycle, forward recursion, first parallel computation of $1VC$ and $1VA$ takes place, and then $1VC$ and $1MV$ are computed in parallel.

It can be seen that the main bottleneck is in the exploitation of maximum parallelism in the second cycle, since it requires more parallel operations than for any other cycle. However, the dependency of TP on T_1 and T_4 by a factor of n (Eq. 47) suggests maximum exploitation of parallelism at the first and fourth cycles. This implies that only the nodes computing ω_i and $\dot{\omega}_i$ in backward recursion and those computing f_i and L_i in forward recursion need a perfect mapping; other nodes can be processed sequentially. The perfect mapping of these nodes requires the capability of performing any two different (or similar) matrix- vector operations in parallel. In this case, T_1, T_4, and T_5 will be minimized while T_2 (slightly) and T_3 (significantly) will be increased. Note that in this case $T_2 > T_1$, which implies that in the first cycle only two matrix- vector operations need to be performed in parallel. Therefore, in Eq. (47), k_1' is minimized while k_2' is increased. This represents an efficient tradeoff between the exploitation of parallelism and the architectural complexity. In fact, performing more operations in parallel, while increasing the architectural complexity, does not significantly reduce TP since k_1' can not be further reduced and only k_2' is reduced. This also explains the poor efficiency of the nonoptimized architecture. With the increase in T_2 and T_3, the value of m in Eq. (54) also increases. However, from an architectural point of view, the exploitation of parallelism at the link level is more difficult than that at the problem level (see Sec. VI.B.). This suggests that one maintain the architectural complexity at the link level to a manageable degree but to increase the number of processors at the problem level.

The minimization of k_1 (and k_2) in Eq. (50) also requires the exploitation of the inherent parallelism in matrix-vector operations. Hence, in order to approach the computation time in Eq. (50), the processor at the link level should be capable of performing any two matrix-vector operations in parallel and exploiting parallelism, according to Eq. (49), in any matrix-vector operation. Therefore, 6 PEs are required for each processor. Knowing the performance of the processor at the link level, i.e., T_1, T_2, and T_3, the number of the processors at the problem level can then be determined from Eq. (54).

For the implemented architecture, $m = 4$, which leads to a total of 24 PEs. This represents more than an order of magnitude reduction in the number of PEs compared with nonoptimized architecture.

D. Models of the Optimized Architecture, Communication, and Synchronization

At the problem level, the architecture is an MIMD with local shared memory and ring topology. Given the limited communication, the MIMD model allows efficient exploitation of large grain and mostly asynchronous parallelism among the link computations. At the link level, in order to perform two matrix-vector operations in parallel, synchronization at basic operation level is required. Also, exploitation of parallelism within basic operation requires synchronization at the primitive operation level. Hence, the model of the architecture at link level is SIMD, which allows efficient exploitation of fine grain parallelism within the link computation. Note that, however, the architecture at link level differs from the classical SIMD in the sense that it should provide synchronization at two levels, i.e., the basic and primitive operation levels.

The communication and synchronization for the architecture at the problem level can be defined based on the flow of data and sequencing of the computation at link level. At the beginning of the computation, all processors are activated by the availability of $\theta_i, \dot{\theta}_i$ and $\ddot{\theta}_i$ (Fig. 2). After the initialization time, all processors except at link 1, enter the wait state. Computation at link $i + 1$ starts by receiving ω_i and $\dot{\omega}_i$ from the processor of link i. Hence, the synchronization between adjacent processors is of data-driven type which allows local synchronization among processors. Communication of ω_i and $\dot{\omega}_i$ can be performed asynchronously through the local shared memory. Communication of $\ddot{P}_i$ is different since both processors of link i and $i + 1$ are activated. However, the analysis of the sequencing of the operations for the optimized architecture can show that $\ddot{P}_i$ is made available before it is needed for the computation of link $i + 1$ [9]. Hence, the transfer of $\ddot{P}_i$ can also be performed asynchronously. With the completion of backward recursion computation, i.e., computation of K_i, the processor of link i enters the wait state. Forward recursion computation starts by the availability of f'_{i+1}. Therefore, the synchronization between the processors of link $i + 1$ and i, in forward recursion, is again of data-driven type and f'_{i+1} is communicated asynchronously. The communication of n'_{i+1} is similar to that of $\ddot{P}_i$ since the sequencing can show that n'_{i+1} is made available before it is needed at link i.

The above analysis shows that synchronization and communication do not cause extra overhead leading to degradation of algorithm performance. Note that simultaneous availability of ω_i and $\dot{\omega}_i$ has led to further overhead reduction.

VI. URRCS ARCHITECTURE

A. Basic Architecture Models and Functions

For the purpose of interfacing to the outside world, URRCS is basically an attached processor which can be interfaced to the bus of an external host as a part of bus memory. The architecture of URRCS consists of an internal host (a general-purpose processor) and several SIMD processors (Fig. 5). URRCS is a modular and expandable architecture; depending on the problem considered, more SIMD processors can be employed to cope with the computational requirements. The host is the control unit of the architecture which handles the interface with the external host, controls the activities of SIMD processors, and performs the necessary input/output operations. The SIMD processors can operate asynchronously and form an MIMD architecture. Hence, from an architectural point of view, URRCS is an MIMD-SIMD architecture.

The host controls the whole architecture by interpreting the instructions sent by the external host through a procedure call. The instruction, which may define a set of computations required for the implementation of a control scheme, is decomposed into a series of computation tasks to be performed by the SIMD processors. The host distributes the data among the SIMD processors and initiates their activities through a procedure call. The activities of the SIMD processors are then carried out independently of the host. The end of the computation is indicated by the SIMD processors to the host, which then transfers the results to the external host. The fact the URRCS is capable of computing various kinematic and dynamic problems allows chaining in the computations or exploiting synergism between kinematic and dynamic computations. This feature is particularly suitable for implementation of the task space control schemes since it reduces the overhead caused by the communication between the internal host and SIMD processors, on one hand, and the internal and external hosts, on the other hand. The SIMD processors are interfaced to the

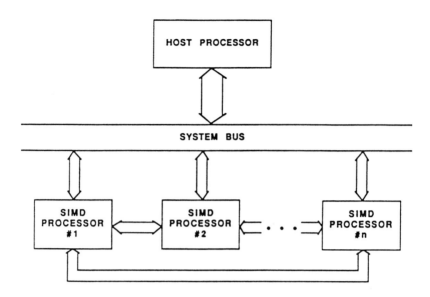

Figure 5.   URRCS Architecture

bus of the host processor as the memory units which allows a high-speed and asynchronous data transfer between the host and the SIMD processors.

The SIMD processors are interconnected through a ring structure which provides a reliable clock distribution as well as high-speed communication among the processors. The communication between adjacent processors can be performed at a rate of one datum per clock cycle. The perfect shuffle topology which is required for exploiting logarithmic parallelism in forward kinematics and Jacobian computation [10] is provided by message passing. The high speed communication reduces the overhead caused by message passing. The basic synchronization mechanism among processors is of the local data- driven type. However, both the fact that all processors are driven by the same clock and the regularity of the computation allow clock-based synchronized communication among processors.

B. SIMD Processor Architecture

The main challenge in the design of URRCS resides in the architecture of SIMD processors. In order to approach the computation time in Eq. (50), parallel computation of the combination of any two matrix-vector operations and exploitation of parallelism in any

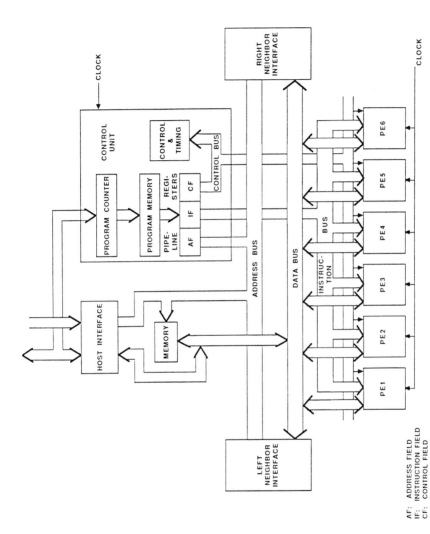

Figure 5.   SIMD parallel processor architecture

AF: ADDRESS FIELD
IF: INSTRUCTION FIELD
CF: CONTROL FIELD

matrix-vector operation are required. Such capabilities demand an extra architectural flex-ibility. The difficulty in the design resides in the fact that SIMD processors usually lack flexibility. Furthermore, the memory organization and data routing among the PEs, which are classical problems in the design of SIMD processors, become even more difficult due to the required flexibility.

The designed processor is a microprogrammable SIMD processor with one Control Unit (CU) and six PEs (Fig. 6). Under the control of the CU, the PEs can perform 6 independent primitive (scalar) operations in parallel, or form two groups of 3 PEs to perform two different or similar basic (matrix-vector) operations in parallel. In computing VA or SV with a group of PEs, parallelism in the operation is exploited by performing an independent addition or multiplication by each PE, but the operations of the 3 PEs within the same group are synchronized by the CU. In computing MV or VC with a group of 3 PEs, parallelism is exploited by performing 3 independent inner products. In this case, the group of PEs is distinguished by its global synchronization and common data sharing. For example, in performing MV, while each row of the matrix is read for corresponding PE, the components of the vector are read for all PEs of the same group (see Appendix 2). Also, in performing two similar or different matrix-vector operations, the operations of two groups of PEs are synchronized by the CU. For other computation problems in robotics, e.g., forward kinematics and Jacobian, matrix-matrix multiplication is required. In this case, under the control of the CU, all six PEs form a single group to exploit parallelism in such an operation. In order to cope with the size of the matrices, the operation is decomposed into two steps; in each step six inner-products are performed in parallel.

This organization provides the required flexibility since the PEs can be regrouped ac-cording to the operation. However, several problems may arise. The first problem is that if individual (or groups of) PEs perform different operations in parallel, then the CU should provide different instructions. The more important problems concern the memory organiza-tion and data path, since different data should be fetched simultaneously or routed among the PEs with a variable pattern. However, the features of the computation can be exploited for solving the above problems and achieving the required flexibility. The first two features are the inherent structure in the matrix-vector operations and, particularly, the small size of the matrices and vectors involved in the computation. Small amounts of data are needed to be fetched and aligned for each matrix-vector operation. The third feature, which is also inherent in all the considered computation problems, is the locality in the computation, i.e., the computation is performed on a limited set of data, which reduces the size of the required memory. Therefore, a cache memory can be used as the basic memory of the SIMD processor, providing very fast memory access. Exploiting the above features, the problem of the flexibility and memory organization is solved by time multiplexing the operations of PEs. The CU is designed to operate several times faster than the PEs, which means that the CU is capable of fetching the data and sending the instructions much faster than the PEs can perform the operations. This solution reduces the hardware complexity since the problems of memory organization and data path can be simply solved by using a single cache memory along with a common data bus. As a result, the architecture of SIMD processor is fairly simple (Fig. 6). This solution also simplifies the programming by reducing the width of CU microcode instruction. The microcode instruction width is 24 bits, with no vertical decoding, which is decomposed into three fields: Address Field (AF) with 8-bit width, Instruction Field (IF) with 3-bit width, and Control Field (CF) with 13-bit width. Consequently, the microcode development and debugging have been quite easy and fast.

The PE employed in the architecture is a simple processor (SN 74516 [40]) capable of performing primitive arithmetic operations (multiply, multiply and add/sub-tract, division, etc.). It runs with a clock frequency of 6 MHz while the primitive operations (except division) require 9 to 12 clock cycles to be completed. It has only one bus, which means that 2 clock cycles are needed for loading the operands. At each clock cycle, one operand along with one instruction can be loaded. The combination of 2 successively loaded instructions defines the type of operation to be performed. The CU runs with the same frequency as the processors which means that, exploiting parallelism in reading data and instructions, all the PEs can be activated within a few clock cycles and can perform different operations in parallel. In Appendix 2, the program for the computation of ω_i and $\dot{\omega}_i$ (which forms the critical path in the overall computation) is presented which further demonstrates the way that CU synchronizes the PEs operations and also the way that different matrix-vector operations are performed.

Practical implementation on the URRCS has shown that the N-E formulation, for a general (without any customization) six DOF manipulator, can be computed in 187 μs. The computation time of the forward kinematics is measured as 60 μs and the forward kinematics and Jacobian ($^{\circ}J_E$) as 75 μs. Such a performance allows the real-time implementation of the most advanced control schemes with a sampling rate greater than 1 KHz.

The main drawback of the architecture developed is that the PEs perform arithmetic operations with a 16-bit fractionary fixed-point format, which means that the variables should be scaled. However, using fractionary arithmetics instead of integer simplifies the problems of scaling and rounding since multiplication does not create overflow and the PEs perform the rounding under the control of the CU (Appendix 2). Nevertheless, this drawback leads to a loss of generality since for each manipulator a different scaling scheme is needed. It should be emphasized that the performance of the URRCS does not result from the speed of PEs or performing fixed- point operations. In fact, the PEs are rather slow since they perform fixed-point multiplication in 1.5 μs. Today, many commercially available floating-point processors are more than an order of magnitude faster in performing floating-point operations, e.g., WTL 1232/1233 [41].

VII. DISCUSSION AND COMPARATIVE STUDY

It is thought to be very unlikely that $P = NC$, where P denotes the class of problems that can be solved sequentially in a time bounded by a polynomial of the input size, i.e., n [32]. NC denotes the class of problems that can be solved by parallel algorithms in $O(\log^k n)$ steps for some constant k. This class of problems is known as Nick's class [33]. It is clear that $NC \subseteq P$. For $k = 1$, the time of $O(\log n) + O(1)$ represents the natural time lower bound.

However, most of the robot kinematic and dynamic problems belong to the class of NC [9]. Furthermore, it is possible to derive parallel algorithms capable of achieving the time lower bound of $O(\log n) + O(1)$ in solving these problems, e.g., the forward kinematics and Jacobian [10], the inertia matrix [11], and the inverse dynamics (the N-E formulation) [17,18]. These results represent a fundamental common property of these problems. They indeed reflect the mechanical structure of the manipulator since, for an open-loop manipulator, these problems can be described by a set of First-Order-Homogeneous-Linear (FOHL) recursive equations which, by using the recursive doubling algorithm [34], can be solved in $O(\log n) + O(1)$ steps. It should be pointed out that although it is shown that the forward dynamics problem also belongs to the class of NC [9,12] but there is not yet any report on an algorithm capable of achieving the time lower bound of $O(\log n) + O(1)$ in solving the problem.

Such parallelism, which is considered as logarithmic parallelism, can be exploited at the highest computational level, i.e., the problem level. However, considering the possible range of the size of these problems, i.e., the number of DOF, the exploitation of only logarithmic parallelism leads to a rather small speedup factor. In this respect, the computation of the inertial matrix represents a remarkable example. The serial computation complexity of evaluating the inertia matrix is $O(n^2)$. In [11], we presented two algorithms which achieve the time lower bound of $O(\log n) + O(1)$ in computing the inertia matrix. The reduction of the computation time from $O(n^2)$ to $O(\log n) + O(1)$ represents a significant unbounded parallelism in the computation. However, for $n = 6$, the best of these two algorithms achieves a speedup factor of 2.4, which is less than the achievable by exploitation of even a limited parallelism in the matrix-vector operations (see Eq. 49).

Our analysis showed that despite its recursive form, the N-E formulation represents a high degree of inherent parallelism which exists at several computational levels and is mostly fine grain. This is also consistent with the fact that the coefficients on the polynomial computational complexity of the N-E formulation are rather large. Hence, in order to achieve a significant speedup, the reduction of the coefficients is of prime importance. This is a general conclusion which is also valid for other kinematic and dynamic problems. In fact, the developed $O(n)$ algorithms for the forward dynamics problem [38,39] encompass even larger coefficients than those of the N-E formulation. However, the reduction of the coefficients requires the exploitation of parallelism at several computational levels which has not been considered in most of the previous proposals.

The MIMD proposals in [13,14,15,16] differ in three main aspects: the architecture, the task graph generation, and the scheduling schemes. With respect to their efficiency for parallel computation of the N-E formulation, the most important aspect is the architecture since it affects the task graph granularity and hence the achievable speedup. For any architecture, the communication overhead is a strong function of its connection dimensionality [35]; the hypercube incorporates the highest connectivity and minimum overhead, while the global shared memory bus connected architecture incorporates the lowest dimensionality and maximum overhead. The overhead imposes a minimum size on the granularity of the task graph (called the critical task size), below which the speedup becomes rather a slow down, i.e., the overhead increases faster than the computation time decreases [36]. The reported speedup factors clearly shown the limited parallelism which can be efficiently exploited by MIMD architectures. In fact, the critical task size of the MIMD architectures, even that of the hypercube, is relatively large, which prevents exploitation of the fine grain parallelism in the N-E formulation.

In [22] and [20], considering MIMD architectures, attempt is made to solve the problem of the critical task size. In both proposals, the large grain parallelism is increased by reducing the data dependency in the computation. In [22] the data dependency is reduced by replacing the propagating data with predicted data. However, the reported simulation results have shown non-negligible error in the computation for relatively fast motion. In [20] the N-E state-space formulation is used, which reduces the data dependency in the computation of angular velocities and accelerations by using Euler angles and their time derivatives. However, this leads to the degradation of the algorithm's accuracy due to the singularity in the orientation representation by Euler angles. Furthermore, the computation of Euler angles and their time derivatives involves time-consuming trigonometric functions and division operations and is inherently serial. As a result, the serial part of the computation is increased, which limits the achievable speed-up. For a six-DOF arm a limited speedup factor of 3.28 is reported.

In [18] an $O(\log n) + O(1)$ SIMD algorithm is developed by transforming the N-E formulation into a set of FOHL recursions. The recursive doubling algorithm is then used to solve the set of equations in a logarithmic time. The computation time of the N-E formulation is derived as

$$T_P = (27\lceil \log_2 n \rceil + 116)M + (9\lceil \log_2(n+1)\rceil + 27\lceil \log_2 n \rceil + 84)A$$

where $\lceil X \rceil$ denotes the smallest integer greater than or equal X. As a result of the transformation, the parallelism at the problem level is increased, leading to an unbounded parallelism. For $n = 6$, the algorithm achieves a speedup of 4.45 (assuming the same time for M and A). Although the speedup increases for redundant arms, a comparison between the above computation time and that of Eq. (50) shows that the algorithm of [18] does not become more the one presented here unless $n \geq 64$. This result clearly confirms our statement about the poor achievable speedup by exploiting only logarithmic parallelism.

The work in [17] is more closely related to that of this paper. In [17] data flow analysis was first applied to the original N-E formulation. Also, an $O(\log n) + O(1)$ algorithm was developed which seems to represent the lower bound (in the sense of the number of operations) in the computation of the N-E formulation. Note that the computation time in [17] is based on the exploitation of maximum parallelism in matrix-vector operations, i.e., Eq. (48). The data flow analysis in [17] was mainly concerned with the regularity of the flow of data. Hence, while the analysis showed the inherent parallelism in the computation, it did not reveal the structure of the computation nor indicated the features of an optimized architecture for the exploitation of parallelism. As a result, a perfect mapping of the computation graph is proposed which shares most of the drawbacks of the nonoptimized architecture of Sec. V.A.

The detailed review of different proposals allows the analysis of our approach with respect to the more general issues regarding the mapping problem in parallel processing. The architecture developed, by combining the features of both MIMD and SIMD architectures, represents a synergistic solution for the efficient exploitation of parallelism in the N-E formulation. In fact, it is not surprising that our attempt, in optimizing conflicting objectives to achieve a better performance, has led to such a solution since synergism is a key factor to designing high performance parallel architectures [37]. The performance of URRCS in the practical implementation of the N-E formulation, and compared with other proposals, shows that synergism can lead to a high speedup factor while minimizing overhead and hardware complexity and providing flexibility. This represents the basis of our approach to efficient computation of different kinematic and dynamic problems in robotics. However, another factor contributing to the flexibility and efficiency of URRCS is the architecture of the SIMD processors which allows efficient computation of various matrix-vector operations. With respect to our discussion in Sec. III, the SIMD processors are optimized for performing matrix-vector operations which are the basic operation of the class of parallel algorithms.

VIII. CONCLUSION

We investigated parallel computation of the N-E formulation. Our hierarchical approach to the mapping problem allowed the analysis of the parallelism at several computational levels along with the architectural features for its exploitation, taking into account different conflicting objectives. Such an approach led to the practical implementation of the N-E formulation and other kinematic and dynamic problems. The analysis and implementation show that it is possible to design and implement parallel architectures capable of providing a complete solution for the computationally intensive problems in real-time manipulator control and simulation.

It should be pointed out that the URRCS was mainly implemented to show the proof of feasibility and to be used as an analysis validation tool. In fact, the implemented architecture, while being only one alternative (perhaps the simplest but certainly the least expensive) among many others, is "over kill" the class of problems considered. However, this architecture can be improved in many aspects. The experience gained in the design and implementation of the URRCS have contributed to the design of a new version of the architecture, called Robot Mathematics Processor (RMP), which is currently under development at the Jet Propulsion Laboratory. In order to achieve even more flexibility and generality, synergism has been considered the key factor in the architectural design. RMP, while performing floating-point operations, is more than an order of magnitude faster than URRCS. From an architectural point of view, it incorporates a more elaborate memory organization and an extensive data path. These features further simplify the microcode development for RMP.

In its minimal configuration, RMP will provide a computing power of 600 MIPS (Million Instruction Per Second) and 300 MFLOPS (Million Floating-Point Operation Per Second). Furthermore, like URRCS, this computing power is tailored to the application domain requirements. Such a performance is expected to provide many unprecedent capabilities, e.g., the faster-than-real-time dynamic simulation capability which is particularly essential for space teleoperation. It will also allow one to attack and explore more computationally intensive problems in robotics, e.g., real-time dynamic trajectory generation and planning. We believe that the performance of and practical results obtained by the RMP and URRCS prove the validity of the parallel processing approach to achieving practical solution for real-time dynamic control and simulation of manipulator.

Acknowledgement

The research described in this paper was carried out by the Jet Propulsion Laboratory, California Institute of Technology, under the contract with the National Aeronautics and Space Administration (NASA).

References

1. A.K. Bejczy, "Robot Arm Dynamics and Control," Jet Propulsion Lab., Tech. Memo 33-669, Feb. 1974.

2. R.P.C. Paul, "Modeling, Trajectory Calculation, and Servoing of a Computer Controlled Arm," Stanford Univ., AI Memo 177 Sept. 1972.

3. J.Y.S. Luh M.W. Walker, and R.P.C. Paul, "Resolved Acceleration Control of Mechanical Manipulators," IEEE Trans. Auto. Control, Vol. AC-25(3), 1980.

4. E. Freund, "Fast Nonlinear control with Arbitrary Pole-Placement for Industrial Robots and Manipulators," Int. J. Robotics Res., Vol. 1(1), 1982.

5. A.K. Bejczy, T.J. Tarn, and X. Yun, "Robust Robot Arm Control with Nonlinear Feedback," Proc. IFAC Sym., Barcelona, 1985.

6. O. Khatib, *Commande Dynamique dans l'Espace Operationnel des Robots Manipulaterus en Presence d'Obstacles*, Ph.D. diss., ENSAE, France, 1980.

7. J.Y.S. Luh, M.W. Walker, and R.C.P. Paul, "On-Line Computation Scheme for Mechanical Manipulators," Trans. ASME J. Dynamic Syst., Measurement and Control, Vol. 120, June 1980.

8. J.M. Hollerback, "A Recursive Lagrangian Formulation of Manipulator Dynamics and a Comparative Study of Dyanmics Formulation Complexity," IEEE Trans. Syst., Man, Cybern,. Vol. SMC-10(11), Nov. 1980.

9. A. Fijany, *Parallel Algorithms and Architectures for Robotics*, Ph.D. Diss., Univ. of Paris XI (Orsay), France, Sept. 1988.

10. A. Fijany and J.G. Pontnau, "Parallel Computation of the Jacobian for Robot Manipulators," Proc. IASTED Intl. Conf. on Robotics and Automation, Santa Barbara, May 1987.

11. A. Fijany and A.K. Bejczy, "A Class of Parallel Algorithms for Computation of the Manipulator Inertia Matrix," IEEE Trans. on Robotics and Automation, Vol. 5(5), Oct. 1989.

12. A. Fijany and A.K. Bejczy, "Parallel Algorithms and Architecture for Computation of the Manipulator Forward Dynamics," Proc. of 3rd Annual Conf. on Aerospace Computational Control, Oxnard, CA, Aug. 1989.

13. J.Y.S. Luh and C.S. Lin, "Scheduling of Parallel Computation for a Computer Controlled Mechanical Manipulators," IEEE Trans. Syst., Man, and Cybern., Vol. SMC-12(2), March 1982.

14. H. Kasahara and S. Narita, "Parallel Processing of Robot-Arm Control Computation on a Multimicroprocessor," IEEE J. Robotics and Automation, Vol. RA-1(2), June 1985.

15. J. Barhen "Hypercube Ensembles: An Architecture for Intelligent Robots," in *Computer Architectures for Robotics and Automation,* Ed. J.H. Graham, Gordon and Breach Science Publisheers, 1987.

16. J. Barhen, *et al.*, "Optimization of the Computational Load of a Hypercube Supercomputer Onboard a Mobile Robot," J. Applied Optics, Vol. 26(23), Dec. 1987.

17. L.H. Lathrop, "Parallelism in Manipulator Dynamics," Intl. J. Robotics Res., Vol. 4(2), 1985.

18. C.S.G. Lee and R.P. Chang, "Efficient Parallel Algorithm for Robot Inverse Dynamics Computation," IEEE Trans. Syst., Man, Cybern., Vol. SMC-16(4), July 1986.

19. D.E. Orin, *et al.*, "Pipeline/Parallel Algorithms for the Jacobian and Inverse Dynamics Computation," IEEE Intl. Conf. Robotics and Automation, March 1985.

20. Y-F. Zheng and H. Hemami, "Computation of Multibody System Dynamics by a Multiprocessor Scheme," IEEE Trans. Syst., Man, Cybern., Vol. SMC-16(1), Jan. 1986.

21. H. Hemami, "A State-Space Model for Interconnected Rigid Bodies," IEEE Trans. Automatic Control, Vol. AC-27, April 1982.

22. E.E. Binder and J.H. Herzog, "Distributed Computer Architecture and Fast Parallel Algorithms for Real-Time Robot Control," IEEE Trans. Syst., Man, Cybern., Vol. SMC-16(4), July 1986.

23. R. Nigam and C.S.G. Lee "A Multiprocessor-Based Controller for the Control of Mechanical Manipulators," IEEE J. Robotics and Automation, Vol. 1(4), Dec. 1985.

24. D.J. Kuck, "Multioperation Machine Computational Complexity," in *Complexity of Sequential and Parallel Numerical Algorithms.* J.F. Traub (Ed.), Academic Press, 1973.

25. K. Kwang, "Advanced Parallel Processing with Supercomputer Architectures," Proc. of the IEEE, Vol. 75(10), Oct. 1987.

26. L.H. Jamieson, "Characterizing Parallel Algorithms," in *The Characteristics of Parallel Algorithms,* L.H. Jamieson *et al.* (Eds.), The MIT Press, 1987.

27. K. Hwang and F.A. Briggs, "*Computer Architecture and Parallel Processing*, Mcgraw-Hill, 1984.

28. R.W. Hockney and C.R. Jesshope, *Parallel Computers*, Adam Hilger, 1981.

29. L. Snyder, *et al.* (Eds.), *Algorithmically Specialized Parallel Computers*, Academic Press, 1985.

30. H.T. Kung, "Why Systolic Architectures?," IEEE Computer, Jan. 1982.

31. S.Y. Kung, "On Supercomputing with Systolic/Wavefront Array Processors," Proc. of the IEEE, Vol. 72(7), 1984.

32. J.S. Vitter and R.A. Simons, "New Classes of Parallel Complexity: A Study of Unification and Other Complete Problems for P," IEEE Trans. on Computer, Vol. C-35(5), May 1986.

33. S.A. Cook, "An Overview of Computational Complexity," Com. ACM, Vol. 26, June 1983.

34. P.M. Kogge, "Parallel Solution of Recurrence Problems," IBM J. Res. Develop., Vol. 18, March 1974.

35. D. Parkinson, "Organizational Aspects of Using Parallel Computers," Parallel Computing 5, pp. 75-83, 1987.

36. C.D. Polychronopoulos and U. Banerjee, "Processor Allocation for Horizontal and Vertical Parallelism and Related Bounds," IEEE Trans. on Computer, Vol. C-16(4), April 1987.

37. G.J. Lipovski and M. Malek, *Parallel Computing: Theory and Comparisons,* John Wiley & Sons, Inc., 1987.

38. R. Featherstone, "The Calculation of Robot Dynamics Using Articulated-Body Inertia," Intl. J. Robotics Res., Vol. 2(1), 1983.

39. G. Rodriguez, "Kalman Filtering, Smoothing, and Recursive Robot Arm Forward and Inverse Dynamics," Jet Propulsion Lab. Publication, 1986.

40. R.W. Blasco, *et al.*, "SN 54/74S516 Co-Processor Supercharges 68000 Arithmetic," Application Note (AN-114), Monolithic Memories Inc., Sunnyvale, CA, 1984.

41. WTL 1232/1233 Floating Point Multiplier and ALU. Weitek Co., Sunnyvale, CA, July 1986.

Appendix 1. Data Flow Analysis

In this appendix, the data flow analysis for the three cases, given by Eq. (42), is presented.

1. $T_1 > T_2$

1.1 Backward Recursion

Let t_{1i} denote the delay between the avialability of ω_i and $\ddot{P}_{i-1}$ at link i. For link 2, $t_{12} = T_1 - T_2$. From Eq. (31),

$$T_{22} = \text{Max}\left((1MV - t_{12}),\ (T_2 - 1VA)\right) + 1VA$$

$$= \text{Max}\left((T_2 - T_1 + 1MV),\ (T_2 - 1VA)\right) + 1VA$$

$$= \text{Max}\left((T_2 - 1VA - 1SV),\ (T_2 - 1VA)\right) + 1VA = T_2$$

Therefore, it follows that for $n \geq i > 1$,

$$t_{1i} = T_1 - T_2$$
$$T_{2i} = T_2$$

Note that, exploiting maximum parallelism, $Av(c_i) = Av(d_i)$. From Eqs. (32)-(34),

$$T_{3i} = \text{Max}\left((1VA + 1SV),\ (1MV + 1VC + 1VA - T_2)\right) + 1VC + 1VA$$

$$= \text{Max}\left((1VA + 1SV),\ (1MV - 1VC - 1VA)\right) + 1VC + 1VA = T_3$$

If $1MV > 1VC + 2VA + 1SV$, then $T_3 = 1MV$ otherwise, $T_3 = 1VC + 1SV + 2VA$. However, the exact value of T_3 does not affect out analysis and we let

$$T_3 = 1VC + 1SV + 2VA$$

It follows that

$$TB_i = T_1 + T_2 + T_3 = 1MV + 3VC + 2SV + 5VA$$

1.2 Forward Recursion

Let t_{2i} denote the delay between availability of n'_{i+1} and f'_i. For link n, f'_{n+1} and n'_{n+1} are available at the same time. From Eqs. (37)- (38),

$$t_{2n} = T_4$$
$$T_{5n} = (1VC + 2VA + 1MV) - (1VA + 1MV) = 1VC + 1VA = T_5$$
$$T_{6n} = TF_n = T_4 + T_5 = 1MV + 2VA + 1VC$$

For link $n - 1$,

$$t_{2n-1} = T_4 - T_5 = 1MV - 1VC$$

Let $1MV > 1VC$. The case $1MV < 1VC$ will be studied later. From Eq. (38),

$$T_{5n-1} = \text{Max} \left(\text{Min} \left((1VC + 1VA - T_4), \; (-t_{2n-1}) \right), \right.$$

$$\left. \text{Min} \left((1VC - T_4), (-t_{2n-1} + 1VA) \right) \right) + 1VA + 1MV$$

$$T_{5n-1} = \text{Max} \left(\text{Min} \left((1VC - 1MV), \; (1VC - 1MV) \right), \right.$$

$$\left. \text{Min} \left((1VC - 1MV - 1VA), (1VA + 1VC - 1MV) \right) \right)$$

$$+ 1VA + 1MV = 1VC + 1VA = T_5$$
$$T_{6n-1} = t_{2n-1} + T_{5n-1} = 1MV + 1VA$$

The last two relations can be easily verified, since $T_{5n} = 1VC + 1VA$ yields that the term $k_{n-1} + P_{n-1} \times f'_n$ is available at the same time as n'_n. Hence, for $n > i \geq 1$,

$$t_{2i} = 1MV - 1VC$$
$$T_{5i} = 1VC + 1VA = T_5$$
$$T_{6i} = T_4 = 1MV + 1VA$$
$$TF_i = T_4 + T_5 = 1MV + 1VC + 2VA$$

2. $T_1 = T_2$

For this case, in backward recursion $t_{2i} = 0$ and $T_{2i} = T_2 = T_1$ for $i \geq 1$. In forward recursion, the data propagation is the same as for the first case.

3. $T_1 < T_2$

3.1 Backward Recursion

With $T_1 < T_2, \omega_i$ is made available before $\ddot{P}_{i-1}$ at link i. Hence, t_{1i} denotes the delay between the availability of $\ddot{P}_{i-1}$ and ω_i (Fig. 3b). It follows that

$$t_{12} = T_2 - T_1$$
$$T_{22} = \text{Max} \left(1MV, (T_2 - 1VA - t_{12}) \right) + 1VA = \text{Max} \left(1MV, \; (T_1 - 1VA) \right) + 1VA = T_1$$

It can be concluded that for $n \geq i \geq 2$,

$$t_{1i} = T_2 - T_1$$
$$T_{2i} = T_1$$
$$TB_i = 2T_1 + t_{1i} + T_3 = T_1 + T_2 + T_3 = 1MV + 3VC + 2SV + 5VA$$

3.2 Forward Recursion

If $1MV > 1VC$, the data propagation and parallel computation time are the same as before. It is unlikely that $1MV < 1VC$, nevertheless the same procedure can be applied to obtain the timing diagram. At link n,

$$T_{5n} = 1MV + 1VA = T_5$$
$$T_{6n} = 1MV + 1VC + 2VA$$

Hence, $K_{n-1} + P_{n-1} \times f'_n$ is made available at the same time as n'_n which yields that

$$T_{6n-1} = 1MV_1VA = T_4$$
$$T_{5n-1} = 1VC + 1VA = T_5$$

Therefore, it can be concluded that for $n > i \geq 1$,

$$t_{2i} = 1VC - 1MV$$
$$T_{6i} = 1MV + 1VA = T_4$$
$$T_{5i} = t_{2i} + T_{6i} = 1VC + 1VA = T_5$$
$$TF_i = T_4 + T_5 = 1MV + 1VC + 2VA$$

Appendix 2. A Sample Program and Timing of the SIMD Processor

In this appendix the program for the computation of ω_i and $\dot{\omega}_i$ by the SIMD processor of link i is presented. The SIMD processor has completed the computation of A_i and is in the wait state. The computation starts by receiving ω_{i-1} and $\dot{\omega}_{i-1}$ from the processor of link $i - 1$. The SIMD processor of link $i - 1$ has a direct access to the memory of the processor of link i and by the end of the data transfer triggers its activities. For the sake of space, only the computation of $\omega_i = A_i^t \omega'_i$ and $\dot{\omega}_i = A_i^t \dot{\omega}'_i$ is presented. ω_i is computed by the group of PEs consisting of $PE1(\omega_{i(x)})$, $PE2(\omega_{i(y)})$, and $PE3(\omega_{i(z)})$. $\dot{\omega}_i$ is computed by the group of PEs consisting of $PE4(\dot{\omega}_{i(x)})$, $PE5(\dot{\omega}_{i(y)})$ and $PE6(\dot{\omega}_{i(z)})$. For each PE the sequence I_1 and I_2 defines the operation to be performed. I_3 and I_4 denote the rounding and write operation, respectively. CC, PE, INS, OP, CON, NOP, and RN stand for clock cycle, processor element, instruction, operand, control, no operation, and right neighbor (processor of link $i + 1$), respectively.

CC	PE	INS	OP	CON
1)	PE1&PE4	I_1	a_{11}	Read
2)	PE2&PE5	I_1	a_{12}	Read
3)	PE3&PE6	I_1	a_{13}	Read

4)	PE1&PE2&PE3	I_2	$\omega'_{i(x)}$	Read
5)	PE4&PE5&PE6	I_2	$\dot{\omega}'_{i(x)}$	Read
6)	NOP	-	-	-
	$\vdots$			
13)	PE1&PE2&PE3	I_3	-	Round
14)	PE4&PE5&PE6	I_3	-	Round
15)	PE1&PE4	I_1	a_{21}	Read
16)	PE2&PE5	I_1	a_{22}	Read
17)	PE3&PE6	I_1	a_{23}	Read
18)	PE1&PE2&PE3	I_2	$\omega'_{i(y)}$	Read
19)	PE4&PE5&PE6	I_2	$\dot{\omega}'_{i(y)}$	Read
20)	NOP	-	-	-
	$\vdots$			
27)	PE1&PE2&PE3	I_3	-	Round
28)	PE4&PE5&PE6	I_3	-	Round
29)	PE1&PE4	I_1	a_{31}	Read
30)	PE2&PE5	I_1	a_{32}	Read
31)	PE3&PE6	I_1	a_{33}	Read
32)	PE1&PE2&PE3	I_2	$\omega'_{i(z)}$	Read
33)	PE4&PE5&PE6	I_2	$\dot{\omega}'_{i(z)}$	Read
34)	NOP	-	-	-
	$\vdots$			
41)	PE1&PE2&PE3	I_3	-	Round
42)	PE4&PE5&PE6	I_3	-	Round
43)	PE1	I_4	$\omega_{i(x)}$	Write local and RN memory
44)	PE2	I_4	$\omega_{i(y)}$	Write local and RN memory
45)	PE3	I_4	$\omega_{i(z)}$	Write local and RN memory
46)	PE4	I_4	$\dot{\omega}_{i(x)}$	Write local and RN memory
47)	PE5	I_4	$\dot{\omega}_{i(y)}$	Write local and RN memory
48)	PE6	I_4	$\dot{\omega}_{i(z)}$	Write local and RN memory
49)	-	-	-	Go RN

At the first step in the computation (clock cycles 1-5), the sequence of I_1 and I_2 denotes the simple multiplication operation. At the second and third steps in the computation (clock cycles 15-33), the sequence of I_1 and I_2 denotes the multiply and add (to the previous result) operation. The last instruction (Go RN) triggers the activities of the processor of link $i + 1$. Note that, ω_i and $\dot{\omega}_i$ are simultaneously available for the computation of links i and $i + 1$. The computation of ω_i and $\dot{\omega}_i$, including their transfer to the local memory and the memory of the processor of link $i + 1$, require 49 clock cycles. With a 6MHz clock frequency, the computation time of ω_i and $\dot{\omega}_i$, i.e., the delay between the activation of the adjacent processors in backward recursion, is less than $12\mu s$. Note that, the SIMD processor is capable of performing 2 MV in parallel and in $8\mu s$.

TECHNIQUES FOR PARALLEL COMPUTATION OF MECHANICAL MANIPULATOR DYNAMICS. PART II: FORWARD DYNAMICS

Amir Fijany and Antal K. Bejczy

Jet Propulsion Laboratory
California Institute of Technology
Pasadena, CA 91109

I. INTRODUCTION

The manipulator forward dynamics problem concerns the determination of the motion of the mechanical system resulting from the application of a set of joint forces/torques which is essential for dynamic simulation. The motivation for devising fast algorithms for forward dynamics computation stems from applications which require extensive off-line simulation capability as well as applications which require real-time dynamic simulation capability. In particular, for many anticipated space teleoperation applications, a faster-than-real-time simulation capability will be essential. In fact, in the presence of unavoidable delay in information transfer, such a capability would allow a human operator to preview a number of scenarios before run-time [1].

The forward dynamics problem can be stated as follows: given the vectors of actual joint position (Q) and velocities $(\dot{Q})$, the external force (F_E) and moment (N_E) exerted on the End-Effector (EE), and the vector of applied joint forces/torques (τ), find the vector of joint accelerations $(\ddot{Q})$. Integratation of the vector of joint accelerations leads to the new values for Q and $\dot{Q}$, and the process is then repeated for the next τ. The first step in computing the forward dynamics is to derive a linear relation (for the given manipulator

CONTROL AND DYNAMIC SYSTEMS, VOL. 40

configuration described by Q) between the vector of joint acceleration $\ddot{Q}$ and the vector of applied inertial forces/torques Γ. Given the dynamic equations of motion as

$$A(Q)\ddot{Q} + C(Q,\dot{Q}) + G(Q) + J^t(q)F_E = \tau \tag{1}$$

and defining the bias vector b as

$$b = C(Q,\dot{Q}) + G(Q) + J^t(Q)F_E \tag{2}$$

the linear relation is derived:

$$A(Q)\ddot{Q} = \tau - b = \Gamma \tag{3}$$

where Q, $\dot{Q}$, $\ddot{Q}$ and $\tau \varepsilon \Re^n$, and $F_E \ \varepsilon \Re^6$ is a combined representation of F_E and N_E. $A(Q) \ \varepsilon \Re^{n \times n}$ is the symmetric, positive definite, inertia matrix, and $J \varepsilon \Re^{6 \times n}$ is the Jacobian matrix (t denotes matrix transpose). The bias vector b represents the contribution due to coriolis and centrifugal terms $C(Q\dot{Q})$, gravitational terms $G(Q)$, and the external force and moment. Hence, in Eq. (3), $\Gamma \varepsilon \Re^n$ is the vector of applied inertial forces/torques. The bias vector b can be computed by solving the inverse dynamics problem, using the Newton-Euler (N-E) formulation [2], while setting the vector of joint accelerations to zero. The computation of the vectors b and Γ, and the integration of computed $\ddot{Q}$ are the common first and last steps in any algorithm for solving the forward dynamics problem.

The proposed algorithms for the forward dynamics problem differ in their approaches to solving Eq. (3), which directly affects their asympototic computation complexity. These algorithms can be classified as:

1. The $O(n)$ algorithms [3]-[6] which, by taking a more explicit advantage of the structure of problem, e.g., by using the Articulated-Body Inertia [3]-[4] and recursive factorization and inversion of the inertia matrix [5]-[6], solve Eq. (3) in $O(n)$ steps without explicit computation and inversion of the inertia matrix.

2. The $O(n^2)$ conjugate gradient algorithms [7,10] which interatively solve Eq. (3) without explicit computation and inversion of the inertial matrix. The conjugate gradient algorithms are guaranteed to converge to the solution in at most n iterations which, given the $O(n)$ computation complexity of each iteration, leads to their overall $O(n^2)$ computational complexity.

3. The $O(n^3)$ algorithms [7] which solve Eq. (3) by explicit computation and inversion of the inertia matrix, leading to an $O(n^3)$ computational complexity.

However, any analysis of the relative efficiency of these algorithms should also include the realistic size of the problem, i.e., the number of Degree-Of-Freedom (DOF), rather than the asymptotic complexity alone. In fact, the comparative study in [3]-[4] shows that the $O(n^3)$ Composite Rigid-Body algorithm is the most efficient for n less than 12. It should be pointed out that the efficiency of the $O(n^3)$ and $O(n^2)$ algorithms has been recently improved [9]-[10]. However, despite these improvements, even the fastest $O(n^3)$ algorithm is far from providing the efficiency required for real-time or faster-than-real-time simulation. This observation clearly suggests that the exploitation of a high degree of parallelism in the computation is the key factor for achieving the required efficiency.

The analysis of the efficiency of the different algorithms for parallel computation is more complex than that for serial computation. In the next section, the three classes of algorithms are analyzed based on their efficiency for parallel computation and it is shown that the $O(n^3)$ algorithms are also the most efficient for parallel computation. However, parallelization of the $O(n^3)$ algorithms represents a challenging problem since it requires the development of parallel algorithms for computation of a set of fundamentally different problems, i.e., the N-E formulation (or, the bias vector), the inertia matrix, the factorization of the inertia matrix, and the solution of triangular systems.

One possible strategy is to first develop efficient parallel algorithm for each specific problem and then design appropriate architecture for each algorithm. However, in addition to the cost issue, this strategy may not lead to an overall efficiency. In fact, the synchronization of different architectures for performing the overall computation and the data alignment and communication among the architectures may result in significant overhead, degrading the overall performance. Another possible strategy is to first consider a specific parallel architecture and then, for each problem, design an algorithm which can be efficiently implemented on the architecture. The first difficulty in this approach is in the choice of the optimal architecture since the algorithmic design is now constrained by the given architectural features which may represent the limiting factor in efficient exploitation of parallelism in each problem. Hence, the choice of architecture is critical and also very complex since it should be made based on the analysis of all the possible algorithms for each problem. The second difficulty is in overhead minimization since proper sequencing and implementation of different algorithms requires data alignment which may lead to excessive data movement within the architecture.

Parallel computation of the $O(n^3)$ algorithms is investigated by Kasahara, Fujii, and Iwata [14] and Lee and Chang [15]. In [14] efficient scheduling schemes for mapping the computation on a rather general purpose multiprocessor architecture is investigated. However, from both algorithmic and architecturalpoint of view, the work in [15] is closer to this paper. In [15] two approaches are developed. In the first approach, an SIMD architecture with n processors interconnected through a generalized-cube network is considered. An $O(log_2 n)$ algorithm for computation of bias vector and an $O(n)$ parallel version of the Composite Rigid-Body algorithm [7] for computation of inertia matrix are developed. A parallel $O(n^2)$ Cholesky algorithm and the $O(n)$ Column-Sweep algorithms are also proposed for the factorization of inertia matrix and solution of triangular systems, leading to an overall computational complexity of $O(n^2)$. However, the main drawbacks of the proposed algorithms are the complexity of the required interconnection network and the communication overhead which results from the excessive data alignment needed for different algorithms. In order to align the data, generated by parallel composite rigid- body algorithm, for parallel Cholesky algorithm a set ordering algorithm is proposed which requires extensive data transfers among the processors and leads to a communication complexity of $O(n^2 \log_2 n)$. As a result, even by using the generalized-cube network which requires $n\lceil \log_2 n\rceil$ gates, the communication complexity, $O(n^2 \log_2 n)$, is still greater than the overall computational complexity of the proposed algorithms, $O(n^2)$, which degrades the performance of parallel computation. In the second approach, in order to further reduce the computational complexity, a systolic algorithm for Cholesky decomposition on a VLSI processor array is proposed. This algorithm reduces the complexity of the factorization, and, hence the overall computation to $O(n)$. However, this reduction is achieved at the cost of extra architectural complexity and excessive communication between the SIMD architecture and the VLSI array.

In this paper, in order to achieve overall efficiency, a both algorithmic and architectural approach for parallel computation of the $O(n^3)$ algorithms is proposed. From an algorithmic point of view, parallel algorithms are developed by carefully analyzing their architectural requirements and not their computational complexities alone. For each problem attempt is made to develop parallel algorithm with simple architectural requirements. Also, for different problems, attempt is made to develop parallel algorithms with as much as possible similar structural properties and hence architectural requirements. As a result, the parallel algorithms for computation of inertia matrix and bias vector and the parallel algorithms for factorization of inertia matrix and triangular linear systems solution have very similar structural features. An architecture for implementation of the algorithms is proposed by taking into account the constraints in practical implementation by using VLSI and WSI (Wafer Scale Integration) technologies. Analysis is done to reduce the complexity of various architectural features, e.g., communication and synchronization mechanisms, without significantly degrading the algorithms' performance.

The computation of inertia matrix, for all practical n, represents the most computationally intensive problem in the $O(n^3)$ algorithm. As a result, the efficiency in parallel computation of inertia matrix significantly affects the architectural choice. A new algorithm for computation of inertia matrix is developed which, though not efficient for serial computation, is highly suitable for parallel computation. This algorithm, compared to the Composite Rigid-Body algorithm [7], not only provides a higher degree of parallelism but also requires much simpler communication and synchronization mechanisms. A parallel version of this algorithm is developed which achieves the time lower bound of $O(\log_2 n)$ with $O(n^2)$ processors. The architectural features required for implementation of the parallel algorithm is then analyzed. It is shown that the algorithm can be implemented on a triangular array of processors with a simple nearest neighbor interconnection without significantly increasing the communication cost. A synchronization mechanism combining both clock-based and data-driven features is also defined for the triangular array. Based on the architecture so defined, efficient parallel algorithms for computation of bias vector, factorization of inertia matrix, and solution of triangular systems are developed. In order to minimize the data alignment overhead, the design of each algorithm takes into account the organization of the data that the algorithm requires within the triangular array.

This paper is organized as follows. In Sec. II parallelism and time and processor bounds in the computation of forward dynamics are analyzed. In Sec. III the new algorithm for computation of the inertia matrix is presented. In Sec. IV parallel algorithm for computation of inertia matrix is developed. In Sec. V mapping the parallel algorithm is analyzed and a triangular processor array is defined as the unique architecture for parallel computation of forward dynamics. In Sec. VI parallel computation of the N-E formulation on the processor array is investigated. In Sec. VII parallel algorithms for solution of the linear system on the processor array are presented. Finally, some discussion and concluding remarks are made in Sec. VIII.

II. PARALLELISM IN FORWARD DYNAMICS: TIME AND PROCESSOR BOUNDS IN COMPUTATION

The analysis of time and processors bounds in parallel computation of a given problem is of fundamental theoretical importance. It can determine the inherent parallelism in the problem and the bound on the number of processors required for exploiting maximum parallelism and achieving the time lower bound. However, in addition to the theoretical importance, it can also provide, as is the case for forward dynamics problem, useful insights for devising more practical and efficient parallel algorithms (in the sense of both computation time and number of processors) for the problem.

Let P denote the class of problems that can be solved sequentially in a time bounded by a polynomial of the input size, n. Also, let NC (for "Nick's Class" [18]) stands for the class of problems that can be solved in parallel in a time of $O(\log^k n)$, for some constant k, with a number of processors bounded by a polynomial of n. For $k = 1$, the time of $O(\log n) + O(1)$ represents the natural time lower bound in the computation. One open question regarding the complexity of parallel algorithms is whether $P = NC$, which is thought to be very unlikely [19] (it is clear that $NC \subseteq P$). The importance of this question resides in its relation to determination of the nature and amount of parallelism in computation. If a problem from P with the serial computational complexity of $O(n^s)$ belongs to NC then an asymptotic speedup of $O(n^s / \log^k n)$ in its computation can be achieved, which indicates an unbounded parallelism and a speedup that grows significantly with the size of problem.

Given an application domain, e.g., robotics, a related question is then whether all the problems from the application domain, representing a subset of P, belong to NC. The answer to this question, which determines the parallelism in the application domain, might be simpler than the general case since, most often, the problems from an application domain exhibit similar algorithmic properties. This is actually the case in robotics since most of the kinematic and dynamic problems belong to the class of NC [8]. Furthermore, it is possible to devise parallel algorithms to achieve the time lower bound of $O(\log n) + O(1)$ in solving these problems [8,11,13,16,17]. In the following, the time and processors bounds in the computation of forward dyanamics by different algorithms are studied.

Using the N-E formulation, the bias vector can be computed in a time of $O(\log_2 n) + O(1)$ with $O(n)$ processors [16]. This implies that the time and processor bounds in the forward dynamics computation are determined by those in the solution of Eq. (3). Note that, with $O(n)$ processors, the integration of the computed joint accelerations can be performed in a time of $O(1)$.

The solution of Eq. (3) by the $O(n)$ algorithms results in a set of first-order nonlinear recurrences which can be represented (at an abstract level) as

$$X_i = \phi_2(x_{i+1})/\phi_1(X_{i+1}) = C_i + \phi(X_{i+1}) \tag{4}$$

where C_i is constant, ϕ_1 and ϕ_2 are polynomials of first and second degree, and $\deg \phi = $ Max $(\deg \phi_1, \deg \phi_2) = 2$. It is well known that the parallelism in computation of nonlinear recurrences of the form of Eq. (4) and with $\deg \phi > 1$ is bounded [20,21]. That is, regardless of the number of processors used, their computation can be sped-up only by a constant factor. This is due to the fact that the date dependency in nonlinear recurrences and particularly, those containing division, is stronger than in linear recurrences [22]. Hence, the parallelism in the $O(n)$ algorithms is bounded, that is, their parallelization leads to the $O(n)$ algorithms which are faster than the serial algorithm only by a constant factor. Note that a rather simple model was used for presentation of the nonlinear recurrences of the $O(n)$ algorithms while they are far more complex than those usually studied in literature, e.g., in [21,22] (see [8] for a more detailed discussion).

For the conjugate gradient algorithms in [7,10], the computation of each iteration, as is shown in [15], can be done in a time of $O(\log_2 n)$ with n processors, leading to the $O(n \log_2 n)$ parallel algorithms. This implies an unbounded parallelism in conjugate gradient algorithms. Asympototically, however, the parallel conjugate gradient algorithms are slower than the best serial algorithms, i.e., the $O(n)$ algorithms, for the problem.

The inertia matrix can be computed in $O(\log_2 n) + O(1)$ steps with $O(n^2)$ processors [8,11,13]. The implication of this result is that it further reduces the analysis of the time and processor bounds in forward dynamics problem to that in a more generic problem, the linear system solution. Csanky [23] has shown that the linear system can be solved in $O(\log_2^2 n)$ steps with $O(n^4)$ processors. This result implies that the linear system solution and hence, the forward dynamics problem, belong to the class of NC. Note that, using Cramer's rule, the linear system can be solved in $O(\log_2 n)$ steps with $O(n!)$ processors [20]. But such a result has neither theoretical nor practical importance.

However, Csanky's algorithm is unpractical since it not only requires too many processors but, more importantly, is numerically unstable [25]. The best stable algorithms for linear system solution achieve a time of $O(n)$ with $O(n^2)$ processors [24, 25]. Hence, parallelization of the $O(n^3)$ algorithms results in the stable $O(n)$ parallel algorithms with $O(n^2)$ processors, which indicates an unbounded parallelism.

The above analysis shows that the forward dynamics problem belongs to the class of NC and that the best known upper bounds on the time and processors are $O(\log_2^2 n)$ and $O(n^4)$, respectively. Hence, theoretically, the parallelism in forward dynamics problem is unbounded. Practically, however, the fastest stable parallel algorithms for its computation are of $O(n)$. With respect to these results two main questions arise as:

1. Given the fact that both the serial $O(n)$ and $O(n^3)$ algorithms result in the $O(n)$ parallel algorithms which one is more efficient for parallelization?

2. How the bounds on the time and processors can be improved?

Let $\alpha_1 n + \beta_1$ denote the polynomial complexity of the serial $O(n)$ algorithms. There is a limited parallelism in both coarse grain and fine grain (in matrix-vector operation) forms in these algorithms [8]. Exploitation of this parallelism leads to the parallel algorithms with polynomial complexity as $\alpha_2 n + \gamma_2 \lceil \log_2 n \rceil + \beta_2$ where, due to the limited parallelism, α_1 is reduced to α_2 only by a small factor. Furthermore, exploitation of both coarse and fine grain parallelism requires additional architectural complexity. For the $O(n^3)$ algorithms, as will be shown, the polynomial complexity of the resulting parallel $O(n)$ algorithm is of the form $\alpha_3 n + \gamma_3 \lceil \log_2 n \rceil + \beta_3$ where α_3 is smaller than α_1 by more than two orders of magnitude. This implies that while the algorithm is asymptotically faster than the serial $O(n)$ algorithms and their parallel versions by a high constant factor, it also provides a much greater efficiency for small values of n. The price to be paid for this efficiency, of course, is an architecture with $O(n^2)$ processors. However, the efficiency of the parallel algorithm and the suitability of the required architecture for VLSI and WSI implementation strongly support the choice of the $O(n^3)$ algorithms for parallelization.

Note that the time and processor bounds were derived based on the $O(n^3)$ algorithms which, asymptotically, are the least efficient. Hence, the parallel algorithm achieving these bounds is not optimal in the sense defined in [26] since the product of its time and processor, i.e., $O(n^4 \log_2^2 n)$, grows much faster than the time of the best serial algorithms, i.e., $O(n)$. Therefore, if the same time upper bound can be derived based on the $O(n^2)$ or $O(n)$ algorithms, then a much smaller processors bound could be expected. The more important issue, however, is the gap between the time of $O(\log_2^2 n)$ and the natural time lower bound of $O(\log_2 n)$. One classical approach for filling this gap is to attempt to reformulate the $O(n)$ algorithms into a set of first-order linear recurrences (see [21,26] for some examples) which can then be computed in a time of $O(\log_2 n)$ with n processors, leading to the optimal bounds on both time and processors. However, such a reformulation needs to be done at a higher level than the basic equations and, so far, there is no report on the feasibility of such a reformulation. In fact, at the present, it seems that the time and processors bounds in the forward dynamics problem are closely related to that in the linear system solution. This observations is further supported by the results in [4,5] in which it is shown that the $O(n)$ algorithms can essentially be considered as a procedure for recursive factorization and inversion of the inertia matrix, that is, a recursive procedure for solution of Eq. (3).

III. A NEW ALGORITHM FOR COMPUTATION OF INERTIA MATRIX

The derivation of the new algorithm used for parallel computation of the inertia matrix is discussed in [8,9,12]. Here, for the   sake of completeness, the algorithm is briefly presented. The main emphasis, however, is to discuss its better suitability for parallel computation over other algorithms.

A. Basic Algorithms for Computation of Inertia Matrix

From Eq. (3) the elements of the inertia matrix can be computed as

$$a_{ij} = a_{ji} = \Gamma_j \tag{5}$$

for the condition given by

$$\ddot{Q}_i = 1 \text{ and } \ddot{Q}_{k \neq i} = 0 \qquad \text{for } k = 1, 2, \ldots, n \tag{6}$$

Two physical interpretations can be thought for the above condition, with each interpretation leading to a distinct class of algorithms as

1. The fist $i-1$ links do not have any motion, that is, they are static, and the accelerations and the forces/torques of the last $n-i+1$ links result from the unit acceleration of joint i. This interpretation leads to the first class of algorithms, designated as the class of Newton-Euler Based (NEB) algorithms. In [7] an algorithm of this class is presented, designated as the Original NEB (ONEB) algorithm, in which inertia matrix is computed by successive applications of the N-E formulation.

2. The last $n-1+1$ links can be considered as a single composite rigid system, since they do not have any relative motion, which is accelerating in space, leading to the exertion of forces and moments on the first $i-1$ static links. This interpretation leads to the second class of algorithms, designated as the class of Composite-Rigid Body (CRB) algorithms. In [7] an algorithm of this class, designated as the Original CRB (OCRB) algorithm, is presented in which the inertia matrix is computed by first computing the center of mass, and the first and the second moment of mass with respect to the center of mass, of a set of composite systems.

In [7] it is concluded that the OCRB algorithm provides a significantly greater efficiency over the ONEB algorithm. In [8,9] an algorithm, designated as the Variant of CRB (VCRB) algorithm, is developed which avoids the redundancies in the OCRB algorithm and achieves a greater efficiency. Note that, however, due to the symmetry of the problem, both interpretations and hence both classes of algorithms should lead to the same results and computational efficiency. In [8,9], it is shown that, by introducing or reducing the redundancy in the computation, the algorithms of the two classes can be transformed to one another and, particularly, to the most efficient one, the VCRB algorithm. Figure 1 shows the relative serial efficiency of and redundancy in different algorithms.

B. Analysis of Parallelism in Algorithms for Inertia Matrix

Although the results presented in Fig. 1 answer the question of the serial efficiency of different algorithms, they do not indicate which algorithm is optimal for parallel computation. For serial computation, removing any redundancy increases the computational efficiency. For parallel computation, however, depending on its impact on the data de-

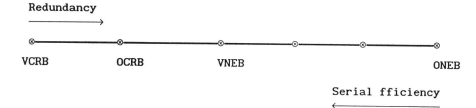

Fig. 1. Comparison of Algorithms for Computation of Inertia Matrix

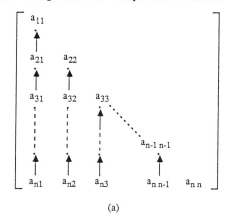

(a)

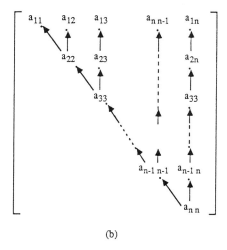

(b)

Fig. 2. Computational Structure of Two Classes of Algorithms

(a) Newton-Euler Based Algorithms. (b) Composite Rigid-Body Algorithms

pendency in the computation, this may increase or decrease the efficiency. The fact that arbitrary algorithms can be developed by introducing or removing different types of redundancy (Fig. 1) represents an additional degree-of-freedom that can be exploited to derive an algorithm which, though perhaps not efficient for serial computation, provides the most suitable features for parallelization. That is, an algorithm that not only achieves the best computation time (in the parallel sense) but also requires the least complex communication and synchronization mechanisms. The derivation of such an algorithm requires a more careful analysis of the two classes of algorithms and the impact of the different redundancies on the data dependency in their computation.

Figure 2 shows, at an abstract level, the computational structure of two classes of algorithms where the arrows indicate the data dependency in (serial) computation of the elements of inertia matrix. From a computational point of view, the CRB algorithms are distinguished by a two-dimensional recursion in their computation (Fig. 2b) whereas the NEB algorithms are distinguished by one-dimensional recursion in computation of each column's elements and the order independency in computation of columns (Fig. 2a).

For the CRB algorithms, due to the two-dimensional recursion, there is data dependency in computation of the diagonal elements as well as each column's elements. Hence, mapping parallel CRB algorithms onto two-dimensional processor arrays requires a rather complex two-dimensional communication and synchronization structures [8,11]. The evaluation of the diagonal elements is more computation intensive and has a stronger data dependency than that of the off-diagonal elements. Consequently, the parallel CRB algorithms are irregular. That is, the subalgorithms for computing the diagonal elements have larger grain size than those for computing the off-diagonal elements [11]. This large grain size limits the maximum exploitable parallelism since the evaluation of the diagonal elements constitutes the major part of the computation. It should be pointed out that the greater serial efficiency of the VCRB algorithm over the OCRB algorithm mainly results from elimination of the redundancy in the computation of the diagonal elements which also reduces the data dependency in the computation. Hence, the VCRB algorithm is more efficient than the OCRB algorithm for both serial and parallel computation [8,11].

For the NEB algorithms, there is data dependency only in the computation of each column's elements and the evaluation of the columns is order independent and can be performed in parallel. Therefore, mapping the NEB algorithms onto two-dimensional processor arrays requires a much simpler two-dimensional communication and synchronization structures than that of CRB algorithms. Furthermore, the evaluation of all the elements requires the same amount of computation, which is less than that of diagonal elements in CRB algorithms, with a uniform data dependency. Hence, the parallel NEB algorithms are more regular and have a finer grain than the parallel CRB algorithms. If all the columns are evaluated in parallel then the computational cost of the parallel NEB algorithms is determined by that of the parallel evaluation of the largest column, the first column. This and the fine grain size in parallel computation of each column imply that the NEB algorithms provide a much higher degree of parallelism than the CRB algorithms while being more regular and demanding much simpler communication and synchronization structures.

Four different types of redundancy can be recognized in the ONEB algorithm. They can be eliminated, respectively, by

1) Choosing a more suitable coordinate frame for projection of the equations.

2) Optimizing the N-E formulation for the condition given by Eq. (6).

3) Using a more efficient variant of the optimized N-E formulation.

4) Introducing a two-dimensional recursion in the computation.

Note that the first redundancy resides in the *extrinsic* equations and results from the choice of coordinate frame for projection of the *intrinsic* equations while the second, the third, and the fourth redundancies reside in the intrinsic equations and are inherent in the formulation. As stated before, by removing all redundancies in the intrinsic equations, the ONEB algorithm can be transformed to the VCRB algorithm. It is important to note that removing any type of redundancy in the ONEB algorithm, as far as it preserves the order-independence property, while does not increase the complexity of required communication and synchronization mechanism, increases its efficiency for parallel computation by reducing the grain size. In this regard, only the removal of the fourth redundancy by introducing a two-dimensional recursion in the computation results in the loss of the order-independence property of the algorithm. In the following, a Variant of the NEB algorithm, designated as the VNEB algorithm, is developed by removing the first three redundancies.

C. A Variant of Newton-Euler Based (VNEB) Algorithm

The VNEB algorithm is presented by its intrinsic equations. This is motivated by the fact that the intrinsic equations provide a suitable abstraction since they result from physical laws which are independent of, and are valid for, any coordinate frame. In this paper, according to the Gibbs notation, vectors are underlined once and tensors (tensors of order 2) twice. Also, in order to simplify and, particularly, unify the derivation of the serial and parallel algorithms, a set of notation, given in Table I and Fig. 3, are used. The VNEB algorithm is then written as

For $i = 1, 2\ldots, n$

For $j = i, i+1,\ldots, n$

$$\underline{\dot{\omega}}(j,i) = \underline{Z}(i) \tag{7}$$

$$\underline{\dot{V}}(j,i) = \underline{\dot{V}}(j-1,i) + \underline{\dot{\omega}}(j,i) \times \underline{P}(j+1,j) = \underline{\dot{\omega}}(j,i) \times \underline{P}(j+1,i) \tag{8}$$

$$\underline{F}(j+1,j,i) = M(j)\underline{\dot{V}}(j,i) + \underline{\omega}(j,i) \times \underline{H}(j) \tag{9}$$

$$\underline{N}(j+1,j,i) = \underline{\underline{K}}(j)\underline{\dot{\omega}}(j,i) \tag{10}$$

For $j = n, n-1,\ldots,i$

$$\underline{F}(n+1,j,i) = \underline{F}(j+1,j,i) + \underline{F}(n+1,j+1,i) \tag{11}$$

$$\underline{N}(n+1,j,i) = \underline{N}(j+1,j,i) + \underline{N}(n+1,j+1,i)+$$
$$\underline{P}(j+1,j) \times \underline{F}(n+1,j+1,i) \tag{12}$$

with $\underline{F}(n+1,n+1,i) = \underline{N}(n+1,n+1,i) = 0$

$$a_{ji} = \underline{Z}(j).\underline{N}(n+1,j,i) \tag{13}$$

Equations (7)-(13) represent an optimized version of the N-E formulation in which, given the conditions described by Eq. (6) and resulting from the exclusion of the velocity-dependent and gravitional terms, the second redundancy of the ONEB algorithm is eliminated. They also represent a more efficient variant of the N-E formulation for the above conditions. This improved efficiency results from avoiding the computation of the linear

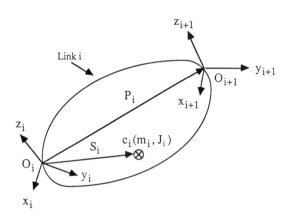

Fig. 3. Link, Frames, and Kinematic and Dynamic Parameters.

a_i, d_i, a_i	D-H parameters of link i.
$Q_i, \dot{Q}_i, \ddot{Q}_i$	Position, Velocity, and acceleration of joint i.
$\tau(i)$	Torque applied at joint i.
$M(i)$	Mass of link i.
$J(i)$	Second moment of mass of link i about its center of mass.
$H(i)$	First moment of mass of link i about point O_i.
$K(i)$	Second moment of mass of link i about point O_i.
$Z(i)$	Axis of joint i.
$S(i)$	Position vector from point O_i to the center of mass of link i.
$P(j,i)$	Position vector from point O_i to the point O_j.
$R(j,i)$	3×3 matrix describing orientation of frame j with respect to frame i.
$\dot{\omega}(j,i)$	Angular acceleration of link j resulting from unit acceleration of joint i.
$\dot{V}(j,i)$	Linear acceleration of link j (point O_j) resulting from unit acceleration of joint i.
$F(k+1,j,i)$	Force exerted on point j due to acceleration of links j through k, i.e., the links contained between points O_j and O_{k+1}, resulting from unit acceleration of joint i.
$N(k+1,j,i)$	Moment exerted on point O_j due to acceleration of link j through k, resulting from unit acceleration of joint i.

Table I. Notation used in derivation of serial and parallel algorithms.

Algorithm	General		n = 6		
	Mul.	Add.	Mul.	Add.	Total
ONEB	$37n^2+15n$	$27n^2+10n$	1422	1032	2454
VNEB	$(39/2)n^2+(195/2)n-95$	$19n^2+55n-66$	1192	948	2140
VCRB	$(9/2)n^2+(231/2)n-181$	$4n^2+88n-137$	644	535	1179

Table II. Comparison of Serial Algorithms for Computation of the Inertia Matrix

acceleration of and the moment acting on the link's center of mass, and the fact that the terms $\underline{H}(i)$ and $\underline{K}(i)$ are constant in link i frame, i.e., frame $i+1$, and can be precomputed.

Note that the VNEB algorithm, like other NEB algorithms, requires n times evaluation of the N-E formulation or its variant. If the equations are projected onto the link frame, as is done in [7], then $O(n^2)$ transformations for the link to link propagation of the variables are required. However, if the equations are projected onto some fixed frame, then only $O(n)$ transformations for projection of the vectors and the tensors are required. This projection scheme eliminates the first redundancy in the ONEB algorithm and leads to a significant reduction of the $O(n^2)$ dependent terms on the polynomial complexity. Although, any coordinate frame can be used for this projection, the EE coordinate frame (frame $n + 1$) is slightly more efficient since the vectors and the tensors of the link n are constant in this frame. Also, though for different reasons, the EE frame is a suitable choice for the parallel algorithm. The computational costs of the ONEB, the VNEB, and the VCRB algorithms are given in Table II where in evaluating the computational cost (for $n = 6$), the time of addition and multiplication is taken to be the same. A comparison of the serial efficiency of these algorithms can show the relative significance of the fourth redundancy with respect to the first three ones. This is an interesting example which indicates the fundamental differences between the serial and parallel algorithms and the analysis of their efficiency. In fact, while the two-dimensional recursion results in a greater serial efficiency of the CRB algorithms over the NEB algorithms, its elimination, which leads to the order independence property of the computation, results in a greater parellal efficiency of the NEB algorithms over the CRB algorithms.

IV. PARALLEL COMPUTATION OF INERTIA MATRIX

A. Time and Processors Bounds in Computation of Inertia Matrix

The serial computational complexity in evaluating the inertia matrix is of $O(n^2)$. No serial algorithm can achieve a better asymptotic complexity since, given n inputs (joint positions), the evaluation of the $O(n^2)$ outputs, the elements of the inertia matrix, requires $O(n^2)$ distinct steps in the computation. Based on the VCRB algorithm, it is already shown that the inertia matrix can be computed in $O(\log_2 n) + O(1)$ steps with $O(n^2)$ processors [8,11]. However, the same bounds on time and processors can be more easily derived based on the VNEB algorithm.

The evaluation of Eqs. (7)-(13), for $i = 1, 2, \ldots, n$, that is, the computation of the columns of the inertia matrix, is order independent and can be performed in parallel. Hence, the time lower bound in parallel computation of the VNEB algorithm is determined by that in computing the first column of the inertia matrix, that is, the evaluation of Eqs. (7)-(13) for $i = 1$. Equations (7)-(13) represent a set of First-Order-Homogeneous-Linear (FOHL) recursions which, for $i = 1$, can be computed in $O(\log_2 n) + O(1)$ steps with n processors, using Recursive-Doubling Algorithm (RDA) [27]. However, parallel evaluation of n sets of FOHL recursions, for $i = 1, 2, \ldots, n$ requires $O(n^2)$ processors. In the following, a parallel version of the VNEB algorithm is presented which not only achieves the time lower bound in the computation, but also, compared to the parallel version of the VCRB algorithm, reduces the coefficiencts on the polynomial complexity.

B. Parallel Algorithm for Computation of Inertia Matrix

As was shown, $O(n^2)$ processors are required to achieve the time lower bound in the computation. For the implementation of the parallel algorithm achieving the time lower bound, a two-dimensional array of $n(n + 1)/2$ processor-memory modules represented as PR_{ij}, for $i = 1, 2, \ldots, n$ and $j = 1, i + 1, \ldots, n$ is considered (Fig. 4 shows the array for $n = 6$). For the parallel algorithm, similar to the serial algorithm, the equations are

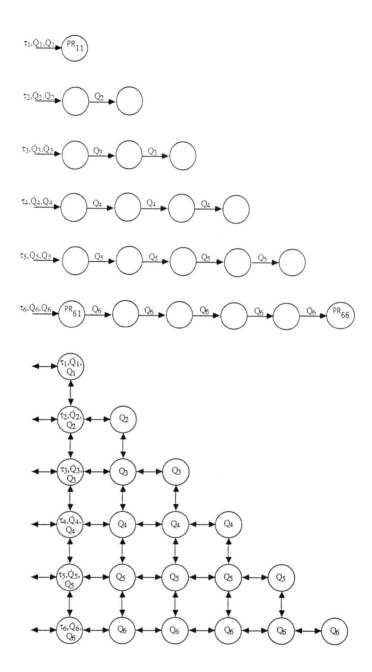

Figure 4. A Two-Dimensional Processor Array. (a) Data Input to Array.
(b) Organization of Initial Data

projected onto the EE frame, i.e., frame $n + 1$. An n DOF all revolute joints manipulator is considered. It is assumed that the joint variables θ_j (or $S\theta_j$ and $C\theta_j$) and constant parameters $S\alpha_j, C\alpha_j,\ ^{j+1}P(j+1,j),\ ^{j+1}S(j),\ ^{j+1}K(j)$, and $M(j)$ reside in the memory of all processors of the jth row. Note that the constant parameter need to be distributed once and before the computation starts. The inputing and distribution of joint variables will be discussed later.

For the parallel algorithm, the elements of the ith column of the inertia matrix are computed by the processors of the ith column of the processor array. The fact that $\omega(j, i) = Z(i)$ for $j = 1, i+1, \ldots, n$, implies a global communication of $Z(i)$ among the processors of the ith column. This requirement can be avoided by introducing two FOHL recurrences as

$$\dot{\omega}(j, i) = \dot{\omega}(j - 1, i) = Z(i) \tag{14}$$

$$P(j + 1, i) = P(j, i) + P(j + 1, j) \tag{15}$$

Eq. (14) does not need any computation while, for the parallel algorithm, the computation of Eq. (15) is required. By computing Eqs. (14)-(15) as a set of coupled recurrences, the terms $Z(i)$ can be considered as the data associated with Eq. (15). Using such a scheme although increases the communication complexity in parallel evaluation of Eq. (15) but will result in the global distribution of $Z(i)$ among the processors of the ith column. The computation of parallel algorithm is then performed as follows.

Step 1:

1) Parallel compute $R(j + 1, j)$ by processors of Row j.

For $j = 1, 2, \ldots, n$

    For $i = 1, 2, \ldots, j$

$$PR_{ji}\ :\ R(j + 1, j) \tag{16}$$

2) Parallel compute $R(n + 1, j)$ by processors of Column i, using RDA.

For $i = 1, 2, \ldots, n$

    For $j = i, i + 1, \ldots, n$

For $\eta = 1$ step 1 until $\lceil \log_2(n + 1 - i) \rceil$, Do $\qquad$ (17)

$$R(j + 2^\eta, j) = R(n + 1, j)$$

$$j + 2^\eta > j + 2^{\eta-1} \geq n + 1$$

$$R(j + 2^\eta, j) = R(n + 1, j) = R(n + 1, j + 2^{\eta-1})R(j + 2^{\eta-1}, j)$$

$$j + 2^\eta \geq n + 1 > j + 2^{\eta-1}$$

$$R(j + 2^\eta, j) = R(j + 2^\eta, j + 2^{\eta-1})R(j + 2^{\eta-1}, j)$$

$$n + 1 > j + 2^\eta > j + 2^{\eta-1}$$

End_Do

3) Shift $R(n + 1, j + 1)$ by processors of Row $j + 1$ to processors of Row j.

For $j = 1, 2, \ldots, n$

$\quad$ For $i = 1, 2, \ldots j$

$\quad\quad PR_{ji} \; : \; R(n + 1, j + 1)$ $\qquad$ (18)

Note that, as the result of the above data transfer, both the terms $R(n + 1, j)$ and $R(n + 1, j + 1)$ reside in the memory of all processors of Row j.

4) Parallel compute $^{n+1}Z(j)$, $^{n+1}P(j + 1, j)$, and $^{n+1}H(j)$ by processors of Row j.

For $j = 1, 2, \ldots, n$

$\quad$ For $i = 1, 2, \ldots, j$

$\quad\quad$ a) $PR_{ji} \; : \; ^{n+1}Z(j) = R(n + 1, j)^j Z(j)$ $\qquad$ (19)

$\quad\quad\quad$ with $^j Z(j) = [0\ 0\ 1]^t$

$\quad\quad$ b) $PR_{ji} \; : \; ^{n+1}P(j + 1, j) = R(n + 1, j + 1)^{j+1}P(j + 1, j)$ $\qquad$ (20)

$\quad\quad$ c) $PR_{ji} \; : \; ^{n+1}S(j) = R(n + 1, j + 1)^{j+1}S(j)$ $\qquad$ (21)

$\quad\quad$ d) $PR_{ji} \; : \; ^{n+1}H(j) = M(j)^{n+1}S(j)$ $\qquad$ (22)

For processors of Row n, Eqs. (20)-(22) do not need any computation since the terms $^{n+1}P(n+1,n)$, $^{n+1}S(n)$, and $^{n+1}H(n)$ are given constant parameters. As the result of the computation of Step 1, all the vectors are projected onto the coordinate frame $n+1$. In the following, the absence of superscripts denotes that the computations are performed in this frame.

Step 2:

1) Parallel compute $\dot{\omega}(j,i)$ and $P(j+1,i)$ by processors of Column i, using RDA.

For $i = 1, 2, \ldots, n$

    For $j = i, i+1, \ldots, n$

        For $\eta = 1$ step 1 until $\lceil \log_2(n+1-i) \rceil$, Do

$$\dot{\omega}(j+2^\eta, i) = \dot{\omega}(j+2^{\eta-1}, i) = Z(i) \tag{23a}$$

$$P(j+2^\eta, j) = P(j+1,i) \tag{23b}$$

$$j + 2^\eta > j + 2^{n-1} \geq n+1$$

$$P(j+2^\eta, j) = P(j+1,i) = P(j+2^\eta, j+2^{\eta-1}) + P(j+2^{\eta-1}, j)$$

$$j + 2^\eta \geq n+1 > j + 2^{\eta-1}$$

$$P(j+2^\eta, j) = P(j+2^\eta, j+2^{\eta-1}) + P(j+2^{\eta-1}, j)$$

$$n+1 > j+2^\eta > j+2^{\eta-1}$$

        End_Do

2) Shift $P(j+1,i)$ by processors of Row j to processors of Row $j+1$.

For $j = 1, 2, \ldots, n$

    For $i = 1, 2, \ldots, j$

$$PR_{ji} : P(j,i)$$

3) Parallel computer $\dot{V}(j,i)$, $F(j+1,j,i)$, $N(j+1,j,i)$ by processors of Column i.

For $i = 1, 2, \ldots, n$

    For $j = i, i+1, \ldots, n$

        a) PR_{ji} : $\dot{V}(j,i) = \dot{\omega}(j,i) \times P(j,i) = Z(i) \times P(j,i)$ (25)

$$b) \ PR_{ji} \ : \ F(j+1,j,i) = \dot{\omega}(j,i) \times H(j) + M(j)\dot{V}(j,i) \tag{26}$$

$$c) \ PR_{ji} \ : \ N'(j+1,j,i) = R(n+1,j+1)\Big(\overset{j+1}{\ } K(j)R(j+1,n+1)$$

$$^{n+1}\dot{\omega}(j,i)\Big) + H(j) \times \dot{V}(j,i) \tag{27}$$

Step 3:

1) Parallel compute $F(n+1,j,i)$ by processors of Column i, using RDA.

For $i = 1, 2, \ldots, n$

 For $j = i, i+1, \ldots, n$

  For $\eta = 1$ step 1 until $\lceil \log_2(n+1-j) \rceil$, Do $\tag{28}$

$$F(j+2^\eta, j, i) = F(n+1,j,i)$$

$$j + 2^\eta > j + 2^{\eta-1} \geq n+1$$

$$F(j+2^\eta, j, i) = F(n+1,j,i) = F(j+2^\eta, j+2^{\eta-1}, i) + F(j+2^{\eta-1}, j, i)$$

$$j + 2^\eta \geq n+1 > j + 2^{\eta-1}$$

$$F(j+2^\eta, j, i) = F(j+2^\eta, j+2^{\eta-1}, i) + F(j+2^{\eta-1}, j, i)$$

$$n+1 > j + 2^\eta > j + 2^{\eta-1}$$

 End_Do

2) Shift $F(n+1,j+1,i)$ by processors of Row $j+1$ to processors of Row j.

For $j = 1, 2, \ldots, n$

 For $i = 1, 2, \ldots, j$

 $$PR_{ji} \ : \ F(n+1,j+1,i) \tag{29}$$

3) Parallel compute $N(n+1,j,i)$ by processors of Column i, using RDA.

For $i = 1, 2, \ldots, n$

 For $j = i, i+1, \ldots, n$

 a) $PR_{ji} \ : \ N(j+1,j,i) = N'(j+1,j,i) + P(j+1,j) \times F(n+1,j+1,i) \tag{30}$

b) For $\eta = 1$ step 1 until $\lceil \log_2(n + 1 - j) \rceil$, Do (31)

$$N(j + 2^\eta, j, i) = N(n + 1, j, i)$$
$$j + 2^\eta > n + 2^{\eta - 1} \geq n + 1$$
$$N(j + 2^\eta, j, i) = N(n + 1, j, i) = N(n + 1, j + 2^{\eta - 1}, i) + N(j + 2^{\eta - 1}, j, i)$$
$$j + 2^\eta \geq n + 1 > j + 2^{\eta - 1}$$
$$N(j + 2^\eta, j, i) = N(j + 2^\eta, j + 2^{\eta - 1}, i) + N(j + 2^{\eta - 1}, j, i)$$
$$n + 1 > j + 2^\eta > j + 2^{\eta - 1}$$

End_Do

2) Parallel compute a_{ji} by PR_{ji}.

For $i = 1, 2, \ldots, n$

For $j = i, i + 1, \ldots, n$

$$PR_{ji} \; : \; a_{ji} = Z(j).N(n + 1, j, i) \tag{32}$$

The computational cost of parallel evaluation of inertia matrix by using the VNEB algorithm, is determined by that of the parallel evaluation of the first column of the inertia matrix. That is, the cost of the n recurrences in Eqs. (17), (23b), (28), and (31) is determined by that of the largest ones, i.e., for $i = 1$, which are of size n. Also, the computational cost of all the $O(n^2)$ terms in Eqs. (16), (19)-(22), (25)-(27), (30), and (32) is determined by the cost of one term since for each column these terms are computed in parallel and the computation for n columns, as will be discussed later, is overlapped. The computational cost of the parallel algorithm is then evaluated as follows.

Step 1: The cost of Eq. (16)-(17) is $4m + (27m + 18a)\lceil \log_2 n \rceil$. Eq. (18) represents a simple data rotation. Eq. (19) does not need any computation. The cost of Eqs. (20)-(22) is $(21m + 12a)$. The cost of this step is then obtained as $(27m + 18a)\lceil \log_2 n \rceil + (25m + 12a)$.

Step 2: Eq. (23a) and (24) represent simple data rotations. The cost of Eqs. (23b), (25) and (26) is $(3a)\lceil \log_2 n \rceil, (6m + 3a)$, and $(9m + 6a)$, respectively. The cost of Eq. (27), by performing the operations from the right hand side for the first term, that is, by performing three matrix-vector multiplications, is $(33m + 24a)$. The cost of this step is $(3a)\lceil \log_2 n \rceil + (48m + 33a)$.

Step 3: The cost of Eq. (28) is $(3a)\lceil \log_2 n \rceil$. Eq. (29) represents a simple data rotation. The cost of Eqs. (30), (31) and (32) is $(6m + 6a), (3a)\lceil \log_2 n \rceil$, and $(3m + 2a)$, respectively. The cost of this step is $(6a)\lceil \log_2 n \rceil + (9m + 8a)$.

Adding the costs of Step 1-3, the computation cost of the algorithm is obtained as $(27m + 27a)\lceil \log_2 n \rceil + (82m + 53a)$.

For the parallel algorithm, using the link frame, similar to the serial algorithm, will decrease the computational efficiency by increasing the coefficient of the $\lceil \log_2 n \rceil$-dependent terms on the polynomial complexity. For the computation performed by processors of each column i, the choice of any fixed frame k, for $n + 1 \geq k \geq i$, leads to the optimal efficiency. In this regard, given the processors organization and the distribution of initial and input data, the choice of EE frame for all columns results in the simplicity and uniformity of the algorithm.

V. ALGORITHM-TO-ARCHITECTURE MAPPING

In previous section, the computational steps and cost, describing the behavioral features of the parallel algorithm, were derived. Considering an abstract processor array, an abstract processor-operation allocation scheme along with some communication primitives were also presented. In order to derive more detailed architectural features, e.g., communication and synchronization mechanisms, required for implementation of the algorithm, mapping the algorithm to architecture needs to be studied. The mapping problem is studied by analyzing the structural features of the algorithm [28,29]. First, the perfect mapping, that is, mapping with minimum communication and synchronization overhead, is studied and the architectural features required for such a mapping are determined. The perfect mapping allows transformation of the abstract processor array into a dedicated or an algorithmi-cally specialized architecture. Then, mapping the algorithm to some different (and less complex) architectures, i.e., nonperfect mapping, is studied to seek efficient design tradeoffs between the algorithmic and architectural complexity.

With respect to the procedures developed for systolic and wavefront arrays [31-33], the perfect mapping represents *a nonconstrained mapping* in which the features of algorithmi-cally specialized architecture are designed to fully support the algorithm. The nonperfect mapping can be considered as the *matching procedure* [32] which is defined as mapping the algorithm to some specific architecture, i.e., a *constrained mapping*, in which the constraints are presented by the features of specific architecture. In our approach, the specific architecture is derived from the algorithmically-specialized architecture. The nonperfect mapping is mainly motivated to reduce the complexity of the architecture and to make it more suitable for VLSI and WSI implementation without significantly increasing the algorithmic complexity in terms of both computation and communication (see also [11] for a similar approach).

The algorithm-to-architecture mapping study also allows a more detailed analysis of practical implementation of the algorithm on the architecture, that is, processor-operation allocation and, particularly, the implementation (synchronization) mode. Note that, Eqs. (16)-(32) provide an abstract space-time representation of the computation in which the indices i and j refer to space (processor) and the index η refers to time (step). Such representation, though valid for computational complexity analysis, does not indicate the practical implementation, or practical space-time mapping, of the computation to the architecture. To see this, note that, these equations might imply that the operation of all the $O(n^2)$ processors are performed in parallel and synchronously. This requires the simultaneous availability of the input data (joint variables) and the intermediate data generated during the computation, for all processors. As will be shown, there is a delay in distribution of the input data among processors of different columns. Furthermore, the computation of Eqs. (17), (23b), (28) and (31) by processors of Column i takes more time than by processors of Column k for $k < i$, which results in the delay in availability of the intermediate data. As will be seen, only the operation of processors of each column needs to be performed in parallel and synchronously while the operations of different columns can be performed asynchronously in an overlapped fashion. Note that, by overlapping the evaluation of different columns, as in their evaluation in parallel, the computational complexity is determined by that of the first column.

A. Data Distribution, Processor-Operation Allocation, Processors Interconnection, and Communication Complexity

The communication complexity in implementing an algorithm on a given architecture is a complex function of the data dependency in computation of algorithm (which defines the communication structure inherent to the algorithm), distribution of the data among processors, processor-operation allocation, and the architecture's interconnection structure. In particular, the distribution of initial data (constant parameters), input data, and dynamic organization of the intermediate data are decisive factors [34,35] since they affect the data alignment required for performing different steps of algorithm and hence the communication overhead.

For the developed parallel algorithm, the input data are only needed for computation of Eq. (16) and do not affect the rest of computation. In order to analyze the distribution of the intermediate data, as a function of the processors interconnection and processor-operation allocation, the data dependency in computation of different equations needs to be studied first. Note that, except for sharing the input data, there is no data dependency in computations performed by different columns. Hence, only the data dependency within the computations performed by each column needs to be analyzed. These computations consist of a set of independent (and local) operations performed by each processors, e.g., Eqs. (16), (19)-(22), etc., and a set of collective (and global) operations for implementing RDA by all processors, e.g., Eqs. (17), (23), (28), and (31). This further implies that only the data dependency in computation of latter equations by using RDA needs to be analyzed.

The communication complexity of RDA in solving a FOHL recurrence of size n is basically $\lceil \log_2 n \rceil t_c$ where t_c is the cost of communicating the basic data of recurrence. However, this communication complexity is preserved if RDA is implemented on an n-processor architecture with a Hypercube augmented with Nearest Neighbor (HNN) interconnection. Also, perfect Shuffle-Exchange augmented with Nearest Neighbor (SENN) interconnection can be used but the communication complexity will be increased by a factor of two since, at each step of computation, first a shuffle to pair the data in adjacent processors and then an exchange of data between adjacent processors is required [36]. The main difference between the two interconnections resides in the distribution of the intermediate results of the computation among processors. To see this, consider the computation of Eq. (17) by processors of Column i. With the HNN interconnection among the processors of Column i, all the intermediate results $R(j + 2^\eta, j)$ as well as the final result $R(n + 1, j)$ are computed by and hence reside in the memory of the processor PR_{ji}. With the SENN interconnection, only the final result $R(n + 1, j)$ is transferred (after the last shuffle) to and hence resides in the memory of the processors PR_{ji} (see [16] for a detailed example). However, only the final results of recurrences in Eqs. (17), (23), (28), and (31) are needed for the rest of the computation since there is no data dependency in terms of the intermediate results among the computation of the different equations. Hence, with either HNN or SENN interconnection, the final result of the computations in Eqs. (17), (23), (24), (29), and (32) reside in the memory of PR_{ji} since they are either computed by or transferred to this processor.

This data organization along with the distribution of constant parameters (Sec. IV.B) guarantees the correctness of the processor-operation allocation given in Eqs. (16), (19)-(22), (25)-(27), (30), and (32), and the fact that no data alignment, in addition to those performed in Eqs. (18), (24), and (29), is required. It can be concluded that both HNN and SENN interconnections lead to the minimization of the communication overhead. However, in order to minimize the communication complexity of the algorithm, the interconnection among the processors of each column should be of HNN.

The distribution of the input data requires a nearest-neighbor interconnection among processors of each row (Fig. 4a). This interconnection, as will be shown, is also used for decomposition of inertia matrix and the solution of resulting triangular systems. Hence, for the perfect mapping of the algorithm, the required interconnection of the processor array is of HNN among the processors of each column (which is not shown in Figs. 4a and 4b) and of nearest neighbor among the processors of each row.

The determination of the data distribution, processor-operation allocation, and processors' interconnection allows the evaluation of the communication complexity resulting from the perfect mapping of the algorithm. Since the operations of different column processors are overlapped, the communication cost of the algorithm, similar to its computational cost, is determined by that in the operations of the first column. Also, since the operations of the processors of each column are synchronized (Sec. V.B), there is no overhead due to communication initiation and the communication cost can be determined based on the number of data.

Let c denote the time for communicating a single datum, i.e., a scalar or a component of a vector or a matrix. As a result of the above processor-operation allocation, the operations in Eqs. (16), (19)-(22), (25)-(27), (30), and (32) are purely local and do not need any communication. The data shifted in Eq. (18) is matrix and those in Eqs. (24) and (29) are vector. The basic data of the recurrences in Eqs. (23), (28), and (31) are vectors and that in Eq. (17) is matrix. Hence, the communication cost of the algorithm is obtained as $(21\lceil\log_2 n\rceil + 15)c$. If the time of m, a, and c is taken to be the same, then it can be seen that the communication cost of the algorithm is much smaller than its computational cost. It is interesting to notice that, unlike many scientific computations [37,38], the parallel algorithm achieves the time lower bound in the computation while remaining highly compute bound. It should be pointed out that by using the SENN interconnection the communication cost of the algorithm is increased to $(42\lceil\log_2 n\rceil + 15)c$, which is still less than its computation cost.

B. Implementation (Synchronization) Mode

The choice of architectural implementation, or synchronization, mode, i.e., SIMD, MIMD, or pipeline, for efficient mapping of an algorithm is a strong function of the algorithm's granularity, data dependency, frequency of the communication, and regularity and determinacy in the computation [28]. The developed algorithm exploits two completely different types of parallelism; in computing different columns of the inertia matrix, which is a coarse grain parallelism with a weak data dependency in computation, and in computing each column, which is a fine grain parallelism with strong data dependency in computation and a rather high frequency of communication. Consequently, the synchronization mode for the processor array needs to be studied at two levels; at local level, for processors of each column, and at global level, for different columns.

For processors of each column, due to the fine grain parallelism, strong data dependency, and high frequency of communication, the SIMD mode is an efficient alternative [28]. The SIMD mode implies that a mechanism for global synchronization of the activities of the processors is needed. This mechanism can be implemented by using a single control unit which provides the global synchronization by issuing instructions in a lock-step fashion for all processors. However, such an implementation leads to extra architectural complexity since for each column a separate control unit and a global control bus are required. Furthermore, the instructions to be executed by each processor require matrix-vector operations and some of them even require a set of these operations, e.g., Eqs. (26)-(27), and (30). Hence, in addition to the global control unit, a local control unit for each processor is also required for properly sequencing the operations and managing the loading and storing the intermediate results from and to the local memory.

If all processors of the same column are driven by the same clock, then due to the regularity and determinacy in the computation, a clock-based synchronization mechanism similar to systolic arrays [30] can be employed. That is, each processor has its own local control unit but all the control units are synchronized by using a global clock. Although, different clocks can be used for different columns, from an implementation point of view, it is more suitable to use a single clock for whole processor array. This further provides the capability of global clock-based synchronization of all processors of the array which is also useful for solution of linear system.

The synchronization of different columns is closely related to the data input mode of the array and the distribution of the input data among the processors. This is due to the data dependency in computation of different columns in terms of sharing the input data or, more precisely, the required global distribution of Q_j among processors of Row j. This synchronization can be efficiently achieved by using a local data driven mechanism similar to wave front arrays [31,32]. The data driven mechanism can be implemented by shifting the data among the processors of adjacent columns. That is, the activity of processors of Column i is triggered by receiving the joint variable Q_k, for $k = i$, $i - 1$, ..., 1, from processors of Column $i - 1$. Each processor PR_{ji} (except PR_{ii}) then, by shifting its data, Q_j, triggers the activity of its right neighbor processor in the same row, i.e., PR_{ji+1}. The synchrony of the processors of Column i and the use of a global clock for the array guarantee the simultaneous availability of the data for and synchrony of the processors of Column $i + 1$. The activity of the processors of Columns i and $i + 1$ can then be performed independently and asynchronously in an overlapped fashion.

This data rotation scheme not only provides a simple and efficient synchronization mechanism but also allows the distribution of the input data among $O(n^2)$ processors to be performed in $O(n)$ steps, while using only n input channels (Fig. 4a). Note that, however, the data input is completely overlapped with the computation of the first column, which further reduces the overhead in parallel computation. This, once again, shows the suitability of the VNEB algorithm for parallel processing. In fact, the order independency in computing the columns of inertia matrix by this algorithm not only provides a high degree of parallelism but also allows the overlapping between computation and communication. It can be concluded that the required synchronization mechanism for perfect mapping of the

algorithm to the processor array is of global clock-based for processors of each column and
of local data-driven for processors of each row.

C. Mapping with Topological Variation

A major drawback of the architecture for perfect mapping of the algorithm is the complexity of the required interconnection among the processors of each column. From an
implementation view point, it is desirable to employ an architecture with a simpler and
more regular interconnection structure, e.g., a processor array with two-dimensional nearest
neighbor interconnection (Fig. 4b). Due to the regularity, modularity, and locality of its
interconnection structure, this architecture is particularly suitable for implementation with
VLSI and WSI technologies. Mapping the algorithm to this architecture represents a nonperfect mapping with topological variation which only affects the communication complexity
of the algorithm. More precisely, it affects the communication cost of RDA in computing
Eqs. (17), (23), (28), and (31).

With a nearest neighbor connection among the processors of each column, the communication structure for RDA can be implemented by rotating the data between adjacent
processors (see also [11]). The time to transfer a data between two processors is proportional to the number of required rotations, which is determined by the distance between
processors. The complexity of such communication scheme can be reduced by exploiting
parallelism since at each step $O(n)$ data rotations can be performed in parallel. At the kth
step in computation of the RDA, the $n2^{k-1}$ required data transfers can be done in parallel,
where each data transfer requires 2^{k-1} data rotations. The communication cost of the RDA
is then obtained as

$$\sum_{k=1}^{\lceil \log_2 n \rceil} 2^{k-1} t_c = (n^* - 1)t_c \tag{33}$$

where $n^* = n$ if $n = 2^m$, and $n^* = 2^m$ if $2^m > n > 2^{m-1}$. Eq. (33) implies that
a communication complexity of $O(n)$ in the computation of RDA can be achieved while
employing a simple nearest neighbor interconnection. Using this scheme the communication
cost of the algorithm is then obtained as $(21(n^* - 1) + 12)c$ which, for $n = 6$, represents
an increase by a factor of about 2 compared with the perfect mapping. Assuming the same
time for m, a, and c, only for $n > 16$ the communication cost of the algorithm becomes
greater than its computational cost. That is, for almost all practical size of the problem, the
algorithm still remains compute bound even with a simple nearest neighbor interconnection.

Note that, however, the local communication can be performed faster than nonlocal one. The speed of local communication can be further increased by using a register-to-register data transfer scheme between adjacent processor. With VLSI and WSI implementation, an even greater speed in the communication can be achieved since it is performed within the chip or wafer. It can be concluded that algorithm still achieves a good performance even when mapped to an architecture with a reasonable complexity and hence a better suitability for practical implementation. In fact, mapping the algorithm to the architecture of Fig. 4b represents an efficient tradeoff between the algorithm's communication and the architecture's interconnection complexity. This also shows, again, the suitability of the VNEB algorithm which, while providing a higher degree of parallelism in the computation, demands a less complex communication structure.

D. Architecture for Parallel Computation of Forward Dynamics

In previous sections, the architectural features required for efficient parallel computation of VNEB algorithm were discussed. Here, these features are summarized. Efficient implementation of other computations required for forward dynamics, i.e., computation of bias vector and solution of linear system, on the same architecture requires additional flexibility and hence additional architectural features which are also discussed here.

The architecture for parallel computation of VNEB lgorithm is an array of $n(n + 1)/2$ processor-memory modules with a nearest neighbor interconnection (Fig. 4b). Each processor-memory module has its own local control unit and program memory. The data input/output is performed through n channels. The synchronization mechanism for processors of each column is of clock-based type and for processors of each row is of data-driven type. The fact that all local control units are driven by the same clock allows the golbal clock-based synchronization (similar to systolic arrays) of all processors (both column and row processors). The local data-driven mechanism can also be extended to all processors. This can be achieved by providing for each processor the mechanism for triggering the activity of its four nearest neighbors by shifting the data.

The capabilty of combining both global clock-based and global data-driven mechanisms results in the extra flexibilty required for efficient exploitation of different types of parallelism. However, in order to efficiently exploit such capability, some considerations in the

design of both architecture and algorithms need to be taken into account. From an architectural point of view, this flexibility requires the design of appropriate local control unit; it
should be capable of switching among three modes, i.e., purely local, global clock-based, and
global data-driven, in performing operations. Furthermore, it should be capable of proper
sequencing of high level local operations such as matrix-vector operations and combination
of these operations. Hence, the local control unit needs a more elaborate and complex design
that those suuggested in theroretical design and analysis of systolic and wavefront arrays.
Note that, however, few practical implementations of systolic arrays, such as Warp [39],
have shown the necessity of a more complex control unit for achieving more flexibility and
generality of the architecture.

From an algorithmic point of view, the use of global clock-based synchronization mechanism requires correct timing of the data while by using the data-driven mechanism this
requirement is replaced by that of correct sequencing of the data [32]. If both mechanisms
are combined and used in the same computation, as in the solution of linear system (Sec.
VII), then both correct timing and sequencing of data should be guaranteed. In order meet
correct timing and/or correct sequencing care should be taken in the algorithmic design.

In the sequel, given the above discussed architectural features, algorithms for parallel
computation of bias vector and complete solution of linear system on the processor array
are developed.

VI. PARALLEL COMPUTATION OF BIAS VECTOR

The bias vector is needed for solution of triangular systems. Hence, its computation
can be overlapped with the computation and factorization of inertia matrix, leading to
an even higher degree of concurrency in computing the forward dynamics. However, the
exploitation of this additional concurrency requires additional processors and a modified
processor array. Given the processor array of Fig. 4, an efficient (and natural) alternative
is to compute the bias vector by processors of the first column. This implies that, by
completion of computation of inertia matrix, the processors of Columns 2 to n enter the
wait state while those of the Column 1 start the computation of the bias vector.

A parallel algorithm for the N-E formulation that achieves the time lower bound of
$O(\log_2 n) + O(1)$ on an SIMD array of n processors is developed by Lee and Chang [16].

This algorithm is also used for computation of the bias vector by setting the vector of joint accelerations to zero [15]. In [16] the parallelism in the N-E formulation is exploited by projecting the equations onto the base frame and transforming them into a set of FOHR, where the computation performed for projection represents the major component in overall computational cost of the algorithm. If the bias vector is computed by the processors of the first column then, by projecting the equations onto the EE frame, this computation can be avoided since it is already performed in Eqs. (16)-(21). The rest of the algorithm of [16], with some minor changes, is reported in Appendix where the absence of superscript denotes that the computation is performed in the EE frame. The computational cost of evaluating the bias vectors by Eqs. (A1)-(A20) is $(15a)\lceil \log_2 n \rceil + (120m + 93a)$ which shows that, by exploiting the synergism between the computation of the bias vector and that of the first column of the inertia matrix, a greater efficiency can be achieved without using extra processors.

Given the distribution of links' parameters and input data, $Q(j)$ and $\dot{Q}(j)$, among the processors of the first column (Fig. 4b), and with a HNN or SENN interconnection, it can be shown, as in Sec. V.A., that the terms $\omega(j,1)$, $\dot{\omega}(j,1)$, $\dot{V}(j,1)$, $F(n+1,j)$ and $N(n+1,j)$ are either computed or transfered to PR_j. This guarantees the correctness of processor-operation allocation in other equations. Note that the rotations performed in Eqs. (A4), (A8), and (A14) are required for data alignment and minimization of communication cost. With a HNN interconnection, the communication cost of the algorithm, which results from the nonlocal operations, i.e., the data rotation in (A4), (A8), and (A14) and computation of FOHL recurrences in (A3), (A6), (A9), (A13), and (A18), is $(15\lceil \log_2 n \rceil + 9)c$ which, again assuming the same time for m, a, and c, shows that the evaluation of the bias vector for all n is compute bound. With the nearest neighbor interconnection among the processors of the first column the communication cost of the algorithm will be increased to $(15n^* - 6)c$. Hence, for almost all pratical n, the algorithm still remains compute bound even with a simple nearest neighbor interconnection.

The clock based mechanism, discussed in Sec. V.B., can also be used for efficient synchronization of the processors. Note that, in addition to the synergism, the computation of the bias vector and the first column of the inertia matrix share common properties, e.g., size of the problem, computational steps, and operations. Fig. 5 shows the organization of

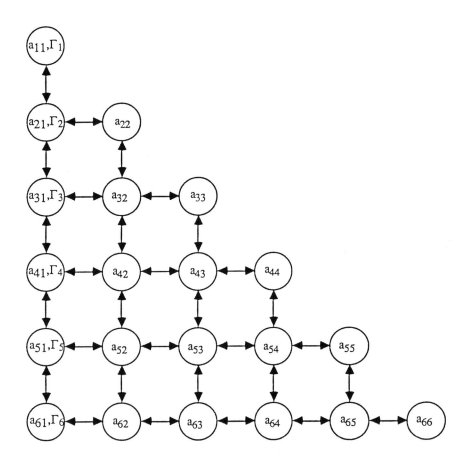

Figure 5. Organization of Data Resulting from Computation of Inertia
Matrix and Bias Vector

data resulting from the computation of inertia matrix and bias vector within the processor array.

VII. PARALLEL SOLUTION OF LINEAR SYSTEM

The linear system is solved in two steps. First the matrix is factored and then the resulting triangular systems are solved. There are numerous reported works on parallel solution of linear system. However, most of these works are concerned with either parallel factorization or parallel solution of triangular systems and less attempts is made to perform both steps within a unique framework and analyze the overall performance. Here, we develop parallel algorithms for complete solution of the system $\ddot{A}Q = \Gamma$ which can be efficiently implemented on the processor array. In order to achieve overall efficiency and minimize various overheads, the algorithmic design is performed based on the distribution of the elements of matrix A and vector Γ within the array (Fig. 5) and the architectural features discussed in Sec. V.D.. In particular, attention is paid to the dynamic organization of data, generated at different steps of the solution, and minimization of overhead due to the data alignment.

A. Parallel Factorization of Inertia Matrix

A symmetric positive definite matrix can be factored by using Cholesky factorization or Square Root-Free Cholesky factorization which is also known as Symmetric Gaussian Elimination (SGE). For the system in Eq. (3) and by using Cholesky factorization, first the matrix A is factored as $A = LL^t$, where L is a lower triangular matrix, and then $\ddot{Q}$ is obtained by solving two triangular systems as $LY = \Gamma$ and $L^t\ddot{Q} = Y$. By using the SGE, first the matrix A is factored as $A = LDL^t$ where L is a lower triangular matrix with 1 as the elements of its main diagonal and D is a diagonal matrix. The solution for $\ddot{Q}$ is then obtained by solving three systems as $LY = \Gamma, DZ = Y$, and $L^t\ddot{Q} = Z$.

The Cholesky factorization requires square root operation for factorization and division for solution of triangular systems. For SGE, the factorization does not require square root operation. Also, the triangular systems $LY = \Gamma$ and $L^t\ddot{Q} = Z$ can be solved without division operation while the solution of diagonal system $DZ = Y$ requires division operation. This implies that, in terms of operation, the algorithms based on the SGE are more regular than those based on the Cholesky decomposition. Note that, also, the square root and division operations are slower than the multiplication and addition operations. Hence, the SGE is considered for the solution of Eq. (3).

Most of the matrix operations can be described by a generic triply nested For loop in terms of three indices i, j, and k. The six permutations of the indices result in six different organizations of computation (known as ijk forms [40]) and hence six different algorithms with different data access and movement patterns. For a given architecture, some permutations might be more efficient than the others due to a less communication overhead.

The factorization of matrix A involves two processes: Elimination (E) and Updating (U). In elimination process, the elements of the factor (s) are computed. In updating process, the already computed elements of the factor (s) are used for updating the remaining elements of matrix A (Figs. x and y). In [40,41] the six different forms of Cholesky factorization and SGE are analyzed in terms of their efficiency for various vector and parallel architectures. These different ijk forms not only represent different data access and movement patterns but also different orderings in elimination and updating processes. A careful analysis of these forms reveals that the kij and kji forms (Figs. 6 and 7) are particularly suitable for a two-dimensional architecture with a limited connectivity, such as the processor array considered here. In both forms once a column of the factor, say L_i, is computed it can be used for updating the remaining submatrix A, say A_{i+1}. This provides a high degree of concurrency since the computation of the next column of the factor, L_{i+1}, can be completely overlapped with the updating of the submatrix A_{i+1}. More importantly, the data communication required for updating can be overlapped with the computations required for elimination and updating.

The choice of kij or kji form for mapping onto the processor array depends on the resulting overhead. The first step in the mapping process is the processor-operation allocation, which is also a decisive factor in the resulting implementation overhead. Given the connectivity of the processor array and the distribution of the elements of matrix A, it is more efficient to assign the computation of each element of the factor $L_{ij}(D_{ii})$ to processor $PR_{ij}(PR_{ii})$. This scheme not only minimizes the communication overhead for factorization but also the organization of the resulting data, L_{ij} and D_{ii}, allows the solution of triangular and diagonal systems to be obtained without any data alignment.

For k = 1 to n-1

E: $D_{kk} = a_{kk}^{(k)}$ (34)

    For s = k+1 to n

    E: $L_{sk} = a_{sk}^{(k)}/D_{kk}$ (35)

    For i = k+1 to n

       For j = k+1 to i

       U: $a_{ij}^{(k+1)} = a_{ij}^{(k)} - a_{jk}^{(k)} L_{ik}$ (36)

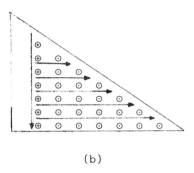

(a)   (b)

Fig. 6. (a) The kij Form of SGE. (b) Data Access for kij Form.
⊕ : Elimination (E). ⊙ : Updating (U).

For k = 1 to n-1

E: $D_{kk} = a_{kk}^{(k)}$ (37)

    For s = k+1 to n

    E: $L_{sk} = a_{sk}^{(k)}/D_{kk}$ (38)

    For j = k+1 to n

       For i = j to n

       U: $a_{ij}^{(k+1)} = a_{ij}^{(k)} - a_{ik}^{(k)} L_{jk}$ (39)

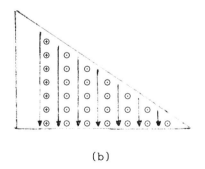

(a)   (b)

Fig. 7. (a) The kji Form of SGE. (b) Data Access for kij Form.
⊕ : Elimination (E). ⊙ : Updating (U).

With this processor-operation allocation scheme mapping kij and kji forms onto the processor array leads to two different data movement patterns. For the kij form, the updated elements $a_{ij}^{(j)}$ need to be communicated to processors of Row i, i.e., PR_{ik} for $k > j$, and to all processors of Column i while the elements L_{ij} need to be communicated to processors of Row i only. For the kji form, the updated elements $a_{ij}^{(j)}$ need to be communicated to processors of Row i while the elements L_{ij} need to be communicated to both processors of Row i and Column i. For both forms, different data move with different velocities through the array. Hence, the correct timing and ordering of the availability of different data for each processor should be guaranteed. This can be more easily and efficiently achieved by using the kij form since for processor PR_{ik} the elements $a_{ij}^{(j)}$ are made available before the elements L_{ij}. This delay can be used by PR_{ik} for communicating the the elements $a_{ij}^{(j)}$ to its neighbor PR_{ik+1} (if $k < i$) or PR_{i+1i} (if $k = i$).

The algorithm for implementation of the kij form on the processor array can be considered as a data-flow algorithm in the sense defined in [42,43]. Each process of the algorithm, i.e, computation of L_{ij} assigned to processor PR_{ij}, consists of sets of instructions. Each set of instructions is executed only when the data it requires are available. This data-flow algorithm can be better described by the activities of processor PR_{ij} as follows.

For $k = 1$ to $j - 1$: Data Communication Cycle

    Wait

   1. Receive $a_{ik}^{(k)}$: $((i + j - 1) + 3(k - 1))c + (k - 1)(d + ma)$

   2. Send Right $a_{ik}^{(k)}$: $((i + j) + 3(k - 1))c + (k - 1)(d + ma)$

   3. Receive $a_{jk}^{(k)}$: $((i + j) + 3(k - 1))c + (k - 1)(d + ma)$

   4. Send South $a_{jk}^{(k)}$: $((i + j + 1) + 3(k - 1))c + (k - 1)(d + ma)$

    Wait

For $k < j$: Updating Cycle

    Wait

   1. Receive L_{ik} : $((i + j) + 3(k - 1))c + (k - 1)(d + ma) + d$

   2. Send Right L_{ik} : $((i + j + 1) + 3(k - 1))c + (k - 1)(d + ma) + d$

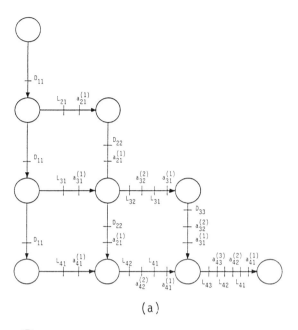

(a)

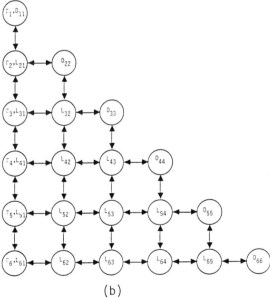

(b)

Figure 8. (a) Data Flow in Kij form of SGE.
(b) Organization of Matrices L and D and vector Γ.

3. Compute $a_{ij}^{(k+1)} : ((i + j + 1) + 3(k - 1))c + k(d + ma)$

Wait

For $k = j$: Elimination

    For $i \neq j$

    Wait

1. Receive $L_{ij-1} : (i + 4j - 6)c + (j - 2)(d + ma) + d$

2. Send Right $L_{ij-1} : (i + 4j - 5)c + (j - 2)(d + ma) + d$

3. Compute $a_{ij}^{(j)} : (i + 4j - 5))c + (j - 1)(d + ma)$

4. Receive $D_{ii} : (i + 4j - 5)c + (j - 1)(d + ma)$

5. Send South $D_{ii} : (i + 4j - 4) + (j - 1)(d + ma)$

6. Send Right $a_{ij}^{(j)} : (i + 4j - 3) + (j - 1)(d + ma)$

7. Compute $L_{ij} : (i + 4j - 3) + (j - 1)(d + ma) + d$

8. Send Right $L_{ij} : (i + 4j - 2) + (j - 1)(d + ma) + d$

    Wait (End of Operations)

    For $i = j$

    Wait

1. Receive $L_{ii-1} : (5i - 6)c + (i - 2)(d + ma) + d$

2. Send Right $L_{ii-1} : (5i - 5)c + (i - 2)(d + ma) + d$

3. Compute $D_{ii} : (5i - 5)c + (i - 1)(d + ma)$

4. Send South $D_{ii} : (5i - 4)c + (i - 1)(d + ma)$

    Wait (End of Operations)

Note that, the exact timing of each instruction is also given where d and ma indicate the time of division and of successive multiplication and addition. As stated above, each set of instructions is executed by receiving the data it requires, that is, the communication and updating cycles are executed by receiving $a_{ik}^{(k)}$ and L_{ik} , respectively, while the elimination process is executed by receiving L_{ii-1} . In addition to correct sequencing, this algorithm can also guarantee the correct timing of data, which is due to the regularity of the computation and flow of data, and the fact that all processors are driven by the same clock. An inspection of the flow of data shown in Fig. 8a (for $n = 4$) can prove the correctness of the timing given above. For example, it can be seen that, in the Communication Cycle the delay between the availability of $a_{ik}^{(k)}$ and $a_{jk}^{(k)}$ is $1c$, and in the Elimination Cycle the term D_{ii} is made available at the same time that the term $a_{ij}^{(j)}$ is computed. In this sense, the algorithm can be considered as a synchronous data-flow algorithm; for each processor a specific set of instructions is executed by receiving the corresponding data but the activities of the neighboring processors are synchronized by using a common clock.

The behavior of the algorithm can be better explained by using the notion of computational front [43]. The computation of each diagonal element D_{ii} generates a computational front which passes through the processors of the remaining subarray, i.e., PR_{jk} for j and $k = i$ to n. The different data of each computational front i generated by D_{ii}, that is, $D_{ii}, a_{ji}^{(i)}$, and L_{ji}, pass through the subarray with a same constant velocity, leading to the correct timing of data. The delay between the generation of front i and front $i+1$, i.e., the delay between the computation of D_{ii} and D_{i+1i+1}, is $5c + ma + d$. Since $n-1$ fronts are needed the computation cost of the algorithm is $(n - 1)(ma + d)$, which corresponds to the time lower bound in SGE, and its communication cost is $5(n - 1)c$. Note that, both the computation and communication costs can be defined based on the timing of the availability of D_{nn}. Figure 8b shows the organization of data resulting from the factorization of matrix A.

B. Parallel Solution of Triangular and Diagonal Systems

The solution of a triangular system is equivalent to the solution of an n-th order linear recurrence. The Column-Sweep [34,44] is an efficient algorithm for parallel solution of the n-th order linear recurrences which achieves the computation time of $O(n)$ with $O(n)$ processors. Here, an efficient implementation of this algorithm on the processor array based on the organization of matrices L and D and the vector Γ is proposed.

The solution of the systems $LY = \Gamma$ and $DZ = Y$ can be obtained by an algorithm which, similar to the factorization of matrix A, involves the two steps of updating and elimination as

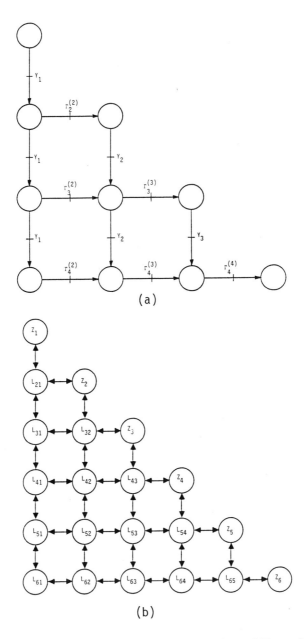

Figure 9. (a) Data Flow in Solution of Lower Triangular and Diagonal Systems.
(b) Organization of Matrix L and vector Z.

For $i = 1$ to n

$$\Gamma_i^{(1)} = \Gamma \text{ (Initialization)} \tag{41}$$

For $i = 1$ to n

$$Y_i = \Gamma_i^{(i)} \text{ (Elimination)} \tag{42}$$

$$Z_i = Y_i/D_{ii} \text{ (Elimination)} \tag{43}$$

For $j = i + 1$ to n $\tag{44}$

$$\Gamma_j^{(i+1)} = \Gamma_j^{(i)} - L_{ji}Y_i \text{ (Updating)} \tag{45}$$

Again, the first step in mapping this algorithm onto the processor array is the processor-operation allocation. Given the distribution of matrices L and D and the vector Γ within the processor array, it is more efficient to assign the computation of $\Gamma_i^{(i)}$ (and hence Y_i) to the processor PR_{ii} since the term Z_i can then be locally computed by the same processor with a simple division. This also eliminates the need for any data alignment for solution of $L^t\ddot{Q} = Z$.

With this scheme, the term Y_i moves through processors of Column i while the term $\Gamma_i^{(j)}$ moves through the processors of Row i (Fig. 9a shows the flow of data for n = 4). The data-flow algorithm for solution of $LY = \Gamma$ and $DZ = Y$ can be described by the activities of the processor PR_{ij} as follows.

For $i \neq j$

   Wait

1. Receive Y_j : $((i - 1) + 2(j - 1))c + (j - 1)ma$

2. Send South Y_j : $(i + 2(j - 1))c + (j - 1)ma$

3. Compute $\Gamma_i^{(j)}$: $(i + 2(j - 1))c + jma$

4. Send Right $\Gamma_i^{(j)}$: $((i + 1) + 2(j - 1))c + jma$

   Wait (End of Operations)

For $i = j$

   Wait

1. Receive $\Gamma_i^{(i)} : 3(i-1)c + (i-1)ma$

2. Compute $Y_i : 3(i-1)c + (i-1)ma$

3. Send South $Y_i : (3(i-1)+1)c + (i-1)ma$

4. Compute $Z_i : (3(i-1)+1)c + (i-1)ma + 1d$

   Wait (End of Operations)

Note that, this algorithm is very similar to the kij form factorization algorithm. It is a synchronous data-flow algorithm in which both correct timing and sequencing of data are guaranteed. This can be easily verified by inspecting the flow of data in Fig. 9a. The computation of each element Y_i generates the ith computational front which passes through the remaining subarray. The different data of each computational front i, Y_i and $\Gamma_i^{(j)}$, pass through the subarray with constant velocities, leading to the correct timing of the data. Both the computation and communication cost of the algorithm can be obtained based on timing of the availability of Y_n and Z_n. The computation cost of the algorithm $(n-1)ma+1d$ and its communication cost is $3(n-1)c$. Note that, since the computation of $Z_i(i \neq n)$ is overlapped with the ith computational front, the contribution of the solution of the systems $DZ = Y$ to the overall computation cost is only $1d$, that is, the cost of Z .

However, the solutions of both $LY = \Gamma$ and $DZ = Y$ can be overlapped with the factorization of the matrix A. The processor PR_{ii} by completing the generation of the ith computational front for factorization can generate the ith computational front for solution of $LY = \Gamma$ and $DZ = Y$. Hence, the ith front of factorization is followed by the ith front of linear system solution. By using this overlapping scheme, the cost of the two linear system solutions is defined based on the cost of the generation of the nth front only, which is $1d$ for computation and zero for communication. Figure 9b shows the organization of the elements of matrix L and vector Z within the processor array.

B. Parallel Solution of Lower Triangular System

The solution of the upper triangular system $L^t \ddot{Q} = Z$ can be obtained by an algorithm which, again, includes the two steps of updating and elimination as

For $i = n$ to 1

$$X^{(n)} = 0 \text{ (Initialization)} \tag{45}$$

For $i = n$ to 1

$$\ddot{Q}_i = Z_i + X_i^{(i)} \text{ (Elimination)} \tag{46}$$

For $j = i - 1$ to 1

$$X_j^{(i-1)} = X_j^{(i)} - L_{ij}\ddot{Q}_i \tag{47}$$

Given the data organization of Fig. 9b, it is more efficient to assign the computation of $\ddot{Q}_i$ to processor PR_{ii}. With this scheme, the term $X_j^{(i)}$ moves through the processors of Column j while the term $\ddot{Q}_i$ moves through the processors of Row i. Hence, the final output of the computation, $\ddot{Q}$, can be made available at the input/output channels in proper order and without any additional overhead since the data output is overlapped with the linear system solution. Again, the data-flow algorithm for solution of the upper triangular system can be described by the activities of processor PR_{ij} as follows.

For $i \neq j$

    Wait

1. Receive $X_j^{(i)}$

2. Receive $\ddot{Q}_i$

3. Send Left $\ddot{Q}_i$

4. Compute $X_j^{(i-1)}$

5. Send North $_j^{(i-1)}X$

    Wait (End of Operations)

For $i = j$

    Wait

1. Receive $X_i^{(i)}$

2. Compute $\ddot{Q}_i$

3. Send Left $\ddot{Q}_i$

    Wait (End of Operation)

The flow of data is shown in Fig. 10a (for $n = 4$) from which the computation and communi-

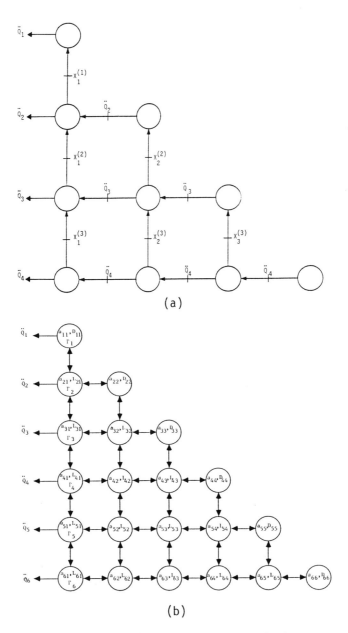

Figure 1o. Data Flow in Solution of Upper Triangular Systems.
(b) Organization of Data and Data Output from Array.

Algorithm		Computation Cost	Number of Processors
Serial	$O(n)$ in [3]	$682n-371$ (4092)	–
	$O(n^3)$ in [7]	$(1/3)n^3+(43/2)n^2+(1549/6)n-113$ (2882)	–
Parallel	$O(n^3)$ in [15]	$5\lceil(n^2-1)/6\rceil+15\lceil(n+1)/2\rceil+5n+$ $74\lceil\log_2 n\rceil+355$ (721)	n
	$O(n^3)$ in [15]	$14\lceil(n+1)/2\rceil+15n+74\lceil\log_2 n\rceil+355$ (722)	$n(n+1)/2+n$
	$O(n^3)$ this paper	$6n+69\lceil\log_2 n\rceil+340$ (583)	$n(n+1)/2$

Table III. Comparison of Serial and Parallel Algorithms

cation costs of the algorithm are obtained as $(n-1)(ma+a)$ and $3(n-1)+1)c$, respectively. This communication cost, as stated before, also includes that data output from processor array which is shown in Fig. 10b.

VIII. DISCUSSION AND CONCLUSION

The computational complexity of serial and parallel algorithms for forward dynamics are presented in Table III in which the time of a, m, and d are taken to be the same. The number in parentheses indicates the cost of the algorithm for $n = 6$. The computational complexity of the algorithm of this paper is obtained by adding the cost of parallel computation of inertia matrix, bias vector, and linear system solution. Similarly, the communication complexity of the algorithm is obtained as $(8n+36n^*-22)c$ which indicates an overall $O(n)$ communication complexity of the algorithm.

It can be seen that the algorithm of this paper achieves a greater computational efficiency over the best algorithm in [15] while demanding a unique architecture with a less number of processors. This indicates both greater speedup and processor utilization efficiency.

If the communication complexity is also taken into account, then the overall performance of the algorithm is even more significant since it achieves a much better communication complexity over that of [15] ($O(n)$ over $O(n^2 \log_2 n)$) while demanding a simple nearest neighbor interconnection structure. Furthermore, the data input/output to and from array is performed through n channels with minimum overhead since they are overlapped with the computation.

Although the parallel algorithm is still of $O(n)$ but the coefficient of n- dependent terms are drastically reduced. In fact, for small n, its complexity complexity is completely dominated by the log n-dependent and the constant terms. For $n = 6$, the algorithm achieves a speedup of about 7 over the serial $O(n)$ and of about 5 over the best serial $O(n^3)$. The speedup significantly increases with the size of the problem, i.e., the number of DOF. For $n = 8$, the algorithm achieves an order-of-magnitude speedup over the serial $O(n)$ algorithm.

In addition to detailed algorithmic analysis, we extensively analyzed the various architectural features required for implementation of the algorithm. Our analysis shows that the proposed architecture is not only highly efficient for parallel computation of forward dynamics but is also very suitable for VLSI and WSI implementations.

APPENDIX. PARALLEL COMPUTATION OF BIAS VECTOR

Given an array of n processor-memory modules, represented as PR_j, $j = 1$, 2, ..., n, and assuming that the first step for projection is already performed as in Eqs. (16)-(21), the algorithm of [16] is given as

Step 2: Compute $\omega(j,1)$, $\dot{\omega}(j,1)$, $\dot{V}(j,1)$ and $\dot{V}_c(j,1)$.

1) Transform acceleration of gravity, $\dot{V}(0)$, to EE frame.

$$PR_1 \ : \ {}^{n+1}\dot{V}(0) = R(n+1,1) \, {}^{1}\dot{V}(0) \tag{A1}$$

2) Parallel Compute $\omega(j,1)$ and $\dot{\omega}(j,1)$.

For $j = 1$, 2, ..., n

a) $PR_j : \omega(j,j) = Z(j)\dot{Q}(j)$ \hfill (A2)

b) For $\eta = 1$ step 1 until $\lceil \log_2 n \rceil$, Do $\qquad$ (A3)

$$\begin{array}{ll} \omega(j, j - 2^\eta) = \omega(j, 1) & 1 \geq j - 2^{\eta - 1} > j - 2^\eta \\ \omega(j, j - 2^\eta) = \omega(j, 1) = \omega(j, j - 2^{\eta - 1}) + \omega(j - 2^{\eta - 1}, 1) & j - 2^{\eta - 1} > 1 \geq j - 2 \\ \omega(j, j - 2^\eta) = \omega(j, j - 2^{\eta - 1}) + \omega(j - 2^{\eta - 1}, j - 2^\eta) & j - 2^{\eta - 1} > j - 2^\eta > 1 \end{array}$$

End_Do

c) Shift $\omega(j - 1, 1)$ by PR_{j-1} to PR_j.

$$PR : \omega(j - 1, 1) \qquad j \neq 1 \qquad (A4)$$
$$PR_1 : \omega(0, 1) = 0 \qquad j \neq 1$$

d) $PR_j : \dot{\omega}(j, j) = \omega(j - 1, 1) \times \omega(j, j)$ $\qquad$ (A5)

e) Parallel compute $\dot{\omega}(j, 1)$ as in (A3). $\qquad$ (A6)

3) Parallel compute $\dot{V}(j, 1)$ and $\dot{V}_c(j, 1)$.

a) $PR_j : \dot{V}(j + 1, j) = \omega(j, 1) \times (\omega(j, 1) \times P(j + 1, j)) + \dot{\omega}(j, 1) \times P(j + 1, j)$ $\qquad$ (A7)

b) Shift $\dot{V}(j, j - 1)$ by PR_{j-1} to PR_j.

$$PR_j : \dot{V}(j, j - 1) \qquad\qquad j \neq 1 \qquad (A8)$$

$$PR_1 : \dot{V}(1, 0) = \dot{V}(1, 1) = \dot{V}(0) \qquad j = 1$$

c) Parallel compute $\dot{V}(j, 1)$ as in (A3). $\qquad$ (A9)

d) $PR_j : \dot{V}_c(j) = \dot{V}(j, 1) + \omega(j, 1) \times (\omega(j, 1) \times S(j)) + \dot{\omega}(j, 1) \times S(j)$ $\qquad$ (A10)

Step 3: Compute $F(n + 1, j)$, $N(n + 1, j)$, Γ_j, and b_j.

For $j = 1, 2, ..., n$

1) Parallel compute $F(n + 1, j)$.

a) $PR_j : F_c(j) = M(j)\dot{V}_c(j)$ $\qquad$ (A11)

b) $PR_j : F(j + 1, j) = F_c(j) \qquad\qquad j \neq n \qquad (A12)$

 $PR_n : F(n + 1, n) = F_c(n) + F_E \qquad j = n$

b) For $\eta = 1$ step 1 until $\lceil \log_2 n \rceil$, Do $\qquad$ (A13)

$$F(j + 2^\eta, j) = F(n + 1, j)$$

$$j + 2^\eta > j + 2^{\eta-1} \geq n + 1$$

$$F(j + 2^\eta, j) = F(n + 1, j) = F(n + 1, j + 2^{\eta-1}) + F(j + 2^{\eta-1}, j)$$

$$j + 2^\eta \geq n + 1 > j + 2^{\eta-1}$$

$$F(j + 2^\eta, j) = F(j + 2^\eta, j + 2^{\eta-1}) + F(j + 2^{\eta-1}, j)$$

$$n + 1 > j + 2^\eta > j + 2^{\eta-1}$$

End_Do

c) Shift $F(n + 1, j + 1)$ by PR_{j+1} to PR_j.

$$PR_j : F(n + 1, j + 1) \qquad\qquad j \neq n \tag{A14}$$

$$PR_n : F(n + 1, n + 1) = F_E \qquad j = n$$

2) Parallel compute $N(n + 1, j)$.

a) $PR_j : {}^{j+1}\omega(j, 1) = R(j + 1, n + 1) \, {}^{n+1}\omega(j, 1)$ \hfill (A15)

b) $PR_j : {}^{j+1}\dot\omega(j, 1) = R(j + 1, n + 1) \, {}^{n+1}\dot\omega(j, 1)$ \hfill (A16)

c) Parallel compute $N(j + 1, j)$. \hfill (A17)

$$PR_j : N(j + 1, j) = R(n + 1, j + 1)\Big({}^{j+1}J(j) \, {}^{j+1}\dot\omega(j, 1) + {}^{j+1}\omega(j, 1)$$
$$\times \, ({}^{j+1}J(j) \, {}^{j+1}\omega(j, 1))\Big) + S(j) \times F_c(j) + P(j + 1, j) \times F(n + 1, j + 1) \qquad j \neq n$$
$$PR_n : N(n + 1, n) = J(n)\dot\omega(n, 1) + \omega(n, 1) \times (J(n)\omega(n, 1)) + S(n) \times F_c(n)$$
$$+ \, P(n + 1, n) \times F_E + N_E \qquad\qquad j = n$$

d) Parallel compute $N(n + 1, j)$ as in (A13). \hfill (A18)

e) $PR_j : b_j = Z(j).N(n + 1, j)$ \hfill (A19)

f) $PR_j : \Gamma_h = \tau_j - b_j$ \hfill (A20)

In the above equations, $\omega(j, i), \dot\omega(j, i)$, and $\dot V(j, i)$ stand for the angular velocity and acceleration of link j, and linear acceleration of point O_j due to the motion of links i through j, respectively. $F(k, j)$ and $N(k, j)$ stand for the force and moment exterted on joint j (point O_j) due to the motion of link k through j. Note that, for sake of simplicity and as in [16], a stationary base (with zero angular velocity and acceleration) is considered. However, the algorithm can be easily modified to include a nonstationary base.

ACKNOWLEDGEMENT

The research described in this paper was performed by the Jet Propulsion Laboratory, California Institute of Technology, under the contract with the National Aeronautics and Space Administration (NASA).

REFERENCES

1. M.H. Milman and G. Rodriguez, "Cooperative Dual Arm Manipulator Issues and Task Approach," Jet Propulsion Lab., Eng. Memorandum 347, Nov. 1987

2. J.Y.S. Luh, M.W. Walker, and R.P. Paul, "On-line Computation Scheme for Mechanical Manipulator," Trans. ASME J. Dyn. Syst., Meas., and Control, Vol. 102, pp. 69-76, June 1980.

3. R. Featherstone, "The Calculation of Robot Dynamics Using Articulated-Body Inertia," Int. J. Robotics Research, Vol.2(2), 1983.

4. R. Featherstone, *Robot Dynamics Algorithms* . Ph.D. Dissertation, Univ. of Edinburgh, 1984.

5. G. Rodriguez, "Kalman Filtering, Smooting and Recursive Robot Arm Forward and Inverse Dynamics," IEEE J. Robotics and Automation, Vol. RA-5, Dec. 1987.

6. G. Rodriguez and K. Kreutz, "Recursive Mass Matrix Factorization and Inversion: An Operator Approach to Open- and Closed-Chain Multibody Dynamics," Jet Propulsion Lab. Publication 88-11, May 1988.

7. M.W. Walker and D.E. Orin, "Efficient Dynamic Computer Simulation of Robotics Mechanisms," Trans. ASME J. Dyn. Syst., Meas., and Contr., Vol. 104, pp. 205-211, 1982.

8. A. Fijany, *Parallel Algorithms and Architectures in Robotics.* Ph.D. Dissertation, Univ. of Paris XI (Orsay), Sept. 1988.

9. A. Fijany and A.K. Bejczy, "An Efficient Algorithm for Computation of the Manipulator Inertia Matrix," J. of Robotic Systems, Vol. 7(1), pp. 57-80, Feb. 1990.

10. A. Fijany and R.E. Scheid, "Efficient Conjugate Gradient Algorithms for Computation of the Manipulator Forward Dynamics," *Proc. of NASA Conf. on Space Telerobotics,* Pasadena, CA, Jan. 1989. 11. A. Fijany and A.K. Bejczy, "A Class of Parallel Algorithms for Computation of the Manipulator Inertia Matrix," IEEE Trans. Robotics and Automation, Vol. RA-5, No. 5, pp. 600-615, Oct. 1989.

12. A. Fijany and A.K. Bejczy, "A New Class of Parallel/Pipeline Algorithms for Computation of the Manipulator Inertia Matrix," In preparation.

13. A. Fijany and J.G. Pontnau, "Parallel Computation of the Jacobian for Robot Manipulators," Proc. IASTED Int. Conf. on Robotics & Automation, Santa Barbara, May 1987.

14. H. Kasahara, H. Fujii, and M. Iwata, "Parallel Processing of Robot Motion Simulation," Proc. 10th IFAC World Congress, Munich, FRG, July 1987.

15. C.S.G. Lee and P.R. Chang, "Efficient Parallel Algorithms for Robot Forward Dynamics Computation," IEEE Trans. Syst. Man Cybern., Vol. SMC-18, no. 2, pp. 238-251, March/April 1988.

16. C.S.G. Lee and P.R. Chang, "Efficient Parallel Algorithm for Robot Inverse Dynamics Computations," IEEE Trans. Syst. Man Cybern., Vol. SMC-16(4), pp. 532-542, July/Aug. 1986.

17. L.H. Lathrop, "Parallelism in Manipulator Dynamics," Int. J. Robotics Res., Vol. 4(2), Summer 1985.

18. S.A. Cook, "An Overview of Computational Complexity," Com. ACM, Vol. 26, June 1983.

19. J.S. Vitter and R.A. Simons, "New Classes of Parallel Complexity: A Study of Unification and other Complete Problems for P," IEEE Trans. Computer, Vol. C-35(5), May 1986.

20. H.T. Kung, "New Algorithms and Lower Bounds for the Parallel Evaluation of Certain Rational Expressions and Recurrences," J. of ACM, Vol. 23, No. 2, pp. 252-261, April 1976.

21. J. Miklosko and V.E. Kotov (Eds.), *Algorithms, Software and Hardware of Parallel Computers*. New york: Springer-Verlag, 1984.

22. L. Hyafil and H.T. Kung, "The Complexity of Parallel Evaluation of Linear Recurrences," J. of ACM, Vol. 24, No. 3, pp. 513-521, July 1977.

23. L. Csanky, "Fast Parallel Matrix Inversion Algorithms," SIAM J. of Computing, Vol. 5, No. 4, pp. 618-623, Dec. 1976.

24. A.H. Sameh and D.J. Kuck, "On Stable Parallel Linear System Solvers," J. of ACM, Vol. 25, No. 1, pp. 81-91, Jan. 1978.

25. A. Bojanczyk, R.P. Brent, and H.T. Kung, "Numerically Stable Solution of Dense Systems of Linear Equations Using Mesh-Connected Processors," SIAM J. of Stat. Comput., Vol. 5, No. 1, March 1984.

26. E.W. Mayr, "Theoretical Aspects of Parallel Computation," in VLSI and Parallel Computation, R. Suaya and G. Birtwistle (Eds.), Morgan Kaufmann Pub., 1990.

27. P.M. Kogge, "Parallel Solution of Recurrence Problems," IBM J. Res. Dev., Vol. 18, pp. 138-148, March 1974.

28. L.H. Jamieson, "Characterizing Parallel Algorithms," in *The Characteristics of Parallel Algorithms*. L.H. Jamieson et al (Eds.), The MIT Press, 1987.

29. L. Synder et al (Eds.), *Algorithmically Specialized Parallel Computers*. Academic Press, 1985.

30. H.T. Kung, "Why Systolic Architectures?," IEEE Computer, Jan. 1982.

31. S.Y. Kung, "On Supercomputing with Systolic/Wavefront Array Processors," Proc. of The IEEE, Vol. 72(7), 1984.

32. S.Y. Kung, "VLSI Array Processors for Signal/Image Processing," in *Parallel Processing for Supercomputers & Artificial Intelligence*. K. Hwang and D. Degroot (Eds.), McGraw-Hill Pub. Co., New York, 1989.

33. S.Y. Kung, *VLSI Array Processor*. Printice Hall, 1998.

34. U. Schendel, *Introduction to Numerical Methods for Parallel Computers*. Chichester, U. K., 1987.

35. B. Lint and T. Agerwala, "Communication Issues in the Design and Analysis of Parallel Algorithms," IEEE Trans. Soft. Eng., Vol. SE-7(2), March 1981.

36. H.S. Stone, "Parallel Processing with the Perfect Shuffle," IEEE Trans. Computer, Vol. C-20, No. 2, pp. 153 161, Feb. 1971.

37. D. Parkinson, "Organizational Aspects of Using Parallel Computers," Parallel Computing 5, pp. 75-83, 1987.

38. D.B. Gannon and J.V. Rosendale, "On the Impact of Communication Complexity on the Design of Parallel Numerical Algorithms," IEEE Trans. Computer, Vol. C-33(12), Dec. 1984.

39. M. Annaratone, et al, "The Warp Computer: Architecture, Implementation, and Performance," IEEE Trans. Computer, Vol. C-36(12), Dec. 1987.

40. J.M. Ortega, "The ijk Forms of Factorization Methods: I. Vectors Computers," Parallel Computing 7, pp. 135-147, 1988.

41. J.M. Ortega and C.H. Romine, "The ijk Forms of Factorization Methods: II Parallel Systems," Parallel Computing 7, pp. 149-162, 1988.

42. P.C. Treleaven et al, "Data-Driven and Demand-Driven Computer Architectures," Comput. Surv. 14, pp. 93-143, 1982.

43. D.P. O'Leary and G.W. Stewart, "Data-Flow Algorithms for Parallel Matrix Computations," Comm. ACM, Vol. 28(8), pp. 840-853, Aug. 1985.

44. D.J. Duck, The Structure of Computers and Computations. John Wiley & Sons, 1978.

INDEX

U

V

W